Microphones

Technology and Technique

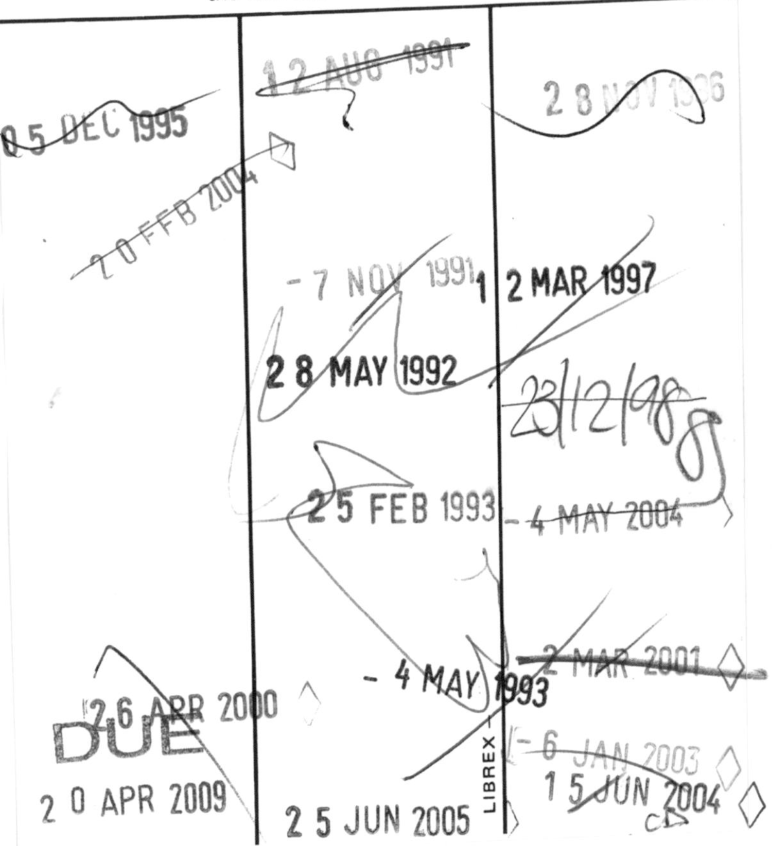

Microphones

Technology and Technique

John Borwick BSc Edinburgh

Focal Press
London and Boston

Focal Press
is an imprint of Butterworth Scientific

 PART OF REED INTERNATIONAL P.L.C.

First published 1990

British Library Cataloguing in Publication Data

Borwick, John
Microphones.
1. Microphones
621.38284

ISBN 0-240-51279-0

Library of Congress Cataloging-in-Publication Data

Borwick, John.
Microphones : technology and technique / John Borwick.
p. cm.
Includes bibliographical references.
ISBN 0-240-51279-0 :
1. Microphone. I. Title.
TK5986.B67 1990
621.382'84--dc20

89-71327
CIP

Laserset by Scribe Design, Gillingham, Kent
Printed and bound in Great Britain by Courier International Ltd, Tiptree, Essex

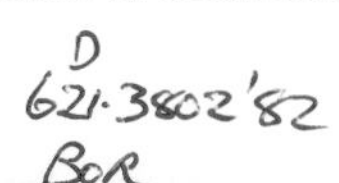

Preface

The title and Contents show that this book is divided into two separate parts, Technology and Technique. This hybrid construction is quite deliberate, and it may reveal my primary motive in writing this book in the first place, which is to pull together these two aspects of microphone knowledge. I believe that the engineers who design, manufacture and market microphones should have a better basic knowledge of the artistic uses to which their microphones are put. On the other hand, it worries me that some recording engineers and producers have only a sketchy idea of how microphones work, and that this may sometimes prevent them from getting the best sound.

By tradition the two mainstream disciplines of science and art are not looked for in a single individual. You are either an engineer or a musician: you cannot be both. You may be able to read a score or a circuit diagram; but not both. Therefore most sound recording, broadcasting and stage-show assignments need at least two people working side by side, yet speaking two different languages. One of them knows all (or nearly all) about the technology while the other knows all about the creative art of making music. This type of partnership can work very well but problems of communication can and often do arise.

In an ideal world, everyone working with microphones – whether primarily labelled a musician, an engineer, an acoustician or a producer – would indeed possess hybrid talents. Well-based technical knowledge of acoustics, electronics and microphone technology should be combined with a well-developed understanding of music and the performing arts. A third essential is an 'educated ear' which responds to and can quickly recognize subtleties in the original sounds and the ways in which these may be modified by the room and the recording/reproducing chain.

A move towards evolving a new hybrid type of individual, at least as far as his or her education is concerned, began more than 40 years ago with the *tonmeister* concept. Special *tonmeister* college courses now recruit bright students who have already demonstrated hybrid potential by securing high-school grades in both science and music. They are then given an in-depth training in both disciplines, plus intensive ear training and hands-on experience with microphones and the latest recording gear.

Very few *tonmeister* graduates will end up in a hybrid role, though successful engineer/producers are to be found, particularly among European recording and broadcasting companies. Most graduates will enter one or other specialization but they invariably find their basic training in the other discipline a practical advantage. Equipment designers, acoustic consultants, architects, studio technicians can all do their own job better if they know what the musicians and producers are trying to achieve, and vice versa.

The *tonmeister* concept can be said to have been born around 1946. In June that year the composer Arnold Schoenberg sent a letter to the Chancellor of the University of Chicago suggesting that the music department should offer classes for 'soundmen' (or, as he would say, *tonmeisters*). He wrote:

> Soundmen will be trained in music, acoustics, physics, mechanics and related fields to a degree enabling them to control and improve the sonority of recordings, radio broadcasts and sound films . . . The student should become able to produce an image in his mind of the manner in which music should sound when perfectly played . . . he will be trained to notice all the differences between his image and the real playing; he will be able to name these differences and to tell how to correct them . . . This can be done and would mean a great advantage over present methods where engineers have no idea of music and musicians have no idea of the techniques of mechanics.

Soon afterwards the very first Tonmeister Institute was opened at Detmold in North Germany. Others have followed, including one at the University of Surrey.

This book is an attempt to help the reader along the way in this process of science/art integration. Part 1 uses very little mathematics and therefore covers only the basic ideas; similarly, Part 2 outlines only some of the microphone balance situations that can arise. The reader is encouraged to study both halves of the book; if you already know a lot about one half, be patient with its simplistic level of treatment and delve more deeply into the other half. I wish you luck.

Acknowledgements

No book of this kind springs complete and fully clothed into the world. It takes its final shape within the author's word processor, helped along the way by influences from the existing literature – whose most relevant papers and books are listed separately among the Further Reading on page 234. It also grows in stature as the author discusses the subject with friends and colleagues, and has here benefited hugely from the assistance freely given by representatives of all the major microphone manufacturers, including AKG Acoustics, Audio-Technica, Beyer Dynamic, Calrec Audio (AMS), C – Ducer, Crown International, Electro – Voice, JVC – Victor, Keith Monks, MB Electronic, Shure Brothers and Sony Broadcast.

John Borwick

Contents

Part 1
The technology

1
Introduction

1.1 Millions of microphones

Today microphones are everywhere. There is a microphone in every telephone, when it is sometimes called the 'mouthpiece' or 'transmitter' to distinguish it from the 'earpiece' or 'receiver'. Most television presenters can be seen to be talking into a microphone which may rise up from the desk in front of them or be pinned to their clothing. Even when we cannot see a microphone in the television picture, we know that either the presenter or artiste has cleverly concealed one (often in association with a tiny radio transmitter so that no cable is needed) or an operator is holding a microphone on a telescopic boom or fishpole, keeping it just out of sight above the frame of the picture set by camera operator colleagues.

In both the above fields, telephony and television, the microphone is the first link in a chain of equipment enabling sounds to be transmitted over long distances. Now, thanks to space travel and satellites, this transmission is no longer confined to terrestrial distances but can reach to the moon and back. Microphones also proliferate in short-distance communications for public address, sound reinforcement and intercom applications – and they supply the signals which are used to cross the barrier of time as well as distance in the field of sound recording.

Put simply, a microphone (Figure 1.1) is a device which converts acoustic energy (received as vibratory motion of air particles) into electrical energy (sent along the microphone cable as vibratory motion of elementary electrical particles called 'electrons'). Once this conversion has taken place, the sound information is freed from the normal acoustical restraints. Sound waves are inherently short lived and can travel only a relatively brief distance before they sink below the limits of audibility. Electric currents, by contrast, can be boosted (amplified) and sent along any required length of wire or they can be added (modulated) onto high-frequency radio waves for 'wireless' transmission all round the world. They can also be stored (recorded) on a variety of different media for subsequent playback (reproduction) at any later date.

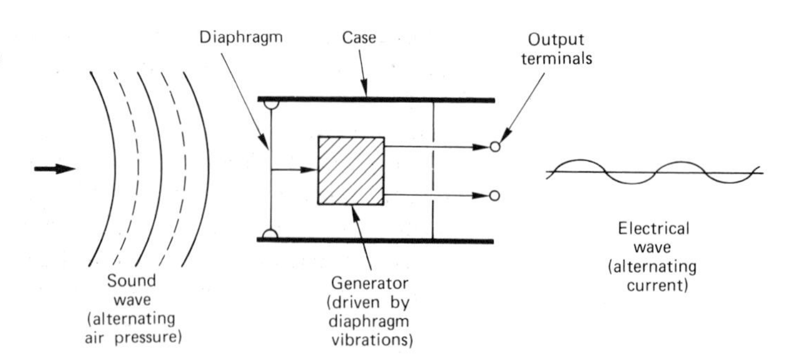

Figure 1.1 A microphone converts acoustic energy into electrical energy

The final stage in the transmission or reproduction process is a reverse conversion of the electric energy back into acoustic energy. The device used for this conversion is called a loudspeaker (or headphone, earpiece, etc.), and generates sound waves which are equivalent, if not actually identical, to the original sound waves which activated the microphone. The generic term for all such energy converters is *transducer*, and the microphone/loudspeaker pairing is just one of a number of reciprocal transducers found in broadcasting and recording. A disc cutting-head, for example, converts electrical energy into mechanical vibration of the stylus which leaves a 'record' of the sound waves engraved into a groove on the master disc surface: the gramophone pickup stylus retraces this waveform and its vibrations are used to generate an equivalent electric current. Similarly the source of the laser light beam which cuts a track of digital pits on the photosensitive surface of a compact disc master converts electrical energy into light energy; the light-sensitive detector in a CD player receives a reflected laser light beam and reverses the transducing process to convert the light energy back into an electric current. Again, a tape recording head converts electrical energy into magnetic energy which is stored as a pattern on the tape; the playback head does the reverse, scanning the magnetic pattern and thereby generating an electric current. (Indeed, many domestic tape machines have a single head which is switched to perform either function as required.)

1.2 The advance of science

Man's understanding of the physics of acoustics and electromagnetism, and his ability to use this knowledge to invent and develop useful mechanisms, has not evolved at a steady rate. It can be taken to have begun in the days of the early Greek civilization around the sixth century BC, when philosophy, science and the arts were studied with equal vigour, often by the same pioneering figures. There then followed about 2000 years of comparative stagnation until the fifteenth century AD, when there was a renaissance of the scientific spirit helped by the spread of printing and the births of such intellectual giants as Leonardo da Vinci (1452–1519) and Galileo (1564–1642), whose pendulum clock was the first accurate means of measuring short time intervals. This period also saw the invention of the microscope and the thermometer. Global exploration awoke an interest in navigational aids such as the compass and the theodolite.

About the same time there was an increased interest in the study of human anatomy, leading in turn to a better understanding of sound propagation and hearing.

The seventeenth century has been called the century of genius, with significant scientific advances made by Galileo as well as Hooke, Newton, Descartes, Kepler, Huygens and Boyle. The nineteenth century brought not only the Industrial Revolution but also an unprecedented upsurge in scientific study and inventive genius. Thus in the past few years we have been celebrating such varied centenaries as the invention of electric light, sound recording, telegraphy and telephony.

Present-day research, invention and development continue at headlong speed, helped by the computer, space travel and the riches and human benefits to be won from global communications and broadcasting. In the audio field, digital processing is moving into all media with important consequences for artistes, producers and engineers alike. The low noise and wide dynamic range available to all users of digital audio has called for some revisions in microphone design and operational techniques.

However, the microphone remains the key to sound quality and so far has been firmly analogue. Future microphones and loudspeakers may indeed incorporate digital converters to provide a completely digitized audio chain between the incident (analogue) sound waves and those that finally reach the ears of the consumer.

1.3 History of discoveries in acoustics

Since prehistoric times, man, in common with his variegated cousins throughout the animal kingdom, has relied on his five senses to keep him informed about his surroundings, to move around safely, to seek out food and shelter and to warn him of danger. Of the five senses, hearing was a relatively late arrival in the evolutionary process but has perhaps become the busiest. Our sense of hearing seems always to be at work, and it has certainly evolved as a uniquely complex and finely tuned mechanism which we are only now beginning to understand properly.

It is therefore hardly surprising that the sense of hearing, and the nature of the sounds that surround us, have interested scientists since the earliest times. As sounds came to be used for communicating (speech) and promoting pleasure (music) the nature of sound and hearing became a necessary study.

Musical instruments certainly existed 5000 years ago, and by about 500 BC Pythagoras had identified the fact that sound is a form of vibration and established a simple numerical relationship between musical notes which sounded well together. He found, for example, that dividing a plucked string at one-half, one-third, etc. of its length produced a pleasing (consonant) interval and thence established pitch ratios for the common musical scale. Later Greek philosophers, including Aristotle, developed theories and practical models showing, for instance, that sound travels through the air as a form of vibration and is reflected from solid surfaces. How the sensation of sound is communicated to the brain remained a mystery for a further 2000 years.

1.3.1 The hearing mechanism

By the middle of the sixteenth century, advances in the tools for scientific research inaugurated a period of accelerated progress in many branches of science. A study of the human anatomy published by Vesalius included a detailed description of the outer and middle ear. This led to examinations of the inner ear by the Italians Fallopio, Eustachian and Corti. Yet it was left to Helmholtz (1821–1894) to establish resonances within the inner ear as the true source of recognition of the pitch (or frequency) of musical sounds. In the present century, Békésy, working in Budapest, developed his travelling-wave theory of pitch recognition, for which he received the Nobel Prize in 1961.

A full explanation of how this motion in the inner ear is converted (transduced) into electricity to stimulate the nerve ends is still awaited. One approach has been to liken the ear's action to that of a microphone, but one that generates pulses of electricity (not simple waveforms) encoded in some way which the brain can decode. There is a similarity between this action and the analogue-to-digital audio signal processing which has come to the fore in recent years.

1.3.2 The propagation of sound

Robert Boyle demonstrated in 1660 that sound needs a medium such as air through which its vibrations can travel. He hung a ringing alarm watch inside a glass jar and, when most of the air had been pumped out of the jar, the ringing became inaudible. Other substances can transmit sounds even more efficiently (and faster) than air – notably water and metals. Incidentally, Boyle also discovered that electric attraction as well as light, unlike sound, can travel through a vacuum.

As early as the middle of the seventeenth century the French mathematician Mersenne measured the time it took for sounds to be reflected over a known distance, and thus calculated the speed of sound as 450 metres per second (1038 feet per second), remarkably close to the figure now known to be correct.

1.3.3 The sensation of sound

In 1681 Robert Hooke studied the relationship between the number of vibrations per second (frequency) and the resulting musical pitch. He used a spinning toothed wheel and held a card against the teeth while running the wheel at various speeds to produce sounds of definite pitch. When he computed the number of card strikes (vibrations per second) he found an exact correspondence between frequency and pitch – though later research has shown that things are not always so simple. Precise estimates of pitch can be affected by various factors, including the intensity of the sound reaching the listener. In 1801 the French mathematician Fourier laid the foundation for much detailed research into waveforms of all kinds when he showed that any sound waveform, however complex, can be reduced to a series of simple sine (single-frequency) waves.

To investigate the relationship between sound intensity (or power) and the perceived loudness, around 1830 Weber studied the scale of 'just noticeable differences' in loudness. Thirty years later Fechner produced an analysis of this intensity/loudness relationship, showing that it was not unlike the frequency/pitch relationship discovered by Hooke and others, and went on to establish similar scales of just noticeable differences for other stimuli and the resulting sensations of the other senses – lights of different brightness or colour, etc. He summed up his findings in what he called 'Weber's Law' in deference to his predecessor but which is now more often referred to as 'Fechner's Law'. This states that as stimuli are multiplied to greater magnitudes, the resulting sensations increase by addition. Thus each time sound intensity is doubled, its perceived loudness goes up in equal steps (the progression is logarithmic rather than linear, as will be explained later). Again, each time sound frequency is doubled, its perceived pitch goes up by a fixed musical interval (which we call an 'octave').

Our ability to identify the direction of a sound source using our ears alone was investigated by Lord Rayleigh. In 1876 he correctly noted that sound coming from one side of the head would reach the nearer ear first and that this sound would also be more intense than that in the more distant ear, at least at high frequencies, because the head casts a 'sound shadow'. This explained the sharper directional hearing acuity for higher frequency sounds but it was not until 1907 that Rayleigh established the phase (i.e. time-of-arrival) differences between sounds reaching the two ears as the mechanism for directional hearing at low frequencies. A third clue to sound-source direction is afforded by the difference in timbre (i.e. proportions of high and low frequencies) at the two ears, and it now seems clear that all three clues contribute to two-eared (binaural) directional location.

1.3.4 Room acoustics

Examination of early Greek and Roman open-air amphitheatres has uncovered a surprising practical knowledge of acoustic behaviour. In particular, numerous sound vessels have been uncovered throughout the seating area, showing that the early builders knew how to employ resonances to reinforce the sounds of the actors' voices (using the principles formulated by Helmholtz many centuries later). Also the shape and layout of the theatres was obviously designed to assist the projection of sounds from the stage to the audience. Much of this is described in a treatise called *The Acoustics of the Theatre*, written by the Roman architect Vitruvius around the year 25 BC.

The behaviour of sound waves in rooms and auditoria was studied in detail by W.C. Sabine in the 1890s. He investigated the build-up and decay (reverberation) of sounds in an over-reverberant lecture theatre at Harvard University. By using a stopwatch he was able to measure the reduction of decay time in seconds when increasing numbers of cushions were placed around the room. From this he developed a relationship between room volume, area of absorbent materials and the reverberation time in a formula which is still used today in planning auditorium acoustics.

1.4 History of discoveries in magnetism and electricity

A parallel pursuit of knowledge into the physics of magnetic effects and the generation and control of electric currents can be traced throughout history. The strange properties of the mineral lodestone (now known to be an oxide of iron) found near Magnesia in Asia Minor were known in ancient times. It had the power of attracting pieces of the same material or iron and would set in a North-seeking direction if suspended. Such a *magnet* can transfer its properties to pieces of steel: a magnetically charged needle, for instance, provides the basis of a navigator's compass. The ends of magnets (poles) were found to possess an opposite (North and South) polarity and it was discovered that unlike poles would attract one another while like poles repelled. There grew up a 'molecular theory' of magnetism, which suggested that all substances capable of being magnetized were made up of tiny particles, each of them a magnet. In the unmagnetized state these tiny molecular magnets are not arranged in any common direction. The act of magnetization consists of aligning the particles so that they act together, producing an accumulated magnetic 'field' in the vicinity.

In the same way, around 600 BC Thales found that pieces of amber (*elektron* in Greek) rubbed with wool could attract light objects, and were then said to be charged with *electricity*. Later, a different form of electrification was identified (for instance, if glass was rubbed with silk) and it was found that the two types of electrification annulled each other if brought together. Gilbert, in *De Magnette* (1600), distinguished between two types of substances, 'electrics' such as amber, glass and resin which attracted a suspended needle after being rubbed and 'non-electrics' such as copper and silver which did not. The former substances we now know as *insulators*, whereas the latter are *conductors* of electricity, and the fact that they allowed the charge to leak away explains why Gilbert's rubbing did not produce the attracting effect. Then in the middle of the eighteenth century Benjamin Franklin gave the names *positive* and *negative* to the two states of static electrical charge and the similarity to the two polarities of a magnet was recognized.

1.4.1 Electric current

In the 1740s the differences between electrical conductors and insulators had been established along with the idea of electricity as being capable of flowing, like a fluid. In 1786 Galvani found that he could make a frog's muscles twitch by touching them with pieces of metal. Then another Italian, Volta, showed that Galvani had in fact been applying an electric 'shock' to the frog muscles through his accidental forming of a junction between two dissimilar metals (copper and iron).

Volta went on to arrange metals in a series of increasing 'positiveness' in static electricity terms and found that, by applying acid at the junctions of a 'pile' of metal pairs, he could produce a lasting state of electrical charge and a continuous current, rather than a brief spark. This was the forerunner of the electric battery, leading to electric bells and other simple devices. Humphry Davy identified the connection between electricity and magnetism and by 1824 Oersted developed the galvanometer for measuring electric currents by means of the magnetic field set up. Ampére showed how this could be used to send messages along wires, deflections of the magnet needle being used to denote letters of the alphabet. This was later improved by Morse and others to operate a buzzer, and the electric telegraph soon spread in popularity.

Ohm reasoned that electricity flowed along a wire much as water flows in a pipe or heat is conducted, and so developed the idea of resistance to current flow. By studying the current to be expected in various wire types and thicknesses from the application of a given 'electromotive force' or 'voltage' he discovered the proportionality propounded in what we now call Ohm's Law:

Resistance = Voltage/Current or $R = V/I$

In the 1830s Faraday showed that current can be induced in a neighbouring conductor whenever a current changes in magnitude or either conductor is moved. He used the idea of lines of force in a medium (ether), the number of lines equalling the number of units of magnetism, and the number of lines cutting through an adjacent conductor determining the magnitude of the current induced. In the event, the inventors moved ahead of the theorists. They soon constructed 'dynamos' (machines which used motion to generate electricity) and the reverse concept, 'motors' (machines which used electric currents to produce motion).

1.4.2 Electromagnetic waves

By the 1860s Maxwell had put together theories to embrace both the electrostatic fields associated with electrically charged bodies and the magnetic fields surrounding current-carrying conductors. He showed that the velocities of propagation of electrical and magnetic disturbances were the same, and were identical to the speed of light. Indeed, it was now realized that light is an electromagnetic disturbance; whether the resulting disturbance waves are visible to the eye depends only on their repetition rate (frequency) falling within that part of the electromagnetic spectrum to which the eye is sensitive.

There then followed the concept of electricity being made up of small (negatively charged) particles to be called 'electrons' (discovered by J.J.Thomson in 1897). A constant flow of electrons along a conductor constituted a current; acceleration or deceleration of electrons caused radiation, and orbital circulation of electrons round positive charges (later established as the model for the atoms of which all matter is made up) gave rise to magnetism.

Nearly 30 years were to pass after Maxwell's studies on electromagnetic radiation before Hertz set up a system capable of generating and receiving his (relatively low-frequency) electromagnetic waves in 1888 – the forerunners of radio or 'wireless' communication. What was needed to enhance the usefulness of telegraphic and radio communication beyond the mere transmission of on/off Morse code clicks was an ability to relay speech and music, and that is where the microphone came in.

1.5 History of the microphone

The word 'microphone' first appeared in 1827 in Wheatstone's description of an acoustic device and was later used by Berliner in

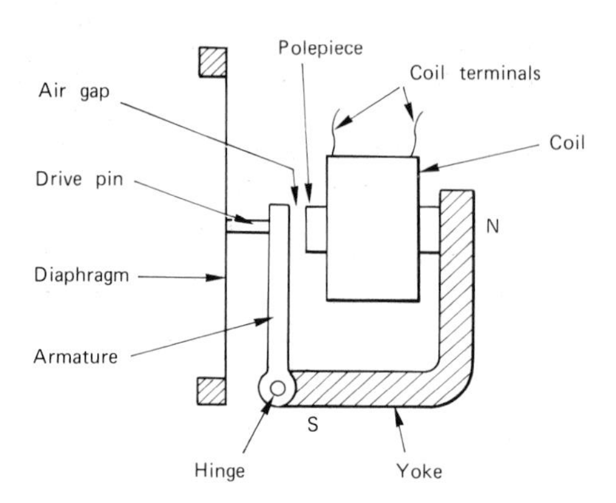

Figure 1.2 Outline diagram of Bell's moving-armature transducer (from the 1876 patent)

1877 and Hughes in 1888 for their 'loose-contact' transducers (later called 'carbon microphones').

Alexander Graham Bell is generally credited with the invention of the first workable microphone (and loudspeaker/earpiece). In 1876, following work on a 'harmonic telegraph' which allowed several Morse code messages to be sent over the same line by transmitting the dots and dashes on notes at different frequencies (musical pitches) to which the receiving device could be selectively tuned, he filed his famous 1876 patent for a 'speaking telegraph' (later to be called the *telephone*). The transducer he described (Figure 1.2) was reversible and so one was to be used at each end of the telegraph wire. It had a battery-energized electromagnet as the source of power with a separate armature connected to a thin diaphragm. When the device was used as a microphone (transmitter) the vibrations of the diaphragm in response to speech air-waves generated an equivalent alternating electric current in the coil. When it was a loudspeaker (receiver) the incoming current in the coil set up a magnetic field which alternately attracted and repelled the armature, causing the diaphragm to vibrate at the original speech frequencies and radiate sound waves.

In the event, Bell's original design proved to be too insensitive but, within a year, he had produced an improved magnetic model which can be seen as the basis for telephone earpieces (though not mouthpieces) for the next three-quarters of a century. Other inventors built and patented alternative transducers designs during 1877, notably Edison and Berliner. The latter's 'loose contact' or 'carbon' microphone is still the form of transducer used for most telephones today. Here the principle is non-magnetic, the diaphragm being backed by a cavity filled with conducting granules (originally metal but later carbon). Vibrations of the diaphragm alternately press and release the granules, thereby varying their resistance and modulating the speech signals onto the direct current (DC) polarizing current supplied by a battery. The addition of a step-up transformer permits very long-distance communication.

Moving-coil microphones, as distinct from Bell's moving-armature model, were described by Siemens and others as early as 1878 but had to await the work of Wente and Thoras in 1931 to become practical and wide-ranging in terms of frequency coverage. Here a circular coil of wire is fixed to the diaphragm and suspended in the air gap between the polepieces of a specially shaped magnet. As the diaphragm responds to incident sound waves, the coil is driven back and forth in the gap and has an equivalent alternating current (AC) generated in it. This design has the distinction in its reverse-action form of providing the basis for the great majority of loudspeakers from the 1930s to today.

A fourth type of transducer, originally proposed by Edison but eventually introduced around 1917 by Wente, relies on electrostatic charges. Basically, it consists of a thin conductive diaphragm separated by a narrow air gap from a solid backplate to form a capacitor. A polarizing voltage is connected across the two plates through a very high resistance so that a quasi-constant electrical charge is established. Motion of the diaphragm causes the charge-carrying ability (capacitance) of the structure to alternate above and below its stationary value and the resulting voltage gives rise to the desired AC signals. A recent variation on this electrostatic or capacitor microphone scheme is the 'electret' microphone, first proposed by Bruno around 1939. This avoids the need for a source of DC polarization by using an electrostatically precharged material (analogous to the magnetically charged state of a 'permanent' magnet) for either the diaphragm or the backplate.

Use of the so-called piezoelectric effect in transducers is a relatively recent development, dating from work on Rochelle-salt crystals by Sawyer in 1931. His 'bimorph' arrangement consists of two slices of piezoelectric material (having the property of developing an electric voltage across the two faces if twisted or displaced) suitably connected to a diaphragm. Once again, diaphragm movement generates the desired AC output without the need for a separate polarizing battery.

Figure 1.3 The first variable-directivity microphone with built-in valve amplifier (Neumann M49, 1953)

A powerful incentive for the development of improved microphones, with less self-generated noise and a wider frequency range to permit the transmission of music as well as restricted speech sounds, arrived with the spread of radio broadcasting. Early transmissions such as Fessenden's in 1900, Marconi's first transatlantic experiments in 1901 and an amazing first broadcast from the Metropolitan Opera with Caruso and Destin in 1910, were indeed primitive in sound-quality terms. However, the invention of vacuum tubes or 'valves' and other refinements led to regular radio broadcasting from about 1920 onwards and intensified microphone research.

Notable broadcasting microphones included the Western Electric double-button carbon and condenser types from 1921, the Kellog double-button carbon types from 1923 and the RCA condenser from 1924. Standards improved from about 1931, when Western Electric introduced their first 600-series moving-coils followed in 1934 by their 77-series, using a divided ribbon to produce the first unidirectional or 'cardioid' microphone, RCA their first ribbon 'velocity' and Brush their Rochelle-salt crystal.

The cardioid directional characteristic proved to be so popular that other designs soon followed. The Western Electric 639A (1939) for example, combined a ribbon and a moving coil in a single case; the Shure Unidyne (1941) used a single moving-coil capsule with an acoustic delay chamber behind the diaphragm to produce the cardioid response. The Neumann M49 (1953) was a double-diaphragm condenser (see Figure 1.3), effectively two cardioids back

Figure 1.4 Prior to 1925 sound recording used an acoustic horn instead of a microphone (EMI photo showing Sir Edward Elgar conducting 'The Symphony Orchestra' in 1915)

to back with a built-in valve amplifier and a potentiometer giving a continuously variable directivity pattern.

Sound recording in the pre-radio valve days had no use for microphones (see Figure 1.4). The sounds were simply picked up acoustically by a large horn having a diaphragm at the apex to which the disc-cutting stylus was attached. The vibratory energy thus received was enough to inscribe the waveform as a spiralling groove on the rotating soft-wax master disc. From 1925, however, the microphones and valve amplifiers developed for radio were introduced to recording studios, giving much greater freedom of choice with regard to layout and balance of the musicians. The electrical signals from the microphone could be passed along wires without serious losses on the way and amplified to give more efficient, and less noisy, disc cutting.

The first electrically recorded gramophone records appeared in 1925 and were initially played on the existing acoustic gramophones. Soon, however, electromagnetic pick-ups were developed whose output signals could be amplified and fed to loudspeakers or headphones. The chain from microphone to final playback was complete.

2
Basic acoustics

A working knowledge of acoustics can lead to a greater understanding of how different microphones operate, how speech and musical sounds are generated and radiated, how best to position microphones and performers for correct musical balance and how to evaluate the sound quality achieved.

2.1 A maths refresher

The use of mathematics has been kept to a minimum throughout this book but a few mathematical ideas have refused to go away. Therefore to save later confusion, the essentials are summarized below in a form which should provide a refresher course for readers who have forgotten their school maths yet not unduly hold up the more recent school graduates who already know it all.

2.1.1 Arithmetic

Perhaps because we are born with ten fingers, we have grown up with a so-called decimal system of numbers. Thus we count objects in units from 0 to nine and then, when we need to express numbers greater than nine, we start with a one again moved one place to the left and followed by 0 to give 10, and so on going from 19 to 20, 99 to 100, etc., e.g.

> The number 385 is made up of three hundreds plus eight tens plus five units

The decimal point separates 'whole' numbers from fractions, which also proceed in steps or divisions of 10, e.g.

> The number 87.46 is made up of eight tens plus seven units plus four tenths plus six hundredths of a unit

A more scientific notation allows numbers which are multiples or submultiples of 10 to be written in shorthand using a smaller 'exponent' or 'index' number after the 10, e.g.

Table 2.1 Multiples and submultiples of units

Factor	*Prefix*	*Symbol*
10^9	giga	G
10^6	mega	M
10^3	kilo	k
10^{-2}	centi	c
10^{-3}	milli	m
10^{-6}	micro	μ
10^{-9}	nano	n
10^{-12}	pico	p

10^2 ('ten squared') means (10×10) or 100
10^3 ('ten cubed') means ($10 \times 10 \times 10$) or 1000
10^{-2} means 0.01
$\sqrt{10}$ ('the square root of 10') equals 3.16 approximately.

This use of indices can be applied to any number: e.g.

$3^2 = 3 \times 3 = 9$
$3^3 = 3 \times 3 \times 3 = 27$

By an extension of this idea, any large number can be rewritten to include a power of 10, e.g.

4 750 000 can be written 4.75×10^6
$0.0035 = 3.5 \times 10^{-3}$

It has become a common practice to derive larger and smaller units for most quantities by adding prefixes which relate to powers of 10. For example, 1 *kilo*metre = 1000 metres, and 1 *milli*metre = one thousandth of a metre. Table 2.1 lists some commonly met multiples and submultiples, and it will be seen that they mainly move in steps of 10^3. The Greek letter μ (mu) is used as the symbol for 10^{-6}. For convenience, Table 2.2 lists the Greek letters which appear in this book.

Table 2.2 Commonly used Greek letters

Greek letter	*Pronunciation*	*Used for*
θ	theta	Angle in degrees
λ	lambda	Wavelength
μ	mu	One millionth (10^{-6})
π	pi	Ratio of circumference of a circle to its diameter (3.14 approx.)
Ω	omega	Ohms

2.1.2 Logarithms

We have seen that $100 = 10^2$, i.e. 'the power to which 10 must be raised to produce 100 is 2'. Look-up tables and pocket calculators with a log function exist which give the power of 10 for any number. These are called 'logarithms', written $\log N = X$, where N is the number and X is the index of 10 which gives the number, e.g.

$\log 1000 = 3$ (since $1000 = 10^3$)

Table 2.3 Examples of logarithms

Number	*Log_{10}*	*Number*	*Log_{10}*
1	0.00	10	1.00
2	0.30	20	1.30
3	0.48	30	1.48
4	0.60	100	2.00
5	0.70	200	2.30
6	0.78	300	2.48
7	0.85	1 000	3.00
8	0.90	2 000	3.30
9	0.95	3 000	3.48
10	1.00	10 000	4.00

Table 2.3 lists some logarithms rounded up to just two decimal places as a brief guide.

A particular convenience of logarithms is that they replace multiplication and division with the simpler arithmetic operations of addition and subtraction, e.g. $3 \times 2000 = 6000$ can be calculated from the table of logarithms as:

$\log 3 + \log 2000 = 0.48 + 3.30 = 3.78$

and the number whose log is 3.78 = 6000.

A logarithmic scale in which equal steps correspond to successive multiples of 10 (or 2) also has the advantage that it accords with the way that the human senses estimate changes in sound intensity, musical pitch, light intensity, etc. As we shall see in Section 2.5, for example, successive doubling of the frequency of a musical tone produces the sensation of equal (octave) steps in musical pitch. Similarly, a scale of equal changes in perceived loudness is found to comprise successive multiplication of the sound intensity by a common factor (see Section 2.7).

2.1.3 Angles and trigonometry

The ratio of the length of the circumference C of a circle to its diameter D is a fixed number, usually represented by the Greek letter π ('pi') (see Figure 2.1(a)), i.e.

$$C/D = \pi \text{ (approximately 3.14 or 22/7)}$$

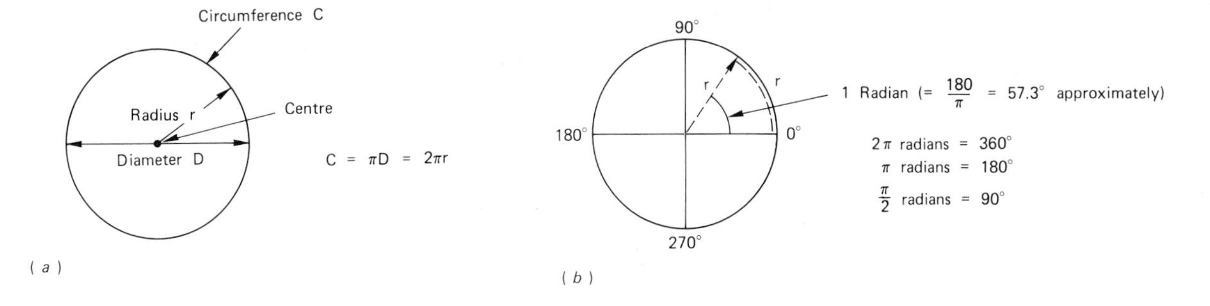

Figure 2.1 (a) The circle defining π as the ratio of circumference to diameter; (b) a circle can be divided into 360° or 2π radians

Similarly, since the radius r is half the diameter, we can write $C = 2\pi r$. Other useful formulae are:

area of a circle $= \pi r^2$
surface area of a sphere $= 4\pi r^2$
volume of a sphere $= \frac{4}{3}\pi r^3$

The circle may be divided into 360° (degrees) (see Figure 2.1(b)), i.e. four quadrants of 90° each. An angle of 90° is called a right angle. Alternatively, the circle may be divided into larger angular units called radians, where 1 radian is the angle subtended at the centre by an arc of length r. Since $C = 2\pi r$, it follows that:

$$2\pi \text{ radians} = 360°$$

The three angles of a triangle add up to 180°. Therefore a right-angled triangle is a special case in which its two remaining angles add

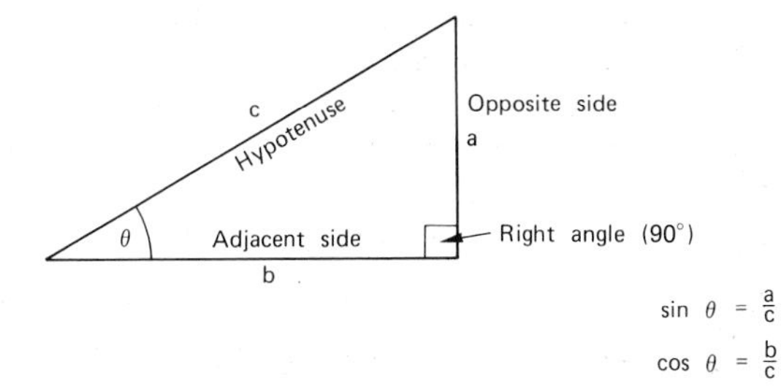

Figure 2.2 The trigonometrical ratios sine and cosine

Table 2.4 Sine and cosine of some useful angles

Angle (θ)	*Sin θ*	*Cos θ*
0°	0	1.0
30°	0.5	0.87
45°	0.7	0.7
60°	0.87	0.5
90°	1.0	0
180°	0	-1.0

up to 90°. The side opposite the right angle is called the hypotenuse. For the angle θ in Figure 2.2 the two most important trigonometrical ratios are the sine and the cosine, where:

$$\sin\theta = \frac{\text{opposite side}}{\text{hypotenuse}} \text{ i.e. } \frac{a}{c}$$

$$\cos\theta = \frac{\text{adjacent side}}{\text{hypotenuse}} \text{ i.e. } \frac{b}{c}$$

Tables exist for all trigonometrical ratios, and the values in Table 2.4 are worth remembering.

2.1.4 Periodic motion

A rotating wheel and the swings of a pendulum are examples of systems performing periodic motion in which the same action is continually repeated. Each repeated action is called a cycle and the time taken for each cycle is one period. The number of cycles performed per second is called the 'frequency', which is expressed in cycles per second (c/s) or, in modern parlance, Hertz (Hz).

Thus in Figure 2.3 the point P is rotating at a uniform speed around the circle at a fixed distance *r* from the centre O. During each revolution or cycle the angle θ will take up all values from 0° to 360°. It is possible to plot a graph of the height PX against the angle of rotation . Starting at A, where θ = 0°, PX is also 0 and may be plotted as the point A′. When P is at the point shown, the angle is θ and P may be plotted as P′ and so on, giving plotted points B′, C′, D′ and

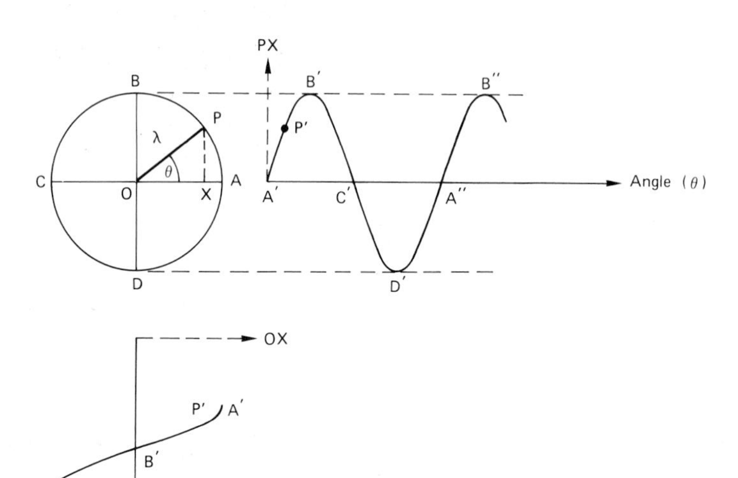

Figure 2.3 Simple harmonic motion: during each cycle of rotation of the point P, the length of the projected line PX oscillates between maximum values above and below the zero line. Plotting PX against the angle θ produces a sine wave. Similarly plotting OX against θ produces a cosine wave

A′′ when $\theta = 90°$, $180°$, $270°$ and $360°$, respectively. Thus the plotted curve A′B′C′D′A′′ represents one cycle of rotation.

Note that $\sin \theta = PX/r$ and so the curve plotted is in fact a graph of $\sin \theta$ against θ and is called a sine wave. Plotting values of OX against θ as in the lower part of Figure 2.3 produces a cosine wave which is exactly the same shape as the sine wave but begins at maximum (since $\cos 0° = 1$) and its cycle ends at A′′. The motion of point X as it oscillates about O between the extremes A and C is called simple harmonic motion (SHM). Note that this motion is typical of the swings of a pendulum or the prongs of a tuning fork.

2.2 How sounds originate

All sounds begin with oscillatory movement of the source. They travel outwards through the air or other media as successive imitative oscillations of layers of particles along the way. They are detected as sympathetic movements of any light structure encountered by the sound waves, such as the eardrum of the listener or the diaphragm of a microphone.

A single shock wave would result from an isolated movement of the source (for example, from a hand-clap). A common pictorial analogy is that of dropping a stone into a still pond when a wave or disturbance is seen to spread outwards in an ever-increasing circle (Figure 2.4(a)). The stone initiated this wave by pushing down a

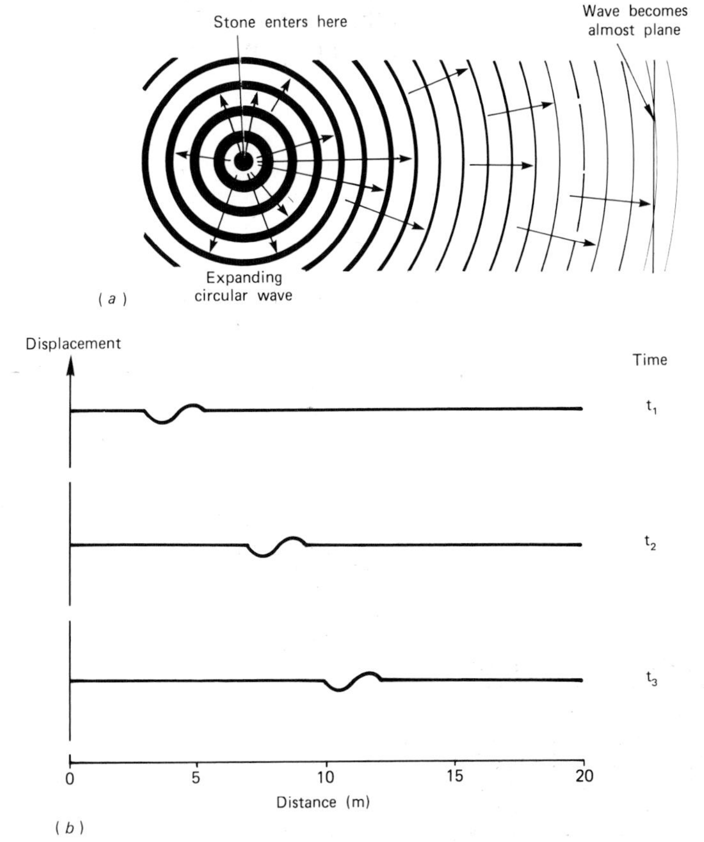

Figure 2.4 Transverse waves on the surface of a pond. (a) The expanding circular wavefront weakens with distance and eventually approximates to a plane wave; (b) a shock wave moves outwards with time

number of elementary water particles. Then, since all particles of a material are linked by elastic restoring forces, this downward motion was progressively relayed outwards through the water (see Figure 2.4(b)). Generally this original wavefront will be followed by a number of up-and-down swings in water level of diminishing amplitude as the gravitational restoring force first causes the depressed particles to return upwards, overshoot the rest position, fall back and eventually reach equilibrium. To create the pond surface analogy of a continuous musical note, it would be possible to plunge a piece of wood up and down in the water at some regular rate when a continuous succession of waves would occur. Thus a cork floating on the water at some distance from the source would be seen to bob up and down in imitation of the source movements.

Already this simple analogy has illustrated several characteristics of sound origination and propagation:

1. The fact that the shock wave travelled outwards in an ever-increasing circle demonstrated that the wave velocity (distance travelled per second) was constant in all directions. Indeed, it would be possible to determine the fixed velocity of the wave, using a stopwatch to measure the time taken to travel a known distance;
2. It also showed that, at least for this simple (small) source, the wave energy was radiated out with equal force in all directions;
3. The fact that the cork moved up and down on the spot proved that it was not the water itself which was moving outwards but only the disturbance;
4. Given a large enough pond, the wave amplitude would be seen to diminish with distance. This dissipation of the wave energy has two causes. First, some of the original energy is expended or absorbed along the way due to frictional losses (converting the mechanical energy into thermal energy). Second, the energy is continually being spread over an expanding wavefront and so exerts less force on individual water particles.

The picture would become confused in a small pond, or where large obstacles existed, because reflections would occur and secondary waves would produce interference patterns. Similarly, using a large source such as a flat plank would produce non-circular (directional) wavefronts.

2.3 Transverse and longitudinal waves

Helpful as they are, pond wave analogies fail to represent a true picture of sound waves since the water surface movements are transverse, i.e. at right angles to the direction of wave propagation. Sound waves in air are said to be longitudinal, i.e. the particles vibrate to and fro in line with the wave direction. Also sound waves naturally travel outwards in three dimensions, so that the wavefront in this simple case of a small 'point' source is a continuously expanding sphere.

In reality, the molecules of a gas such as air are continually in a state of rapid random motion, thus exerting pressure on any object immersed in it. The particles also possess mass and so are attracted

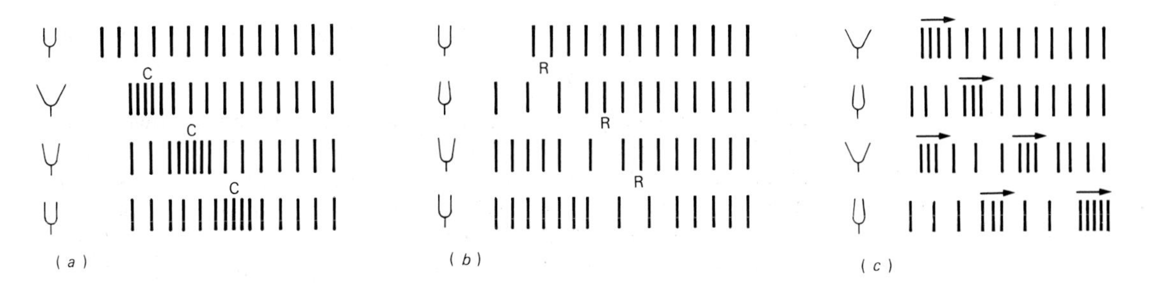

Figure 2.5 Longitudinal sound waves radiated by a tuning fork. (a) When the prong swings to the right, a compression wave C travels outwards; (b) a swing back to the left similarly produces a rarefaction wave R; (c) continuous motion of the tuning fork generates an alternating pressure wave

by the earth's gravitational pull. The resulting barometric pressure at sea level is about 100 000 Pascals (1 Pa = 1 newton per square metre) or 14.7 lb/in^2. When the right-hand prong of the tuning fork shown in Figure 2.5(a) swings to the right, the displacement of adjacent air particles will produce a crowding together (compression) which will travel outwards as shown. When the prong swings to the left (Figure 2.5(b)) there is an opening out of the air particles (rarefaction). As the tuning fork continues to vibrate (Figure 2.5(c)) – performing SHM as described in Section 2.1.4 – the alternating compressions and rarefactions form a moving pattern in which the air pressure will oscillate above and below the steady barometric value at the same frequency and following the same sine law as the tuning fork. It is this pressure wave to which eardrums and microphone diaphragms are sensitive. The swings in air pressure corresponding to audible signals are almost incredibly small. The quietest sound detectable by average ears represents a pressure of only 0.00002 Pa (20 μPa) and is associated with a particle amplitude of just one tenth of the diameter of a hydrogen molecule. This SHM is radiated through the air as a progressive longitudinal sound wave. There is naturally an incremental time delay in the vibrations of successive layers of air particles with distance measured out from the source.

To represent the longitudinal wave of Figure 2.5 as a sine curve, the individual displacements of air particles can be drawn as distances above and below the datum line to represent displacements to right and left, respectively (see Figure 2.6(a)). This produces a sine wave similar to that in Figure 2.3 and a clearer display of the sinusoidal variation of particle displacement is obtained in Figure 2.6(b) simply by increasing the vertical scale.

Of course, each cycle of displacement values is repeated along the direction of wave propagation and the distance occupied by one cycle (e.g. from crest to crest) is called a *wavelength*, λ. The point which a particle or wave has reached in its cycle is called its phase. Another definition of wavelength is therefore 'the distance between successive points in a wave which are in phase'. Phase differences are stated in degrees with reference to the 360° cycle, so that a crest, for example, is 180° out of phase with the next trough.

Particles at A, L and M are momentarily at their normal rest position, but it will be seen that L is at a point of compression (highest pressure) because air particles on each side of L are displaced towards it. By contrast, A and M are points of rarefaction, and it is possible to deduce from this the curve of Figure 2.6(c) which represents the sinusoidal rise and fall along the wave of air pressure

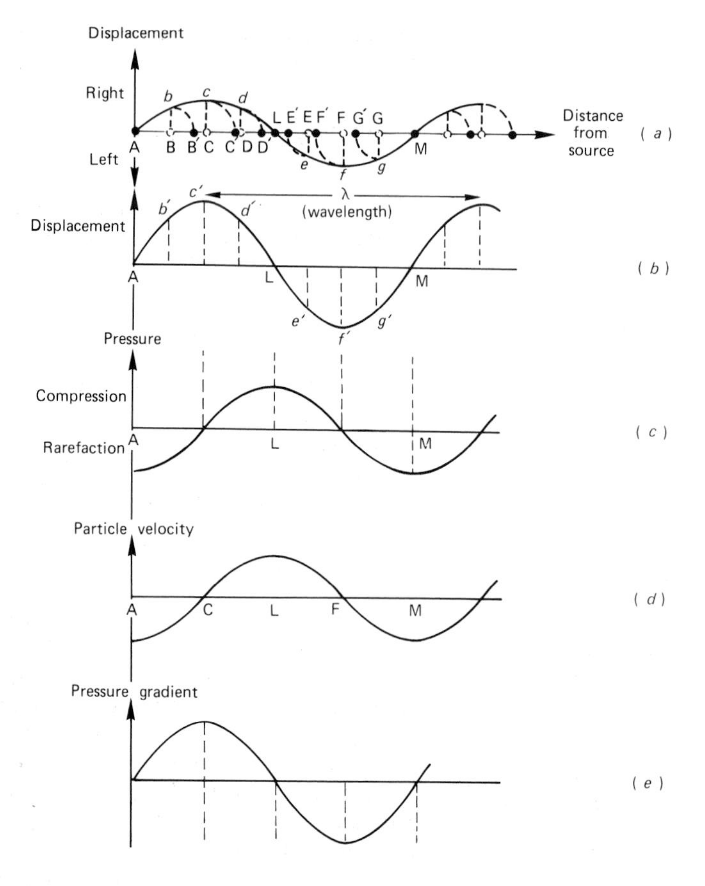

Figure 2.6 Representation of a longitudinal wave as a transverse wave. (a) Right and left particle displacements are drawn as distances above and below the zero line; (b) the vertical scale is increased for clarity and the wavelength is shown as the distance between successive points which are in phase (e.g. crests of the wave); (c) pressure variation; (d) particle velocity variation; (e) pressure gradient variation

with respect to normal atmospheric pressure. Note that displacement and pressure are effectively 90° out of phase.

Similarly, by inspection we see that the velocity of individual particles varies sinusoidally. It is zero at points of maximum displacement C and F, and a maximum at points of zero displacement A, L and M, giving the curve shown in Figure 2.6(d). Particle velocity is therefore in phase with pressure. In each cycle, the particle moves from its rest position out to the right, back out to the left and then to the centre. Calling the maximum displacement or amplitude a, the particle travels a distance $4a$ in each cycle. Its average velocity, i.e. the distance travelled per second, is therefore given by:

$$v = 4af$$

where v = distance travelled per second,
a = amplitude, and
f = frequency.

It follows that in a *constant-velocity* system amplitude is inversely proportional to frequency, and in a *constant-amplitude* system velocity is proportional to frequency. It also appears that the rate of change of pressure, or the pressure gradient, is in phase with the displacement (see Figure 2.6(e)). We may note in passing that pressure gradient is related to the steepness of the slope, which

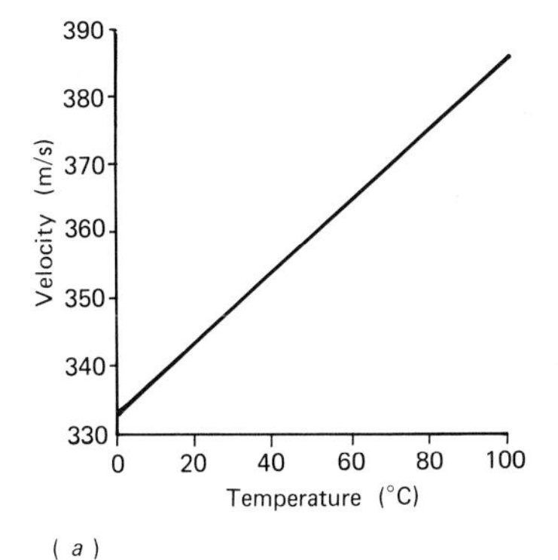

(a)

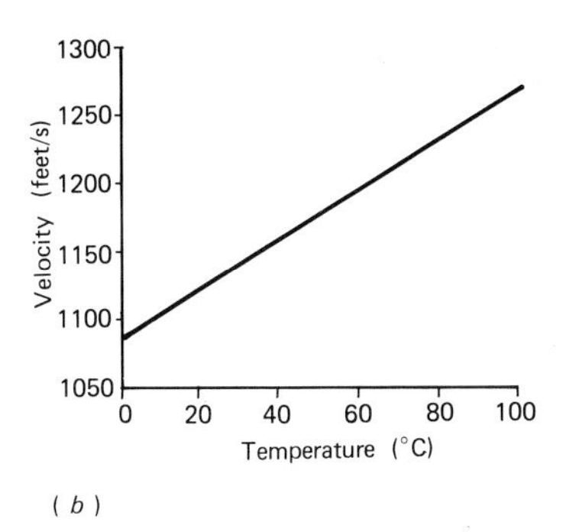

(b)

Figure 2.7 Velocity of sound in air as a function of temperature: (a) metres per second; (b) feet per second

would, for example, be halved if the wavelength were doubled, and so we can state that pressure gradient is proportional to frequency.

2.4 Velocity of sound

The velocity c at which sound waves travel through any medium is determined by the medium's elasticity and density. For dry air at 0°C the accepted figure is:

$$c = 331.45\ \text{m/s}$$
$$\text{or } c = 1087.42\ \text{ft/s}$$

This is independent of pressure, so that sound travels at the same speed at any altitude, everything else being equal. However, c is dependent on both temperature and humidity. The rise in velocity with temperature is shown in Figures 2.7(a) and 2.7(b), which are derived from the expressions:

$$c = 331.45\ \sqrt{(1 + t/273)}\ \text{m/s}$$
$$\text{and } c = 1087.42\ \sqrt{(1 + t/273)}\ \text{ft/s}$$

where t is the temperature in degrees Celsius.

The presence of moisture has the effect of reducing air density (the opposite of what we might have expected) and therefore causes an increase in sound velocity. Humidity also has an effect on the absorption of sound energy in air. The fact that molecular turbulence (air friction) reduces the intensity in a sound wave has already been mentioned, along with the steady fall in intensity (by 3 dB for each doubling of distance) due to continual expansion of the area of the wavefront. Compared with dry air at 20°C, increasing humidity has very little effect on low-frequency sounds but progressively increases the absorption at frequencies from 2 kHz upwards, particularly in the common region between 10% and 40% RH. These variations in sound velocity and absorption with climatic conditions explain many of the problems met in tuning musical instruments, microphone balance and sound-reinforcement applications to be discussed later. There is also, of course, the added complication that strings and the bodies of instruments physically expand as the temperature increases, causing further problems with tuning.

2.5 Frequency and musical pitch

The number of cycles per second (written Hertz or Hz) performed by a vibrating structure or wave is called its frequency f. Defining the wave velocity c as the distance travelled per second, it follows that, since the wavelength λ is the distance betwen successive crests in the wave:

$$c = f \times \lambda$$

The range of frequencies to which the human ear is sensitive is usually taken to be 20–20000 Hz. Therefore based on a nominal value of c at 21°C = 344 m/s (or 1129 ft/s), Figure 2.8 shows the variation of wavelength with frequency over the audible range. The figure also shows typical frequency ranges for various sound media.

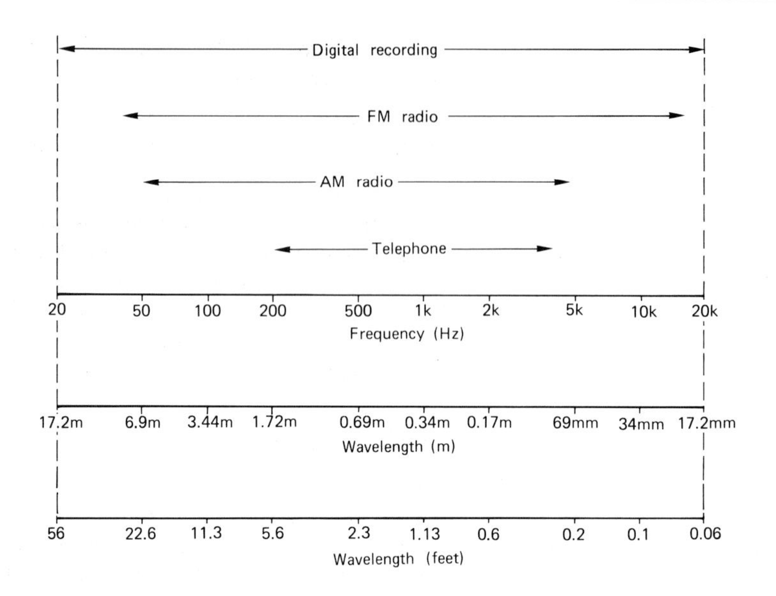

Figure 2.8 Wavelength in air for sounds at frequencies between 20 Hz and 20 kHz (based on c = 344 m/s) and showing the typical bandwidths of various sound media

The frequency of a sound is directly related to its pitch on the musical scale. The standards organizations have set International Concert Pitch by making 440 Hz the recommended frequency for middle A (in the treble clef). Musicians through the centuries have chosen different tunings, and there are national differences even today despite the trends towards worldwide standardization brought about by broadcasting and recording.

Figure 2.9 Frequency and musical pitch: the range of fundamental frequencies and overtones or harmonics for various instruments and voices

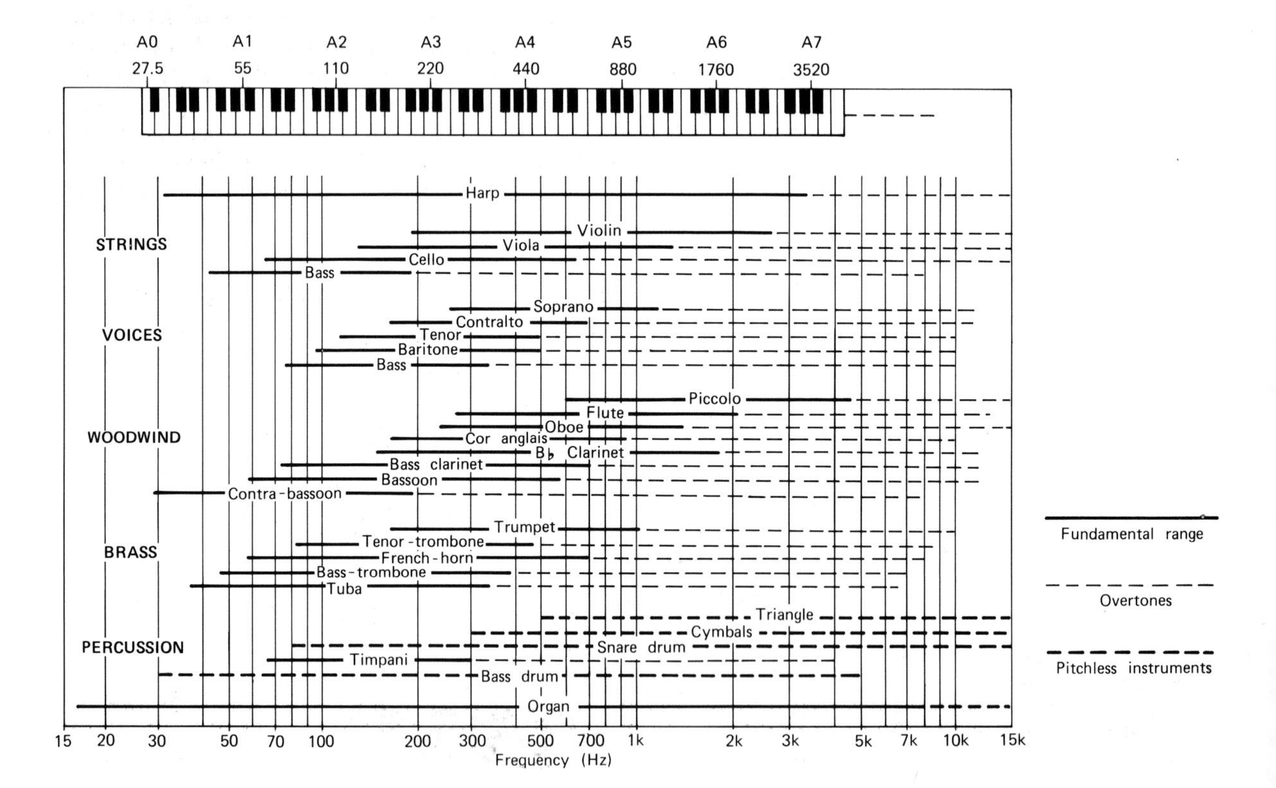

The octaves of A encompassed by the notes on a piano form a series of frequencies as follows (shown also in Figure 2.9):

A0	A1	A2	A3	A4	A5	A6	A7
27.5	55	110	220	440	880	1760	3520

Note that the frequencies for this series of 'equal' pitch intervals are in geometric progression with a common ratio of 2 (moving up one octave in pitch always means doubling the frequency). A logarithmic scale (with the common ratio of 10) is therefore preferred in graphs and response curves throughout the field of audio engineering since this accords equal importance to octaves (and other pitch intervals) in the same way as the human ear does.

2.6 Harmonics

Most sound sources do not in fact vibrate in just one mode, or radiate sounds at a single frequency. Several modes of vibration tend to be set up simultaneously, generating a plurality of sound frequencies. The tuning fork is an exception, and radiates a pure sine wave at the particular frequency fixed by its exact weight and shape. This is why a tuning fork makes such a reliable reference to which other instruments can be tuned. Complex structures such as church bells may have one dominating vibrational mode which gives the bell its recognizable pitch on the musical scale, but this is accompanied by almost randomly unrelated overtones.

The vibrating strings and air columns of the majority of musical instruments tend to generate a family of related frequencies (see Figure 2.10). The lowest frequency is caused by vibrations of the whole string or air-column length. This is called the *fundamental frequency* and fixes the musical pitch of the complex note. Partials or overtones are simultaneously generated as shown due to break-up of the vibrating system into shorter lengths. These overtones are called *harmonics* and form a series of simple multiples of the fundamental frequency. The fixed or stationary points in these

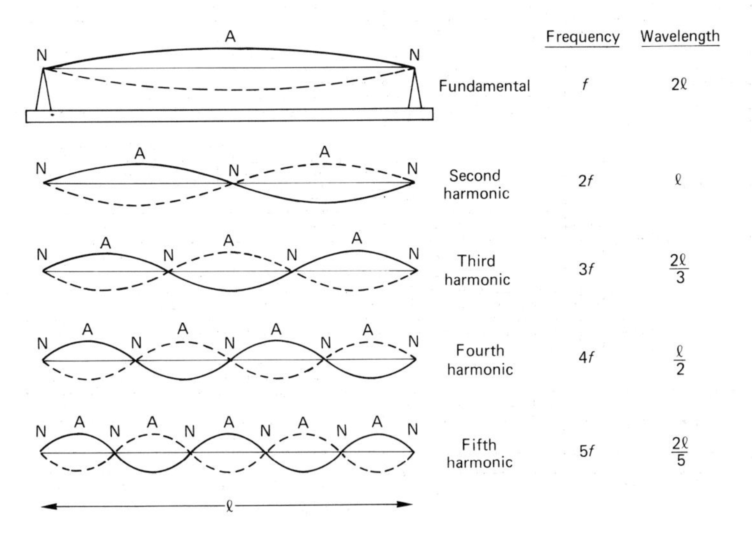

Figure 2.10 Harmonics: how a vibrating string simultaneously generates a family of frequencies which are simple multiples of the fundamental frequency

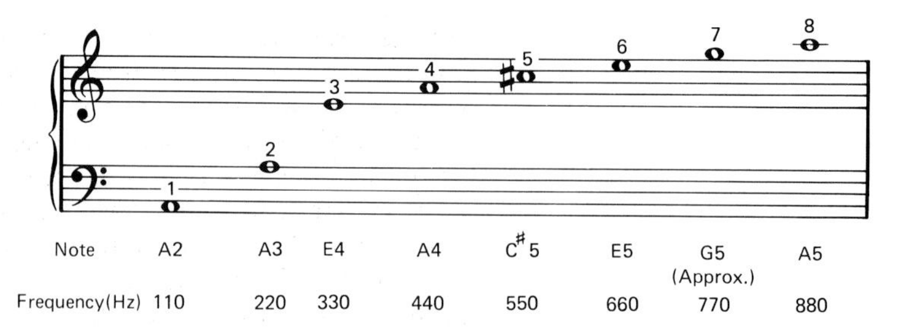

Figure 2.11 The first eight harmonics of the note A2

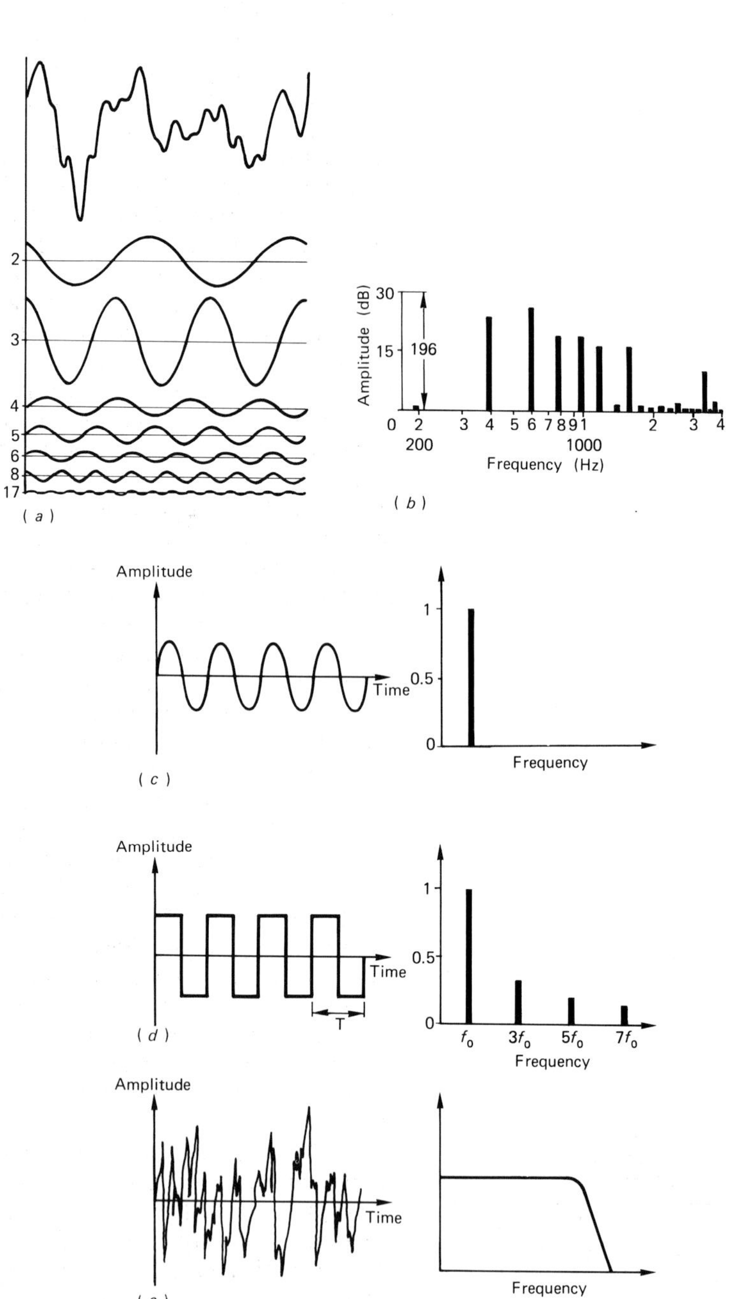

Figure 2.12 Waveforms and spectra. (a) The complex waveform of a violin playing the note G3, with the sine-waves corresponding to the various harmonic frequencies; (b) the equivalent spectrum or frequency distribution in terms of relative amplitudes (after Seashore); also the waveforms and spectra for (c) sinewave, (d) square wave and (e) random noise

standing wave patterns are called *nodes* and the points of maximum amplitude are *antinodes*. It will be seen that the number of antinodes or loops defines the harmonic number. The quantity and relative strengths of the harmonics produced by a given instrument largely determine its characteristic timbre, enabling us to distinguish a violin from an oboe, for example, even when they are playing the same note. Figure 2.11 shows in musical notation the first eight harmonics of the note A2. In practice, the seventh and higher odd-numbered harmonics do not fall exactly on notes of the scale and tend to introduce a dissonant effect.

The way in which the presence of individual harmonics can influence tonal quality was summarized by Jeans, who suggested that the second harmonic (the octave) adds clearness and brilliance but nothing else; the third harmonic adds brilliance plus a certain hollow or nasal quality (e.g. clarinet); the fourth harmonic adds yet more brilliance, even shrillness; the fifth harmonic adds a rich horn-like quality; the sixth harmonic adds a delicate nasal shrillness; the seventh, ninth and higher odd harmonics introduce roughness, harshness or dissonance (see below).

The waveform of a sound containing more than one frequency is arrived at by simple addition of the sinusoidal waveforms of the separate frequencies. Figure 2.12 illustrates the complex waveform of a violin playing the note G below middle C (196 Hz) and indicates the analysis of this into its separate harmonic frequencies: (a) as sinewaves and (b) as a spectrum (frequency plot of relative amplitudes). The waveforms and spectra corresponding to (c) a sinewave, (d) a square wave (which contains only odd-numbered harmonics) and (e) random noise are also shown.

When two musical notes are present together, the effect may be to various degrees pleasant (consonant) or unpleasant (dissonant). This effect is naturally subjective but its origin is found to depend on the arithmetic simplicity of the ratio of the two fundamental frequencies. Thus if the two frequencies are identical, i.e. in the ratio 1:1 (unison), there is perfect consonance and the waves will either reinforce or partially cancel each other depending on the relative phase between the waves reaching the ear of the listener. The next most consonant interval is the octave, frequency ratio 2:1, then the perfect fifth (3:2), the perfect fourth (4:3) and the major third (5:4).

Part of the reason why two frequencies not having a simple ratio produce dissonant effects is that non-linearity in the ear's response creates sum and difference frequencies which are either inharmonic or introduce annoying beats – the throbbing effect (at the difference frequency) heard when two notes of almost the same frequency are sounded together. In practice, when musical notes are sounding together, each comprising a fundamental frequency plus harmonics, it is the coincidence or otherwise of the harmonic frequencies which largely determines the degree of consonance or dissonance (see Figure 2.13).

The major and minor scales of Western music have been built up from the most consonant intervals, as shown in Table 2.5. This simple tuning forms what is called the Just Diatonic scale and produces maximum sonority when pairs of notes or chords are sounded. However, this natural tuning suffers from the disadvantage that modulation between keys is restricted. Since about the mid-1700s,

Table 2.5 Frequency ratios for the Just Diatonic scale

C	D	E	F	G	A	B	C
1	$\frac{9}{8}$	$\frac{5}{4}$	$\frac{4}{3}$	$\frac{3}{2}$	$\frac{5}{3}$	$\frac{15}{8}$	2

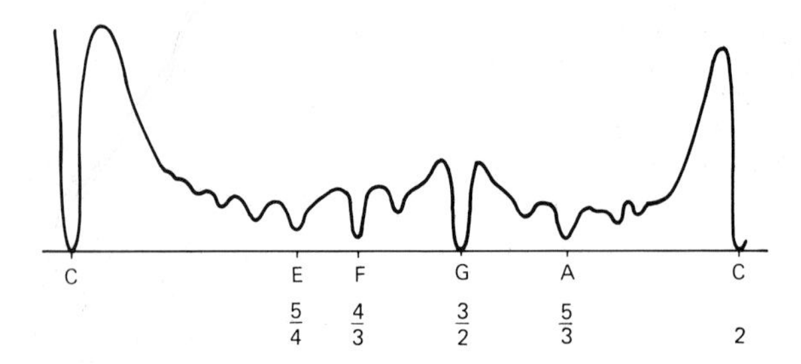

Figure 2.13 Dissonance: the degree of dissonance produced when various notes within the octave co-exist with the note C (after Helmholtz)

therefore, when free key modulations became of paramount importance, keyboard instruments have generally been tuned in what is called Equal Temperament. This involves slight mistuning of all notes other than the octaves, with a consequent loss in sonority, to divide the octave into 12 equal intervals called tempered semitones. The frequency ratio for each semitone is theoretically $\sqrt[12]{2}$ or 1.0595, but piano tuners work to a routine of counting beats to establish the correct spacings between pairs of notes within each octave.

When singers or string players perform without the accompaniment of keyboard or other fixed-tuning instruments they may consciously or unconsciously revert to using the natural pitch intervals, with a more harmonious and resonance-reinforced result.

2.7 Intensity and loudness

In the same way that the frequency of a sound can be related to the subjective impression of pitch as judged by a listener, a relationship obviously exists between the strength of a sound wave – its intensity – and the impression of loudness that it produces.

The initial amplitude of the vibrations of the source is, of course, directly related to the amount of energy (power) applied. Striking, bowing or blowing harder will produce greater amplitudes and higher radiated power (measured in watts). In the case of the simple point source radiating spherical waves, this power is evenly distributed over the surface of a continually expanding sphere. Sound intensity is defined as the rate at which energy is passing through unit area of the wavefront, say 1 m^2. Therefore at a given distance r from the source the intensity (assuming no losses) will equal the total power of the source W divided by the area of the wavefront. The surface area of a sphere is $4\pi r^2$ and so:

$$\text{Intensity} = \frac{W}{4\pi r^2}$$

This is the well-known inverse square law (also applicable to light waves, radio transmissions, etc.), and tells us that the intensity of a spherical wave falls off as the square of the distance (see Figure 2.14). At greater distances from the source the curvature of the wavefront decreases, ultimately approaching the characteristics of a plane wave, and the rate of reduction in intensity for further short distances falls considerably.

The human ear responds to a phenomenal range of sound intensities. At 1000 Hz, for example, an intensity of only 10^{-12} W/m^2 is just audible (at the so-called 'threshold of hearing') for an average

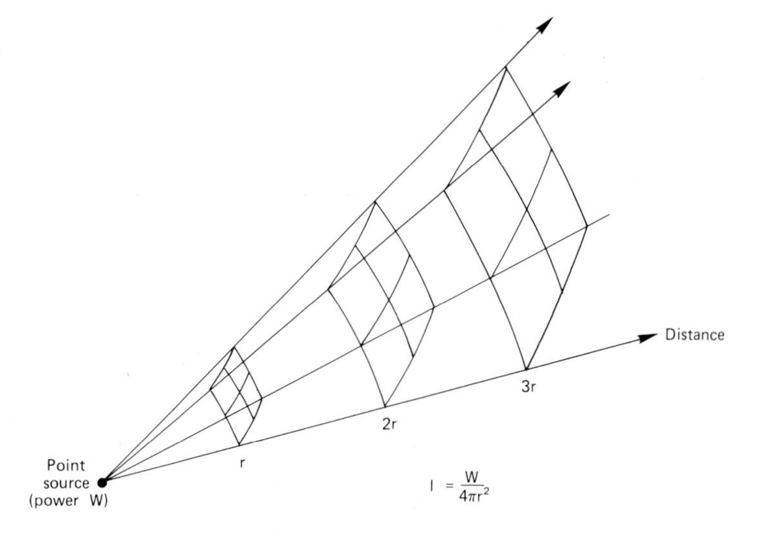

Figure 2.14 Inverse square law. In a spherical wave the area of the wavefront increases as the square of the distance, and so intensity falls off as the square of the distance (i.e. −3 dB for each doubling of distance)

listener whereas the highest intensity that can be comfortably tolerated ('threshold of feeling') is about 10 W/m^2. This corresponds to a ratio of 10000000000000 (ten trillion) or 10^{13}:1. The intensity scale used by audio engineers is logarithmic, providing more manageable numbers and according neatly with the manner in which the ear estimates equal steps in perceived loudness as resulting from equal ratios or multiplicative steps in intensity. The logarithmic unit used is the Bel (after Alexander Graham Bell), representing a factor of 10, there being 13 Bels of intensity change over the 10^{13}:1 audible range. In practice, a smaller unit is more useful, the decibel (dB); there are 10 dB to a Bel and thus about 130 dB between the thresholds of hearing and feeling.

These two thresholds form the lower and upper limits of the well-known Fletcher–Munson curves and the more recent Robinson and Dadson curves shown in Figure 2.15. The curves are 'equal loudness' contours averaged for a large group of people who were asked to indicate the intensity level at which tones at various frequencies sounded 'equally loud' compared with a 1000 Hz tone, which was reproduced on a rising scale of intensities in 20 dB steps (or '20 phons' in the equivalent scale of 'loudness'). It will be seen that the human ear is far from being equally sensitive to sounds at all frequencies. Sensitivity is highest at around 2–5 kHz but falls off considerably at lower frequencies and also, more irregularly, at higher ones.

To express the power, intensity or sound pressure for a sound of a given force, the word 'level' is added and the figure quoted is then the power level, intensity level or sound-pressure level (SPL) with respect to an agreed datum value. This is usually taken as the just audible (threshold) value at 1000 Hz, i.e. 10^{-12} W for power, 10^{-12} W/m^2 for intensity and 20 μPa (or 2×10^{-5} N/m^2) for sound pressure. The intensity level (IL) in decibels for a sound of intensity I is therefore defined by

$$\mathrm{IL} = 10 \log(I/I_o)$$

where I_o is the reference intensity 10^{-12} W/m^2. Since intensity is proportional to the square of pressure, the SPL in decibels is given by

Age 20
Age 60
Phons
120
100
80
60
40
20
0
Threshold of hearing
Level (dB)
Frequency (Hz)
Pressure (Pa)
140
120
100
80
60
40
20
0
-20
20
50
100
200
500
1k
5k
20k
200
20
2
2 x 10^-1
2 x 10^-2
2 x 10^-3
2 x 10^-4
2 x 10^-5

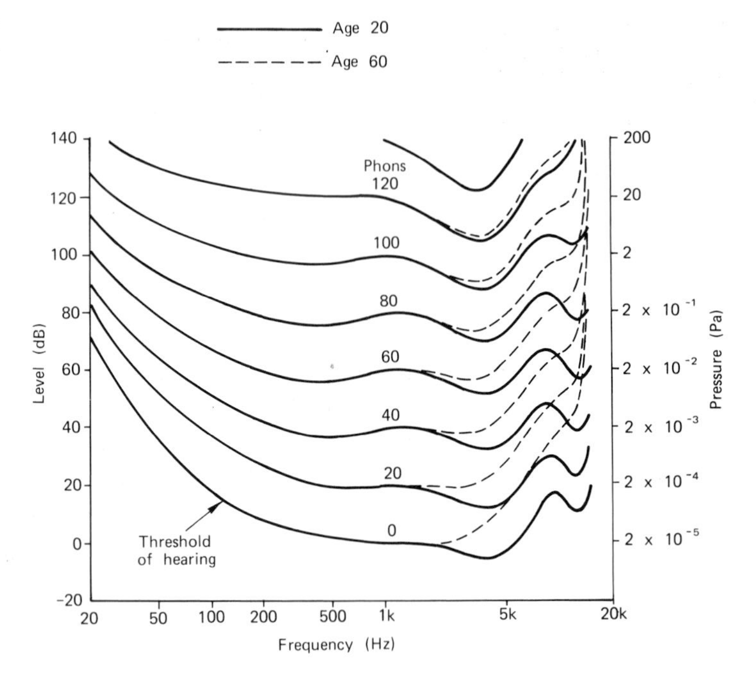

Figure 2.15 Curves of equal loudness with reference to the subjective loudness at 1000 Hz, also showing the typical hearing loss at age 60 (after Robinson and Dadson)

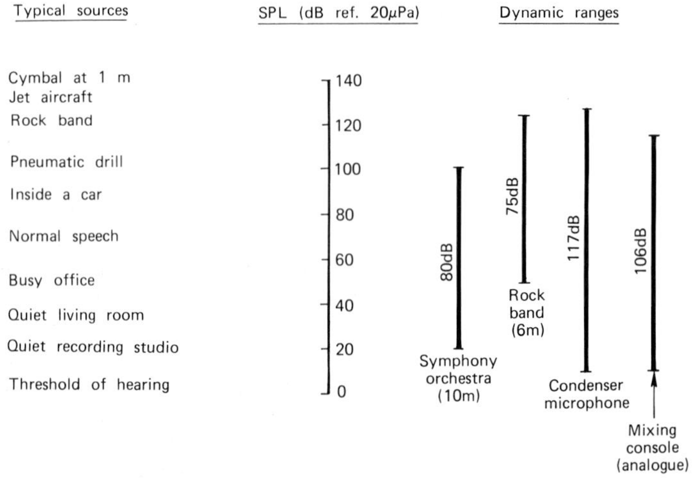

Typical sources
SPL (dB ref. 20μPa)
Dynamic ranges
Cymbal at 1 m
Jet aircraft
Rock band
Pneumatic drill
Inside a car
Normal speech
Busy office
Quiet living room
Quiet recording studio
Threshold of hearing
140
120
100
80
60
40
20
0
80dB
Symphony orchestra (10m)
75dB
Rock band (6m)
117dB
Condenser microphone
106dB
Mixing console (analogue)

Figure 2.16 Peak SPL for some typical sound sources and environments, plus typical dynamic ranges

$$\begin{aligned} \text{SPL} &= 10 \log (P^2/P_o^2) \\ &= 20 \log (P/P_o) \end{aligned}$$

The ratio between the highest and lowest levels generated by a sound source, or to which a broadcasting or recording system can respond, is called the *dynamic range*. Typical dynamic ranges are shown in Figure 2.16 alongside a scale indicating the peak SPL for some common types of sound or environment.

2.8 Resonance

Many vibrating structures (and this most clearly includes all 'tuned' musical instruments) possess at least one natural mode of vibration.

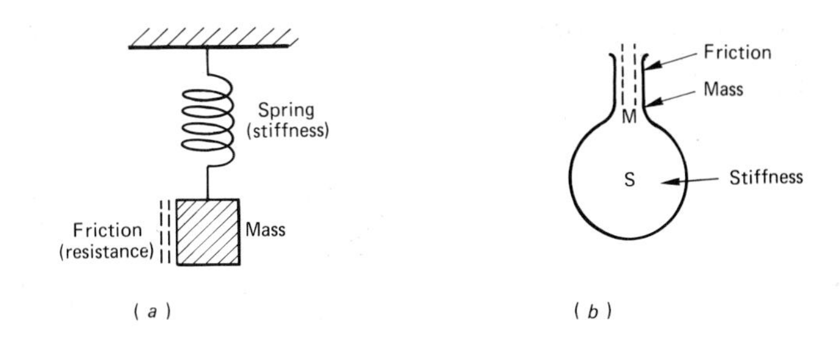

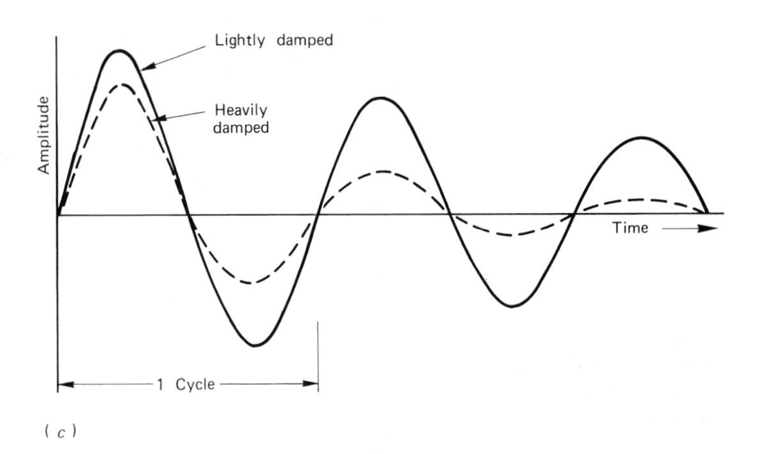

Figure 2.17 Simple resonant systems, showing (a) a mechanical vibratory system; (b) an acoustic (Helmholtz) resonator; and (c) the decreasing swings in amplitude following a shock excitation for both lightly and heavily damped systems

This occurs at the frequency for which the moving energy of the mass is exactly balanced by the energy stored in the elastic restoring force or stiffness (the opposite of compliance). When such a mechanical system (Figure 2.17(a)) is shock excited into free vibration, it performs a simple harmonic motion as described in Section 2.3, in which the applied energy alternates between being entirely associated with the moving mass at nodal points of maximum velocity and entirely stored in the stiffness (or elastic compression) at the antinodes, where velocity is zero and displacement from the position of rest is at a maximum. Figure 2.17(c) shows the result of such a shock excitation, and the frequency of this natural mode of vibration is called the resonant frequency.

In the general case of a vibrating system the resonant frequency is given by:

$$f = \frac{1}{2\pi}\sqrt{\frac{s}{m}}$$

where s = stiffness and m = mass. It will be seen that either decreasing the stiffness or increasing the mass lowers the resonant frequency. There is a clear analogy with tuned electrical circuits, to be discussed in Section 3.7.5. Analogous acoustic oscillatory systems can also be used to control microphone response or room acoustics. Figure 2.17(b) shows a typical acoustic resonator, usually referred to as a Helmholtz resonator. The mass of air in the neck can be considered as vibrating against the stiffness (reciprocal of compliance) of the volume of enclosed air.

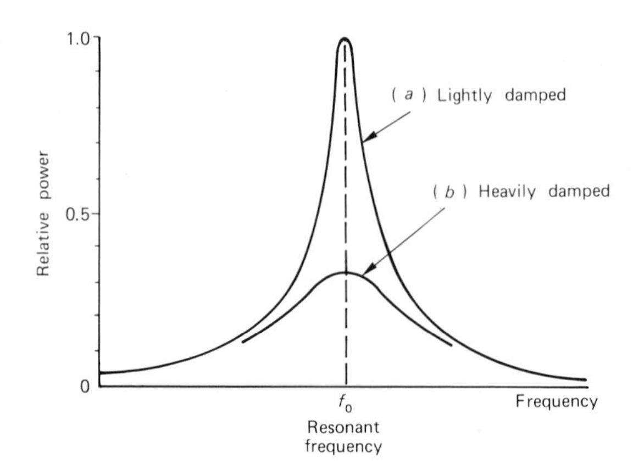

Figure 2.18 Resonant response of a vibratory system when driven at different frequencies

The amplitude of the shock-excited vibration shown in Figure 2.17(c) progressively reduces with time. An obvious part of this energy loss is due to sound radiation, but there will also be a significant dissipation of energy due to internal or external friction or resistance – often called damping. The solid-line curve in Figure 2.17(c) illustrates the case where damping is low, as for a tuning fork held in the hand. The broken-line curve shows the more rapid decay in amplitude when energy is dissipated more rapidly, for example when the stem of the tuning fork is placed on a table or box, or the amount of friction or resistance is increased.

In systems where separate portions of the total mass are capable of independent vibratory modes, more than one resonant frequency may exist. This is the way that harmonic frequencies are set up, for example, in vibrating strings or air columns (as was shown in Figure 2.10).

Apart from the free vibrations discussed so far, there are many examples of forced vibrations where one system is being driven not by a single shock excitation but by an oscillatory force at some particular frequency. In general, the driven system will vibrate at the frequency of the driver but with restricted amplitude. However, if the driver is tuned through the resonant frequency of the driven system the velocity will rise to a peak and then fall away, as shown in Figure 2.18. The maximum value of the velocity at the resonant peak and the sharpness of the tuning are greater if the system is lightly damped (a). On the other hand, the presence of considerable damping can produce a broadly tuned system which will respond to a wide band of frequencies (b).

In a few cases, sharp tuning is a desirable feature (for example, in the tuned resonator tubes mounted below the bars of a xylophone). More often, broad tuning is the aim, as in the design of microphone diaphragms, loudspeaker cones and cabinets. Here designers aim to produce the chosen (usually flat) frequency response characteristics by juggling with the relative values of the three controlling properties – stiffness, mass and resistance (i.e. damping) – depending on which feature they want to be independent of frequency – displacement, velocity or acceleration.

These three types of driven system are said to be stiffness-controlled, resistance-controlled and mass-controlled, respectively (see Figure 2.19):

Figure 2.19 The three types of driven system used in microphone design. (a) Stiffness control with f_o placed at the high end of the frequency range; (b) resistance control with f_o at a central frequency; (c) mass control with f_o at the low end

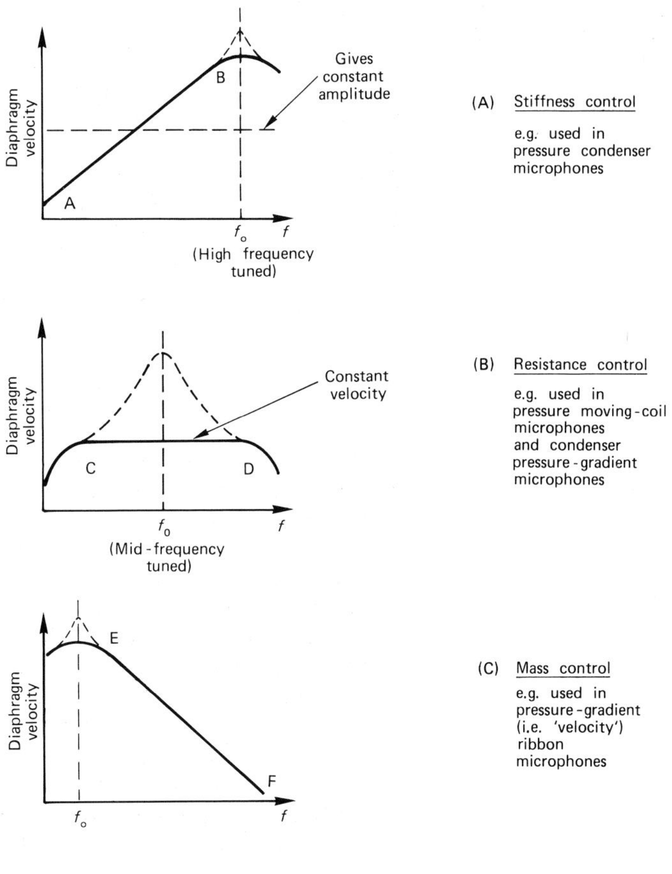

Note: Diaphragm velocity $v = 4\ af$ (where a = amplitude)

(*a*) if v is proportional to frequency we have a 'constant amplitude' system.

(*b*) if v is the same at all frequencies we have a 'constant velocity' system.

(*c*) if v is inversely proportional to frequency we have a 'constant acceleration' system.

1. *Making the stiffness large*, so that f_o is higher than the band of frequencies for which a flat response is required, produces an upward-sloping velocity characteristic AB which gives constant amplitude at all frequencies within the required range. This is the approach used, for example, in designing pressure condenser microphones where the output voltage is proportional to diaphragm displacement.
2. *Making the resistance relatively large* produces the situation of a well-damped, broadly tuned system. The flat velocity characteristic CD will then extend on either side of the system's resonant frequency, which is deliberately placed near the centre of the required band. This approach is appropriate for moving-coil microphones, where the output voltage is proportional to diaphragm/coil velocity.
3. *Making the mass relatively large* produces a falling velocity characteristic EF for which the acceleration is broadly independent of frequency. This is the approach chosen for pressure-gradient operated microphones (e.g. ribbon). Note that all

systems are resistance controlled in the vicinity of their resonant frequency, but resistance must be kept high if a flat velocity response is required in that region. Well above and below resonance, the system is either mass controlled or stiffness controlled.

So far, the driving system has been assumed to be more powerful or massive than the driven system. When systems coupled together are of roughly equal mass, the resulting vibrations may shift to some intermediate frequency.

2.9 Source directivity

The sound source principally discussed so far has been the simplest imaginable – a dimensionless 'point' source or tiny pulsating sphere. This is clearly unattainable, but is approached in practice by any source which is physically small compared with the wavelength in air for the particular frequency being radiated. A main characteristic of a point source is that it is non-directional, i.e. it radiates spherical waves, as discussed in Section 2.2 (see Figure 2.20(a)).

The radiation behaviour of practical sound sources can be deduced by imagining them as being made up of numerous point sources, the pressure at any point in the sound field being the sum of the pressures produced by the individual sources. An interesting situation arises if we consider two spaced point sources of equal strength radiating the same frequency but vibrating 180° out of phase (Figure 2.20(b)). Two spherical waves will be sent out into the air as shown by the solid and broken-line circles. Since the two waves are in antiphase, it follows that they exactly cancel each other at all points along the axial lines 90° and 270°, while the combined effect is to produce enhanced (additive) energy radiation in the two directions 0° and 180°. This illustrates a basic principle of sound wave transmission in any medium that, when more than one wave-train is set up, each wave continues to travel outwards independently. However, in the overlap region where more than one wave exists, an interference pattern is created by the simple addition of the forces acting on the individual particles in the medium.

This type of double radiator in antiphase is called an acoustic doublet or dipole and produces the familiar figure-of-eight radiation pattern shown shaded in Figure 2.20(b). Examples of practical

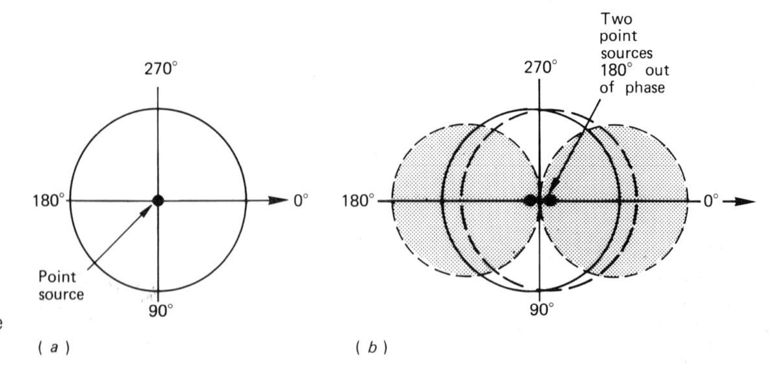

Figure 2.20 Source directivity. (a) A single point source generates a spherical wavefront; (b) two spaced point sources 180° out of phase produce bidirectional radiation with zero signal at right angles to the line joining the two points

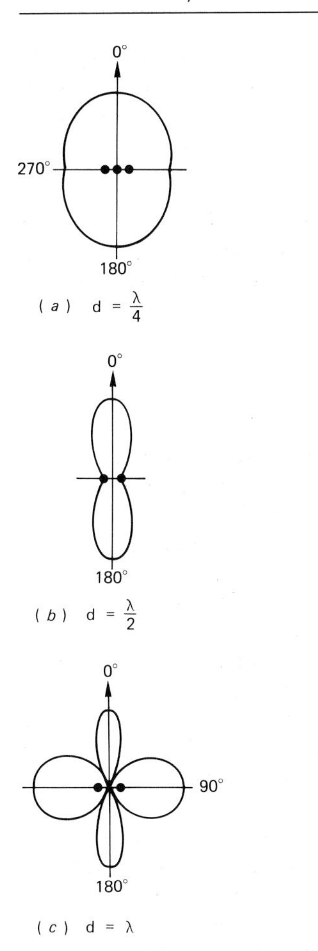

Figure 2.21 Radiation patterns for two in-phase point sources separated by a distance equal to (a) quarter wavelength; (b) half wavelength; (c) wavelength

dipoles include the H-type or frame radio or television aerial and an open-backed loudspeaker (such as the Quad Electrostatic or indeed any loudspeaker drive unit mounted on an unenclosed flat baffle). Dipole-like radiation also takes place into half-space from a single point source placed close to a wall, since an 'image' source in antiphase appears on the other side of the wall.

A different radiation model is obtained if two point sources vibrating in phase are imagined to be set at various distances apart. Figure 2.21 illustrates the directivity patterns for two in-phase point sources spaced apart by λ/4, λ/2 and λ.

By an extension of this model of two closely disposed point sources it is possible to consider an array of many point sources arranged in a straight line. By vibrating in phase, the array would act as a 'line' source capable of concentrating the radiation into the form of an expanding cylinder with the line array at the centre. This assumes a line of infinite length. In practice, the radiation pattern depends on the ratio of line length:wavelength and generally consists of a major axial lobe with several minor side lobes. All such sources in free space will generate a three-dimensional expanding wavefront – which can be visualized by imagining the directivity pattern rotated about its 90°/270° axis through the full 360° angle.

Finally, if an infinite number of imaginary line sources are placed together in the same plane a source of rectangular plane waves will result. With practical dimensions, the radiation patterns will take the forms illustrated in Figure 2.22.

As mentioned in Section 2.7, a small area of a spherical wavefront at some considerable distance from the source approximates to a flat or plane surface. The radiation beyond this point in any chosen direction may then be likened to that of a plane wave, as just described. Plane waves are generated, for example, by a vibrating piston at the entrance to a long tube. The special feature of a plane wave is that, unlike the spherical wave, the wavefront area does not continually expand except for any effects at the fringes. Sounds are therefore focused or beamed to travel with very little attenuation over long distances.

The key to the directional properties of any practical sound source is its physical size in relationship to the wavelength. If the source is much smaller than the wavelength, it may be taken to generate spherical waves: if it is much larger, then something approaching a beamed plane wave is the result. Remembering that sound wavelengths in air extend from about 17 mm at 20000 Hz to 17 m at 20 Hz,

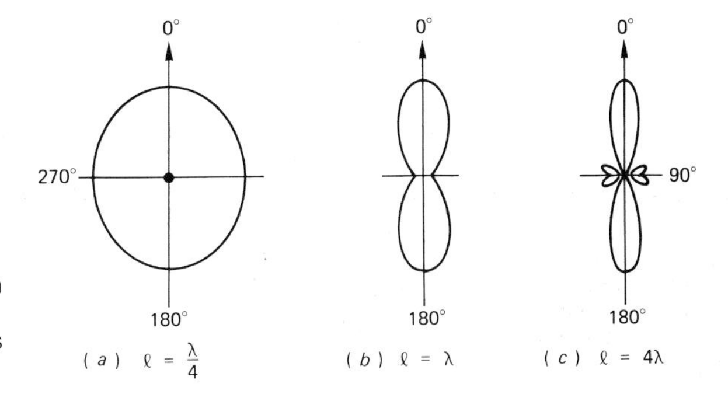

Figure 2.22 Radiation patterns for a plane rectangular sound source when the length equals (a) quarter wavelength; (b) wavelength; (c) four times the wavelength

we see that most sources and certainly all musical instruments are intermediary in size and will tend to be more directional when playing their highest notes and less directional for low notes. This will be discussed in detail in Chapter 8.

The directivity D of any source is defined as the ratio of the intensity at a given distance on the principal (or reference) axis to that which would result from a point source of the same power. The directivity index DI is this ratio expressed in decibels: i.e. $DI = 10 \log D$, e.g. for a pulsating sphere $D = 1$ and $DI = 0\,\text{dB}$ and for a hemispherical source mounted on a rigid baffle $D = 2$ and $DI = 3\,\text{dB}$.

2.10 Reflection and standing waves

A common example of an interference pattern set up by two independent sound waves occurs when a plane wave (for simplicity) meets a totally reflecting boundary at right angles and is reflected back along its original path. The superposition of the incident and reflected waves at successive intervals of a quarter-cycle is illustrated in Figure 2.23. It produces a pattern of alternating nodes and antinodes forming a stationary or standing wave with twice the amplitude of the original wave. Naturally, the wall behaves as a nodal

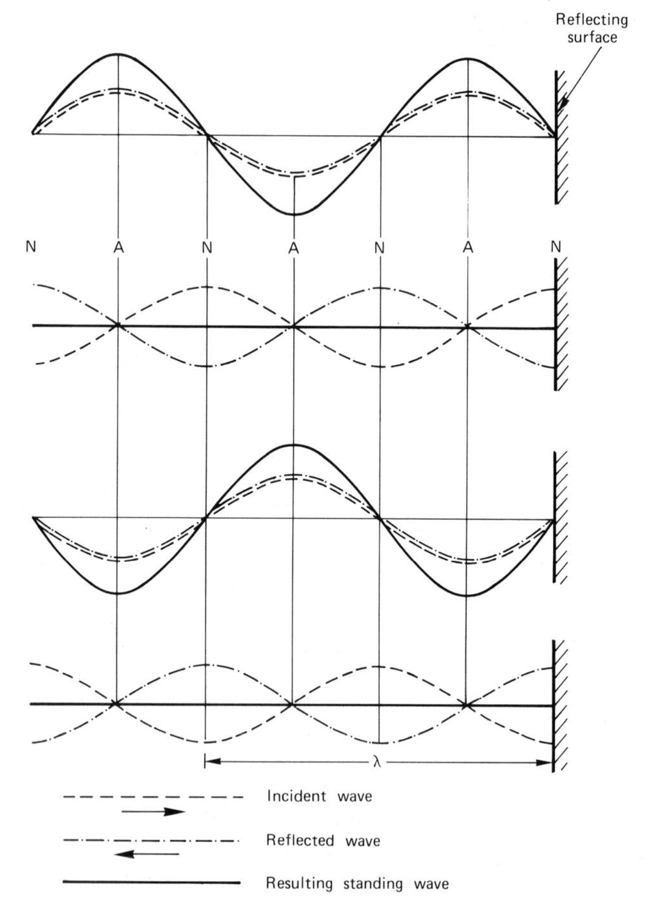

Figure 2.23 Standing wave formed when a sound wave is totally reflected back along its original path, shown at time intervals of one-quarter cycle

point (of zero displacement of the adjacent air particles) and other nodes occur at multiples of $\lambda/2$ from the wall where the two waves are either at zero amplitude or in antiphase (therefore cancelling). Similarly, the antinodes or loops occur at intermediary points $\lambda/4$, $3\lambda/4$, $5\lambda/4$ from the wall. If some of the incident energy is absorbed or transmitted into the wall, the standing-wave pattern will still occur, but the amplitude will be less than $2a$ at the antinodes and will not quite fall to zero at the nodes.

This standing-wave effect on steady tones can be heard or picked up by a microphone in the vicinity of any wall surface, and must clearly be borne in mind when placing musicians or microphones near a reflecting surface. It also means that a microphone diaphragm experiences a greater driving force, and produces a higher output voltage when sounds arrive along the axis and are reflected by the diaphragm. This effect is called pressure doubling. It occurs only when the microphone dimensions are large compared with the wavelength, and so results in a boost at high frequencies (see Section 4.3). The standing-wave diagram in Figure 2.23, of course, serves to illustrate the natural modes of vibration of the stretched strings and air columns of musical instruments (to be discussed in Chapter 8).

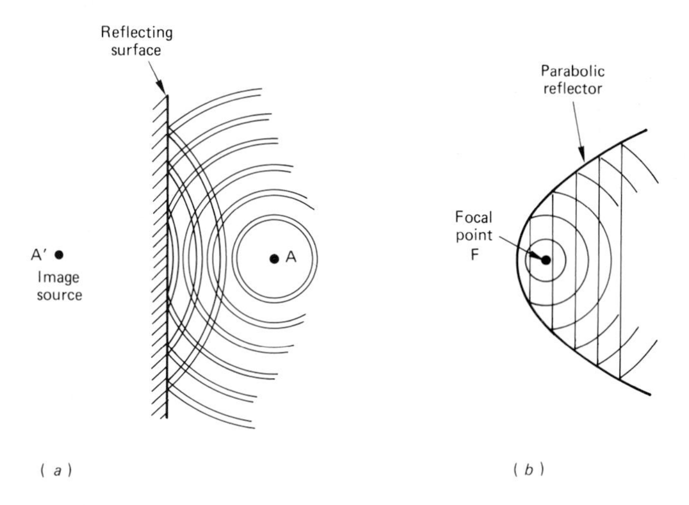

Figure 2.24 Reflection of a spherical wave (a) from a flat wall (mirror); (b) from a parabolic reflector, producing plane waves on the axis

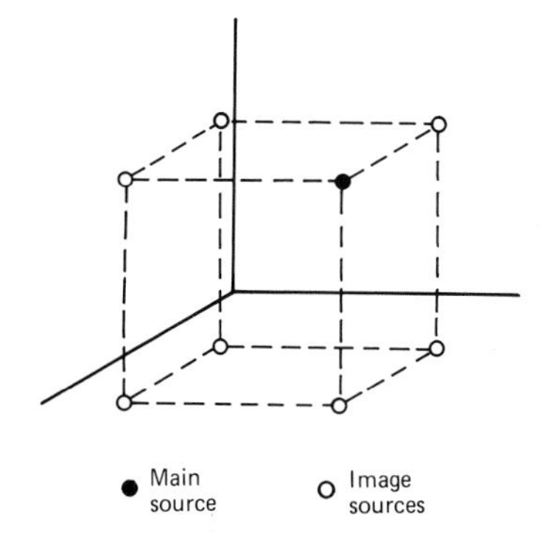

Figure 2.25 Creation of seven image sources by placing a sound source near the corner of a room

It is often helpful to use optical analogies to explain the behaviour of sound waves, at least for short wavelength sounds. Figure 2.24(a) shows by way of example that reflections of the spherical waves from a source A result in a reverse spherical wave whose origin is a secondary 'image' source mirrored at the same distance behind the reflecting surface as A is in front of it. Similarly Figure 2.24(b) illustrates that placing a sound source at the focal point F of a (relatively large) parabolic reflector produces a forward-directed beam of plane waves. The analogy with car headlamps is obvious. Other analogies can be used to explain the reflection of sound waves arriving at oblique angles, when the angle of reflection equals the angle of incidence, the setting up of multiple images by reflection at room corners (Figure 2.25) and the 'flutter echo' type of standing wave often heard between parallel walls in rooms and studios.

The reciprocity between many source and 'receiver' mechanisms is also in evidence in the parabolic diagram of Figure 2.24(b). If, instead of placing a sound source at the focal point of a parabola, we imagine a plane wave stream arriving along the axis of a parabolic reflector, the energy received over the whole surface will be focused back on the focal point exactly in phase. This is the principle of the parabolic reflector microphone which gives high sensitivity along the axis, while sounds arriving off-axis are not reflected in phase and so are suppressed. Such parabolic microphones are popular for recording distant sources in relatively noisy surroundings, e.g. birdsong and bat-on-ball at cricket matches (see Chapter 4).

2.11 Comb filter effect

In the general case, when a sound wave meets a reflecting surface at other than normal (0°) incidence the wave is reflected obliquely. The angle of reflection equals the angle of incidence and the reflected wave may be represented diagrammatically as though it originates from an image source S1 as shown in Figure 2.26(a). For a listener or microphone at point M there will be interference between the direct wave and the reflected wave. The resulting pressure at any instant will depend on the time delay (phase shift) due to the extra distance travelled by the reflected wave.

It will be clear that for certain frequencies this extra distance will equal one or more times the wavelength λ so that the two waves will arrive at M exactly in phase and add, i.e. reinforce each other. Similarly, there will be a series of frequencies for which the waves arrive at M exactly 180° out-of-phase and will cancel. This is illustrated in Figure 2.26(b) for the 'worst-case' situation of a perfectly reflecting surface so that peaks of twice the direct-wave amplitude and troughs falling to zero occur respectively at octave

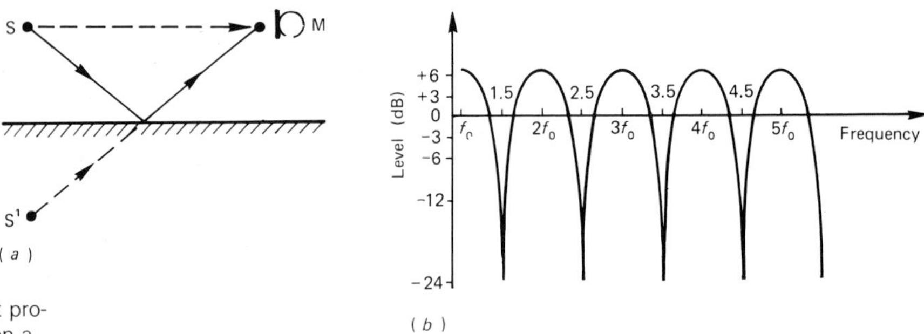

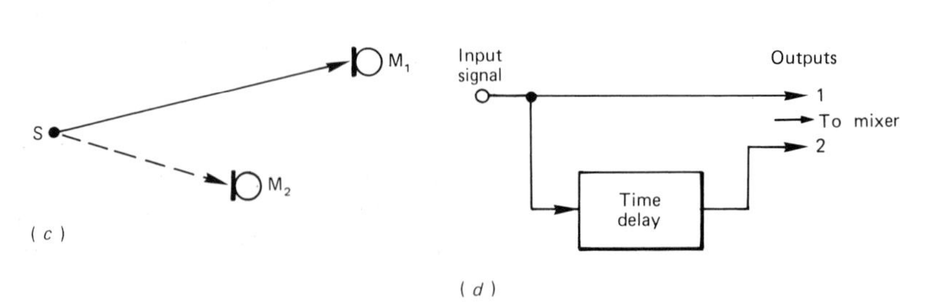

Figure 2.26 Comb filter effect produced by interference between a sound wave and a delayed version of itself. (a) Microphone M receives direct wave from source S and a delayed wave from image source S^1; (b) frequency response shows alternating peaks and troughs resembling the teeth of a comb; (c) a similar effect is produced with microphones M1 and M2 at different distances from source S; (d) the effect can be introduced deliberately using a time-delay network

spacings. The resemblance to the teeth of a comb has caused this interference pattern produced by two waves (one a delayed version of the other) to be called the 'comb filter' effect.

It is not always possible to hear the odd phasey effect and sound coloration which this produces when listening to sounds near a wall or large obstacle. This is due to our ability, using two ears, to distinguish between sounds arriving from different directions. However, the distortions are easily identified if one ear is covered, and placing a microphone near a wall or screen is certain to run into problems.

Practical situations which arise very often include reflections from table tops, lecterns and polished wood or stone floors, as discussed in Section 4.6.7 in connection with the boundary-type microphone. A similar problem affects all situations in which two or more microphones pick up a sound source from different distances (see Figure 2.26(c)). Then the electrical outputs, when mixed, will be subject to the same kind of comb filter additions and cancellations at specific frequencies, causing disturbing distortion. This subject occurs in later discussions on spaced microphones for stereo (Section 5.4) and multi-microphone balances generally (Section 10.3). Figure 2.26(d) illustrates how a comb filter response might be introduced deliberately in a synthesizer by routing part of the signal through a time-delay network.

2.12 Echoes and reverberation

A familiar example of sound wave reflection is observed when sounds reach the ear after reflection over a fairly long distance so as to produce a repeat or echo of the original. In fact the human ear cannot resolve and identify a separate sound echo unless the original wave and the reflected version are separated by a time interval of at least 1/15th of a second (60 ms). Reflections over shorter distances are perceived as simply reinforcing and prolonging the sound. Since sound travels at 344 m/s it follows that the minimum path length difference for a separate echo to be heard is about 344/15 = 23 m.

A similar calculation will predict the time interval to be expected when a nearby sound source is reflected from a large surface, say, 150 m distant:

Total distance travelled by wave = $(2 \times 150) = 300$ m
Therefore time interval = $300 \div 344 = 0.87$ s.

In a room or concert hall, unlike the open air, sounds suffer multiple reflections from the enclosure walls with the result that a complex and diffuse soundfield is created. There is a unique pattern of growth and decay when a source of sustained sounds is placed in such an enclosure (see Figure 2.27(a)). The growth and decay generally follow an exponential (logarithmic) law, as shown.

When the source begins to radiate sounds, the intensity measured at any point will increase by small increments as successive reflections arrive from the walls, floor and ceiling, until an equilibrium value is reached (when the rate of sound energy loss due to absorption just equals that being supplied by the source). Again, when the source is

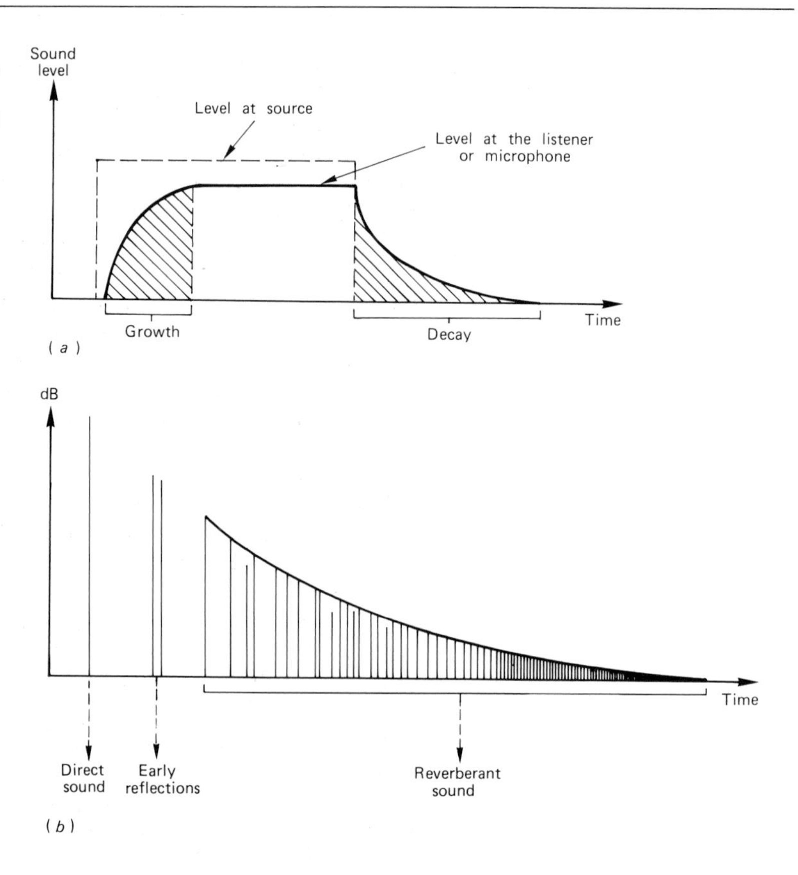

Figure 2.27 Multiple reflections in a room or concert hall. (a) The curve of growth and decay; (b) the sequence of direct sound, early reflections and reverberation observed for a single impulse sound

switched off, the sound intensity does not fall to zero immediately but fades more or less gradually, depending on the acoustic characteristics of the room.

This prolongation of sounds in an enclosure is called reverberation (see Figure 2.27(b)). Ever since the researches of W. C. Sabine around 1900, a principal yardstick for describing the acoustic properties of an auditorium has been the reverberation time (RT). This is defined as the time taken for sounds to decay by 60 dB (i.e. to one-millionth of their original intensity). Sabine discovered that RT was approximately proportional to the volume of the room and inversely proportional to the total absorption present in the boundary materials, furnishings and the audience.

That RT will increase with volume is only to be expected, since the greater path lengths between reflections reduce the rate of boundary absorption. Similarly, increasing the amount of absorption must decrease RT as more energy is dissipated at each reflection. This is summarized in the Sabine formula:

$$\mathrm{RT} = \frac{0.16V}{A}\ \mathrm{s}$$

where V = room volume in cubic metres and A = total absorption in m^2-sabins (a unit representing $1\,\mathrm{m}^2$ of perfectly absorbing surface, e.g. an open window). This simple formula does not take into account room shape and the distribution of absorptive materials, yet it serves a useful purpose, particularly in later refined versions such as that of Norris and Eyring.

Assuming that the room dimensions are large compared with the wavelength, the sound is taken to be completely diffuse after a large number of reflections, i.e. the average sound density is the same throughout the room and all directions of propagation are equally probable. The presence of reverberant energy produces a gain in intensity which is approximately proportional to RT. Therefore a long RT (live acoustics) is desirable in music making, for example, to assist in making weak sounds audible throughout the enclosure. By contrast, where maximum clarity is a requirement (e.g. for speech) a short RT will give less masking and blurring of sounds (dead acoustics). In practice, a compromise value is desirable to provide both high intelligibility and a measure of loudness reinforcement for distant listeners.

Practical materials and structures do not absorb incident sound energy equally at all frequencies. Defining the absorption coefficient of a given material as the fraction of the sound energy absorbed would give a perfect absorber a coefficient of 1 at all frequencies. Table 2.6 illustrates the wide range of coefficients met in practice and suggests that proper tailoring of the acoustics of a given room or auditorium will call for careful disposition of a number of absorber types to bring RT down to the target value at all frequencies for the intended application. Typical RT/volume curves for various enclosures are shown in Figure 2.28.

Included in Table 2.6 are typical absorption coefficients for the audience area, empty and occupied. Clearly, this is a significant factor which must be taken into account when designing concert

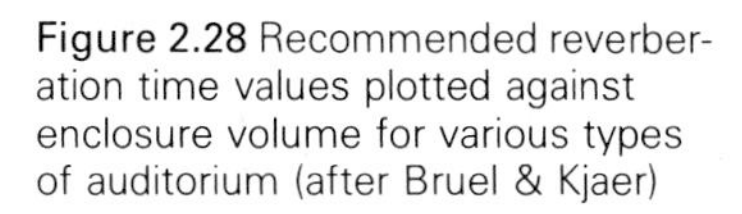
Figure 2.28 Recommended reverberation time values plotted against enclosure volume for various types of auditorium (after Bruel & Kjaer)

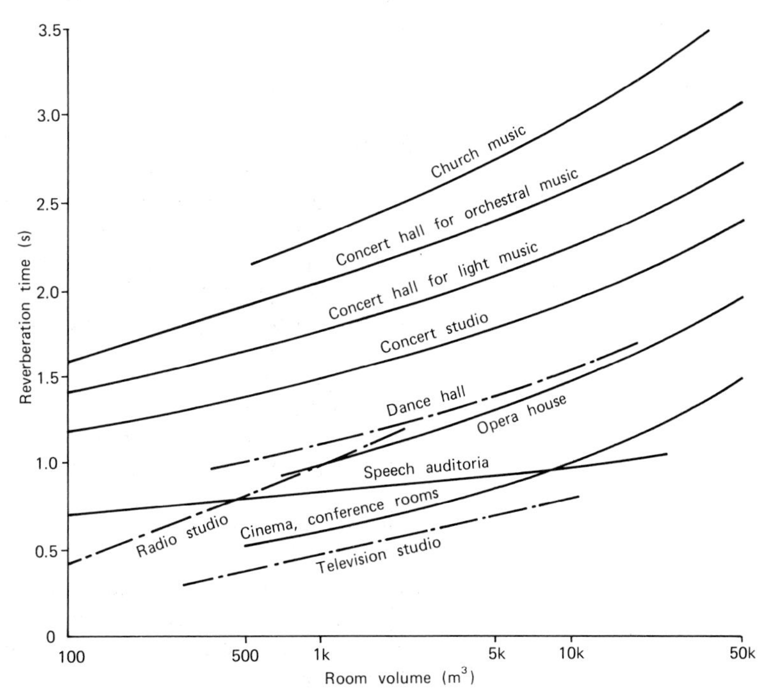

Table 2.6 Variation of absorption coefficient with frequency

Surface	*Frequency (Hz)*			
	125	*500*	*1000*	*4000*
Occupied audience area	0.45	0.84	0.95	0.84
Cloth seats, empty	0.44	0.76	0.87	0.70
Heavy curtains	0.14	0.57	0.72	0.62
Wooden floor	0.14	0.09	0.08	0.07
Acoustic tile	0.08	0.55	0.68	0.70

halls, etc. It also explains the often considerable decrease in reverberation experienced at a concert compared with conditions when rehearsing in the empty hall. Later designs of auditorium seats aim to make the RT more nearly equal for empty, full and half-full hall conditions by using constructions which present approximately the same absorption characteristics whether the seat is occupied or empty.

The very effective absorption of sound by a seated audience means, when scattering effects are also considered, that very little of the sound energy falling on the seating area is reflected. It is also an example of how absorption tends to be very unevenly distributed in some situations, so that the simple formulae of Sabine and others must be regarded as giving only broad guidance to an enclosure's acoustic properties. It also explains why recording engineers often find that layers of distinct change in tonal balance can be found in a hall or studio and produce marked shifts in sound balance when the microphone is moved only a few centimetres.

A specially troublesome situation arises in small rooms when repeated reflections between parallel wall surfaces (including floor and ceiling) can set up pronounced resonance effects at specific frequencies related to the room dimensions. The room therefore tends to reinforce certain frequencies in the same way as an open organ pipe or a violin string; the walls act as pressure antinodes, and the wavelength for the fundamental resonance is equal to 21. There are also tangential and oblique resonant modes resulting from repeated reflections involving four or six surfaces, respectively. The general formula allowing all simple room resonance frequencies to be calculated is:

$$f = \frac{c}{2}\sqrt{\left[\left(\frac{p}{L}\right)^2 + \left(\frac{q}{W}\right)^2 + \left(\frac{r}{H}\right)^2\right]}$$

where c is the velocity of sound, L, W and H are the room length, width and height, and p, q and r are integer numbers 0, 1, 2, etc.

Figure 2.29 shows how the sound pressure would be found to vary along the length of a room (i.e. putting q and $r = 0$) with perfectly reflecting end walls when the first three resonant mode frequencies are sounding – corresponding to $p = 0$, 1 and 2 (wavelengths of $2L$, L and $0.66L$). In a room for which $L = 3.44$ m, for example, the first three lengthwise room resonances (or eigentones) would occur at 50, 100 and 150 Hz. The presence of room resonances has the effect of colouring the sounds, making speech or music boomy. The problem is mainly associated with small rooms in which the principal

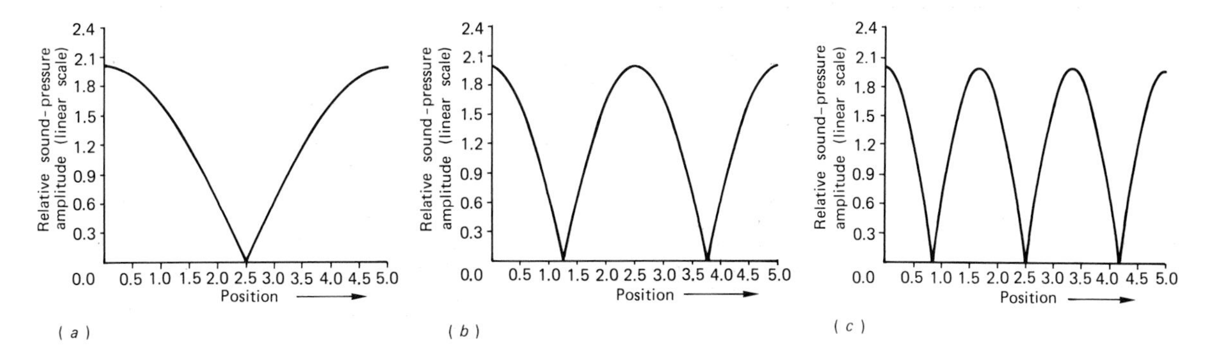

Figure 2.29 Room resonances, showing sound-pressure variation along a room when the length equals (a) 0.5 wavelength; (b) wavelength; (c) 1.5 wavelength

resonance frequencies are spaced far apart and fall well within the audible range. It can be reduced by adding absorption on at least one of the facing surfaces. Of course, it is aggravated in a room having two or more of its dimensions equal or in a simple ratio, since this duplicates some or all of the resonance modes.

2.13 The soundfield

We have seen that, when a continuous source is radiating sound into a room, two types of field are in operation, the direct field produced by direct radiation from the source and the reverberant or diffuse field made up of numerous reflected waves from walls, etc. This is illustrated in Figure 2.30 for three types of room – reverberant, semi-reverberant and 'dead'. Placing a microphone very close to the source (the near field) will pick up an unpredictable signal which can vary considerably with position and is not greatly dependent on the room characteristics. At greater distances, the direct field may be taken to decrease in SPL (sound-pressure level) at a rate of 6 dB per doubling of source/microphone distance (the inverse square law). In the 'far field', therefore, the diffuse field, which is almost uniform throughout the room, will tend to dominate the microphone input. The distance at which the contributions to the effective SPL from both the direct and the diffuse fields are equal is a characteristic of the particular room, and is called the room radius or critical distance. It

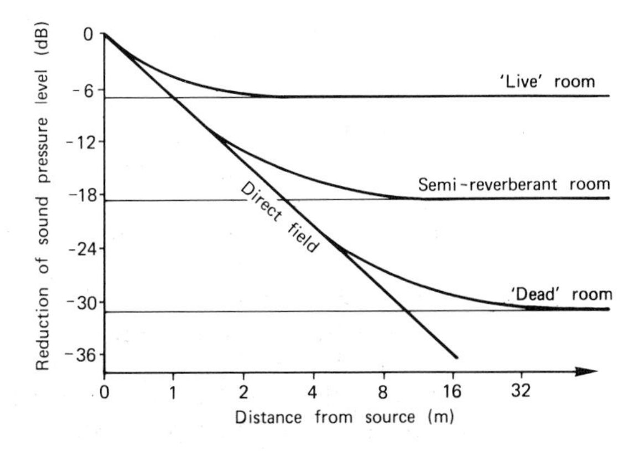

Figure 2.30 Comparing the direct field, in which SPL falls 6 dB for each doubling of distance (inverse square law), with the diffuse field in which SPL remains at a uniform level depending on the reverberation time

is related to the room volume and the reverberation time, and may be calculated from the expression:

$$r = 0.057\sqrt{\frac{V}{T}}$$

where r = room radius (m), V = room volume (m^3), and T = reverberation time (s).

The progressive decrease in the ratio of direct-to-reverberant sound with microphone (or listener) distance is a vital key to balance technique, and will be discussed in detail in Section 7.1.5.

2.14 Diffraction

Just as happens with light waves, sound waves tend to be scattered by or bend round the edges of any obstacles in their path, instead of being simply reflected. This process is called diffraction or shadowing. An obstacle which is much larger than the wavelength is found to cast a 'sound shadow', with a fringe of 'half shadow' due to diffraction (Figure 2.31). A relatively small obstacle, on the other

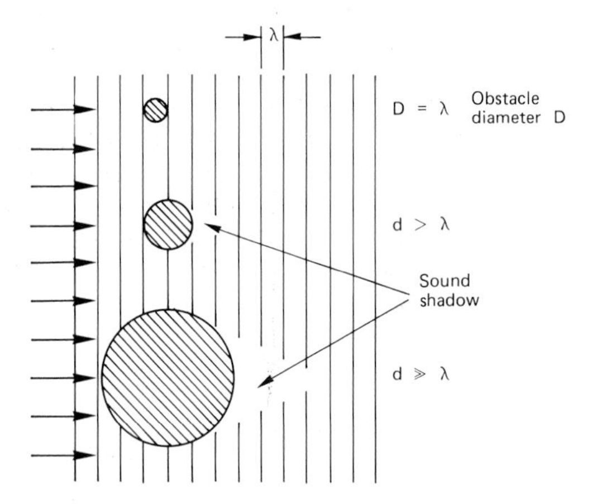

Figure 2.31 Diffraction: the formation of a sound shadow at high frequencies where the obstacle is large compared with the wavelength

hand, casts practically no shadow and the wavefront travels onwards undisturbed. With the general run of obstacle shapes and sizes, the effect of diffraction is to set up a complex interference pattern of shadows and 'bright spots', producing varying pressures over the obstacle surface and in the near field. This has important consequences for the designers and users both of audio sources (musical instruments and loudspeakers) and microphones.

2.15 Diffraction and human hearing

Diffraction of sound waves arriving at the head of a listener has a critical effect on hearing acuity at different frequencies and contributes a great deal to the ways in which a listener estimates the direction from which sounds are coming. Regarded simply as an obstacle, the human head is a slightly misshapen sphere of

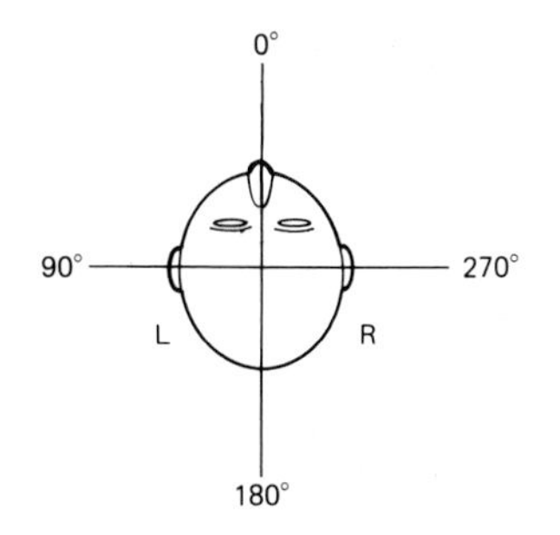

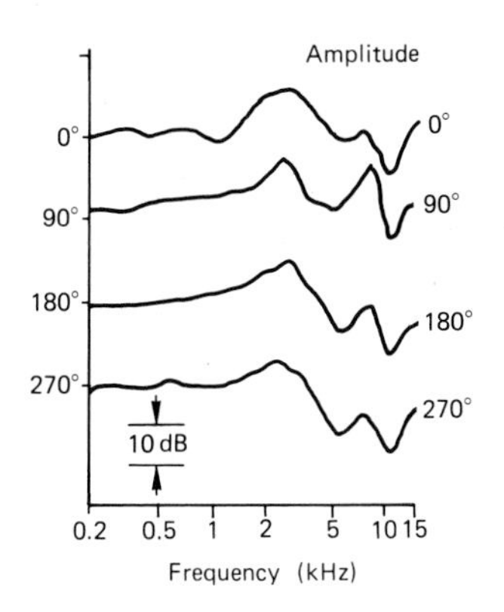

Figure 2.32 Diffraction effects introduce marked variations in the frequency response at the ear for sound arriving at different angles

approximately 21 cm diameter, with an ear at each side. It therefore interferes very little with sound waves at low frequencies below about 1 kHz (wavelength 34 cm).

However, diffraction begins to be important at higher frequencies when standing waves and therefore pressure doubling are set up at the point of incidence, with a sound shadow affecting the parts of the head distant from the source.

Considering first the horizontal plane through the ear centres and measuring angles with respect to the frontal axis, Figure 2.32 shows typical frequency responses measured at the entrance to the left ear canal for sounds incident at 0°, 90°, 180° and 270°. It will be seen that the response is consistent at low frequencies and indeed very nearly independent of the angle of incidence. At high frequencies, however, pressure doubling produces, on average, the expected 6 dB rise at 90° while there is a marked level loss at 270°, where the source is on the 'wrong' side of the head as far as the left ear is concerned.

When both ears are considered, for binaural as opposed to monaural listening, it will be appreciated that definite interaural amplitude differences (IADs) will cause the left and right ears to send quite different signals to the brain for sounds incident at any angle other than 0° or 180° – at least for frequencies from about 1250 Hz (wavelength 27 cm) upwards. Thus IAD is a significant clue to the angular location of sound sources, as will be described in Section 2.17.

The wide fluctuations in level within the high-frequency band are due to outer ear (pinna) geometry and, of course, being a fixed characteristic, are largely ignored or compensated for by the brain. They include a broad peak of up to 12 dB around 2–3 kHz and a sharper trough around 8–10 kHz, irrespective of incident angle. A shallow dip appears at 1200 Hz for frontal incidence due to shoulder reflections.

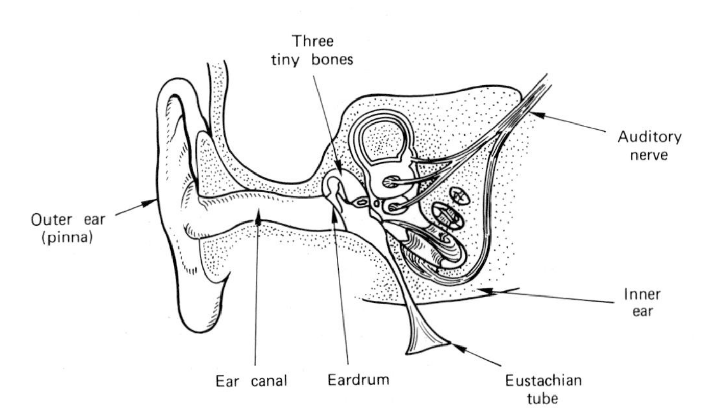

Figure 2.33 The human ear. Sounds arriving at the outer ear are funnelled into the ear canal to set the eardrum into vibration. This energy is transmitted by a lever system of three tiny bones to the inner ear where it is converted into electrical impulses carried by auditory nerves to the brain

Without going into ear anatomy in any detail, it can be stated that further fluctuations in frequency response (but not in directional hearing effects) are introduced as sounds make their way from the outer ear along the ear canal to the eardrum (a thin membrane closing the inner end of the canal) and thence via an impedance matching network of tiny bones to the inner ear and, after conversion to electrical impulses, to the brain (see Figure 2.33). The ear canal itself is a narrow tube about 25 mm long by 7 mm wide and introduces

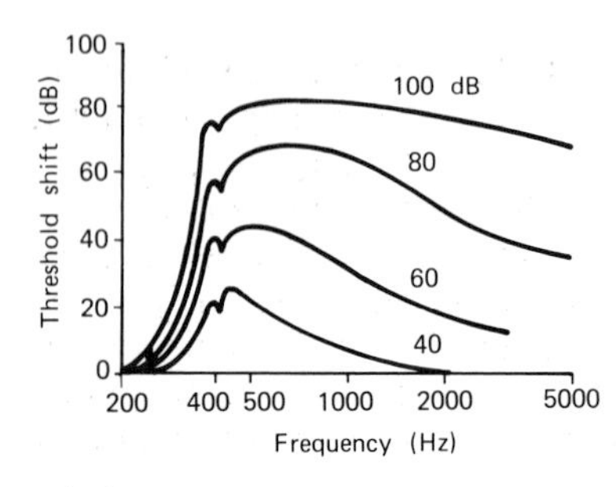

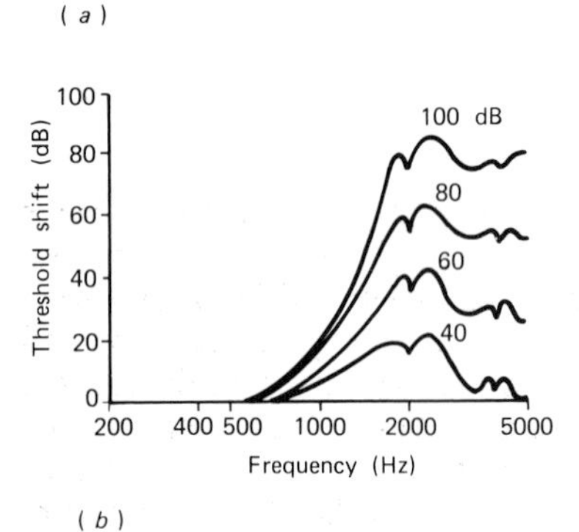

Figure 2.34 Masking of one pure tone by another: the shift in audibility threshold for various levels of masking tone at (a) 400 Hz and (b) 2000 Hz

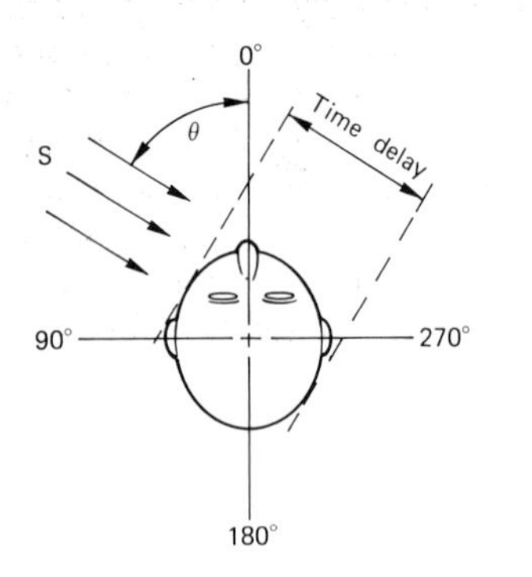

Figure 2.35 The interaural time delay is zero for a source S at 0° and proportional to sin θ, reaching a maximum of about 0.8 ms at 90°

a quarter-wave (closed-pipe) resonance rise at the eardrum of about 10 dB at 3 kHz. When this is added to the 12 dB peak revealed in Figure 2.32, the greater overall sensitivity in the 3 kHz region illustrated earlier in Figure 2.14 is explained. It may also be mentioned that the action of the ear is essentially non-linear. For example, the eardrum vibrations are simply air loaded on outward swings but must bear against the bone assembly leading to the inner ear on inward swings. This non-linearity leads to the production of spurious harmonics even when a pure tone (fundamental) is heard (see the reference to dissonance in Section 2.6). There is also a kind of reflex limiter action affording some protection against unduly loud sounds. This starts to operate at about 80–90 dB SPL but has a rather slow reaction time of 10 ms, and so is ineffective against very sudden sharp impulses.

Finally, mention should be made of masking, the well-known effect which makes it more difficult to hear one sound in the presence of another. In general, a masking tone or noise band has a greater obscuring effect on higher frequencies. This is illustrated in Figure 2.34, which shows the effective upwards shift in the audibility threshold level at all other frequencies in the presence of a masking tone at 400 Hz or 2000 Hz and at various masking levels.

2.16 Interaural time delay

Another important consequence of head dimensions and the interaural spacing (of approximately 21 cm) in binaural listening is that off-axis sounds arrive at the more distant ear after a small interaural time delay (ITD). In Figure 2.35 it is clear that sounds arriving from source S at some angle θ to the median plane through the listener's head will reach the left ear earlier than the right. The ITD will increase in proportion to sin θ and reach a maximum at 90°. Regarding the head as a sphere of diameter 21 cm, the maximum path difference from ear to ear will be $\pi D/2$ = 33 cm. Taking the speed of sound to be 344 m/s, this gives a maximum ITD of 0.9 ms, which is close to the 0.8 ms found by experiment for real non-spherical heads.

Note that this ITD clue to source direction works only for low frequencies, i.e. when the wavelength is large compared with the head dimensions, and in effect produces interaural phase differences. When the wavelength is 38 cm or less (900 Hz or above) there is a degree of phase ambiguity, and it becomes uncertain which ear signal is leading. The confusion is complete at 1250 Hz, when the phase shift is 180°. However, there is evidence that the brain then shifts its attention away from the fine detail of complex sounds to concentrate on the ITD of the leading edge or 'attack' of the delay envelope, enabling transient sounds of less than a few milliseconds' duration to be located.

Certainly, ITD provides a vital location cue at low frequencies where, as was shown in the previous section, IAD is practically non-existent. An ITD of only 0.03 ms can be detected, corresponding to a path length difference of only 1 cm. This relates to an angle shift of a mere 3° at the front, but angular discrimination decreases to about 7.5° at the side. Up to about 800 Hz, where the maximum ITD corresponds to half a wavelength (180°), the ITD introduces an identifiable phase 'lag' at the more distant ear, enabling the brain to locate the sound source to left or right of centre accordingly.

Recent research has discovered that the brain analyses the information sent to it by the two ears in a highly complex manner. For example, paths of communication exist, not previously identified, between the two halves of the brain and from the brain back to the ears. The explanations given here are therefore an oversimplification, but they should serve as a pointer to the more important directional hearing mechanisms.

2.17 Directional hearing

The two previous sections have indicated the principal mechanisms providing directional cues in human hearing, interaural amplitude difference (mainly effective at high frequencies) and interaural time delay (mainly effective at low ones). To these may be added a third directional cue, an interaural timbre or spectral content difference, arising from shifts in the relative amplitudes of the overtones or harmonics present in complex sounds. This third cue and various others such as bone conduction, visual cues and head movement are brought into the argument since simple IAD and ITD alone cannot explain some of the subtleties of human directional hearing.

So far, only directional hearing in the horizontal plane has been considered in a left–right sense. If the discussion is extended to include front–back hearing and sound incidence in all three dimensions, a number of limitations and confusions emerge. To begin with, circumstances can arise in which it becomes difficult to decide whether a source on axis is in front or behind the listener (0° or 180°) or indeed at any angle of elevation in the median plane (vertical through the centre of the head). Assuming identical ears, this becomes a (monaural) function of small spectral differences. Blauert has shown that the perceived median plane direction is associated with particular frequency bands, quite regardless of the actual source direction, as follows:

Front: 260–550 Hz 2.5–6 kHz
Rear: 700–1800 Hz 10–13 kHz
Above: 7–10 kHz

Again, if only the ITD is being considered, we should note that a given ITD could arise from any point on an arc or so-called 'cone of confusion', as shown in Figure 2.36.

One situation in which confusion in the directional hearing mechanism might be expected to arise is in determining the direction of a sound source in a reverberant environment. In the diffuse field the intensity at the ear more remote from the source may be about the same as that at the nearer ear, or even greater, due to random reflections (or the presence of standing waves or sound-reinforcement louspeakers). However, research by Haas demonstrates that the brain identifies the source direction from the first sound received: provided there is a time delay of between 5 and 35 ms, the direct sound can be up to 8 dB weaker than the reflected sound and still be correctly identified.

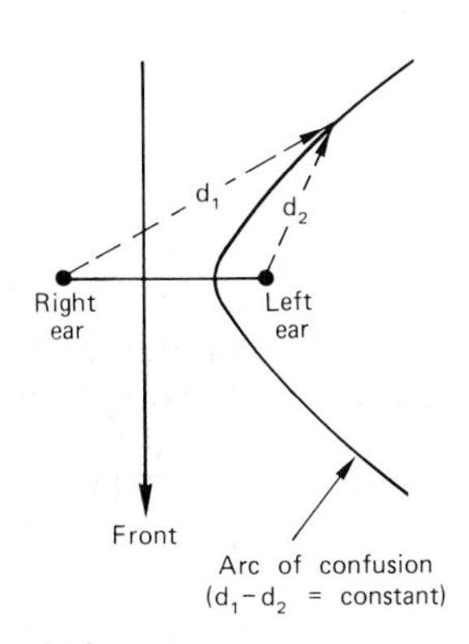

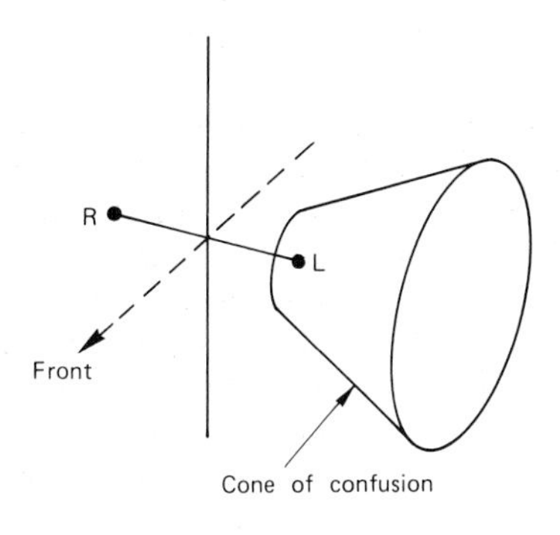

Figure 2.36 The same ITD, proportional to $(d_1 - d_2)$, will exist for a source located anywhere on (a) an arc and (b) a three-dimensional cone, making it difficult to decide the true direction based on ITD alone

3
Basic electronics

3.1 Molecules and atoms

The notion of air as being made up of small particles was used in the previous chapter to describe the mechanism of longitudinal transmission of sound waves. In fact all material substances whether in the solid, liquid or gas state are composed of extremely small particles known as *molecules*. A molecule is the smallest part of any substance which can retain its characteristic properties. All molecules are in a state of continuous motion, an important difference being that in a solid object they are relatively close together and so the strong attractive forces between them enable the object to retain its shape. In a liquid they are more widely spaced and so the liquid will take up the shape of any container into which it is poured. In a gas the even greater molecular spacing allows the gas to expand freely or contract in response to the ambient pressure conditions imposed.

A molecule is capable of subdivision into elementary electrochemical units called *atoms*. When the molecules of a substance comprise only one kind of atom the substance is called an element. Substances whose molecules contain atoms of more than one element are called chemical compounds. At present just over a hundred elements have been identified.

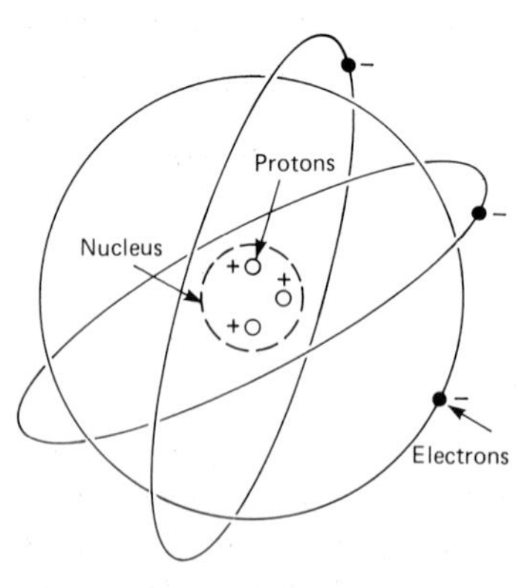

Figure 3.1 Model of the atom as an electrically stable system comprising a heavy positive nucleus surrounded by orbiting electrons

Each atom is conceived as resembling a tiny solar system (a model now somewhat modified by wave mechanics theory) with small negatively charged particles (*electrons*) orbiting like the Earth and other planets around a very much heavier, positively charged nucleus acting as the sun (Figure 3.1). This nucleus contains the same number of positive particles (*protons*) as there are electrons, together with a generally equivalent number of neutral particles or *neutrons*.

Hydrogen is the simplest element, its atoms consisting of just one electron and one proton. Then comes helium with two electrons, two protons and two neutrons and so on up the atomic series to nobelium, with 102 of each. The electrons stack up in a series of orbital shells of fixed radii, and the energy contained in an atom depends on its particular electron arrangement.

3.2 Static electricity

The electrons in the outer peripheral shell of some elements are relatively loosely bound to the atom, partly because of the presence of negatively charged electrons on the shells between them and the positive nucleus, and can migrate between adjacent atoms or be removed by the application of a suitable force. This force can be frictional, for example, as when a glass rod is briskly rubbed with a piece of silk and some of the electrons in the glass will transfer to the silk. This sets up a charge of static electricity in each object, the silk being negatively charged and the glass being 'ionized' or given a positive charge due to the temporary deficiency of electrons.

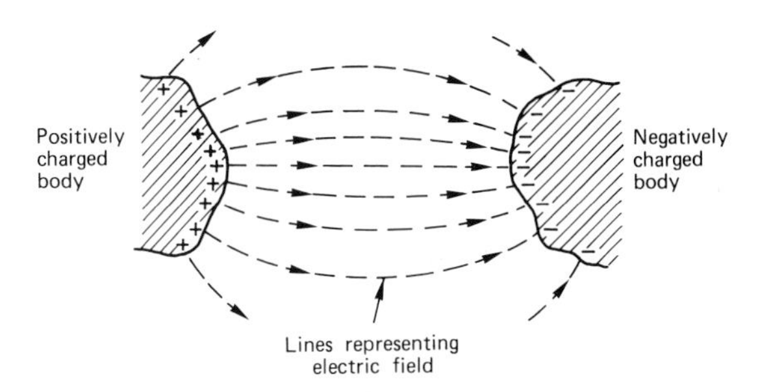

Figure 3.2 A static electric field is set up between two oppositely charged bodies

Figure 3.2 shows two oppositely charged bodies a small distance apart. Since work must have been done to remove electrons from their normal orbits, the system shown must possess potential energy. This reveals itself through (1) the difference of electrical potential (pd) and (2) a mechanical force of attraction between the two objects. It is visualized as an electric field of imaginary lines of force as shown. The general rule relating to electrically charged bodies may be stated as follows: 'Like charges repel and unlike charges attract one another.'

With time, the charges will tend to leak away, as the excess electrons find a path back to the positive object, and both objects will return to their neutral state. This process could be accelerated by bringing the objects slowly together until the potential difference was sufficient to cause the electrons to jump the gap and produce a spark discharge. Alternatively, the objects could be joined by a wire or thread of electron-conducting material, as considered in the next section.

3.3 Electric current

In certain substances the outer-shell electrons are relatively easily made to flow or migrate in a specific direction when an electrical charge or potential difference (pd) is applied at two points. In Figure 3.3 a pd is applied at opposite ends of a wire of such a substance (labelled positive and negative) and the resulting electromotive force (emf) causes a flow of electrons away from the negative terminal and towards the positive one.

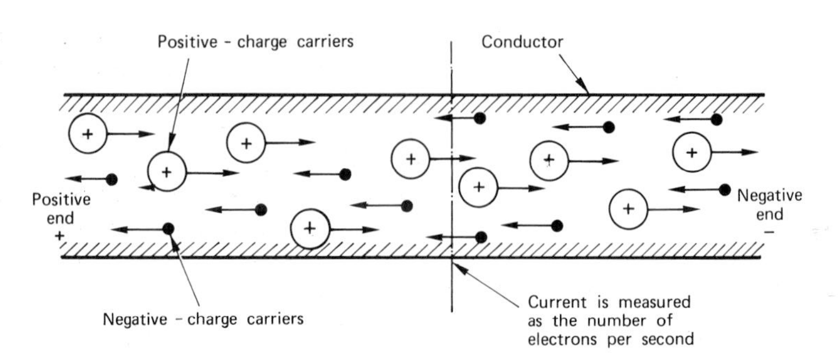

Figure 3.3 Applying a potential difference across the ends of a conductor results in a flow of electrons towards the positive end and a smaller contrary flow of positively charged particles

This organized flow of electrons is called an *electric current* and substances which permit easy current flow are called *conductors*. Among the best conductors are silver, copper and gold in that order. The number of electrons flowing past a certain point per second defines the rate of current flow (symbol I) which is measured in amperes (or amps). One amp corresponds to a flow of one unit of charge (Coulomb) per second, which is 6.24×10^{18} electrons.

It may be noted in passing that a small amount of contrary flow of positively charged particles may take place (e.g. of small water drops or dust particles), but the principal current flow is of electrons. By convention, current is regarded as flowing from positive to negative, which is in the opposite sense to the actual flow of electrons. As a further comment, it may be noted that, despite the fact that a current starts to flow as soon as an electric field or potential difference is applied across the ends of a conductor, the actual movement of electrons consists of a relatively slow drift with a velocity of only a few millimetres per second.

In contrast to the category of substances just described as conductors, there exist numerous materials whose electrons are relatively tightly bound to their parent atoms so that electron flow is strongly resisted. Instead an applied electric field sets up a condition of strain within the atoms causing them to present negative and positive charges on opposite faces. These substances are called *insulators*, and can be used to act as a barrier to or restrict the flow of currents or charges. Effective insulators include rubber, plastics and dry air.

The flow of an electric current is accompanied by a number of side effects, including the redistribution of charges, heating of the conductor (and the radiation of light when the conductor is a thin filament as in an electric lamp), chemical changes and the generation of a magnetic field.

3.4 Resistance

The flow of electrons is impeded to some extent by the molecular lattice in even the best conductors. This opposition to current is called resistance, and results in some of the electrical energy being progressively converted to heat along the current path. In the familiar analogy, comparing the flow of current along a conductor with the flow of water in a pipe, electrical resistance is seen as the equivalent of frictional resistance in the pipe. The rate of water flow is

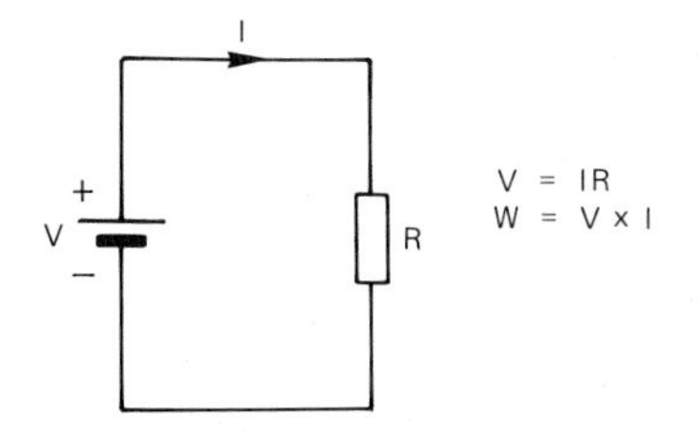

Figure 3.4 Ohm's Law. Applying a potential difference V volts to a circuit of resistance R ohms produces a current of I amperes, where $V = IR$

proportional to the pressure difference between the ends of the pipe, and frictional resistance causes the water pressure to fall uniformly along the pipe.

The analogous quantity to water pressure is the pd, measured in volts, and this is similarly proportional to the rate of flow of current and falls uniformly along the conductor. Thus doubling the applied voltage will double the current, while doubling the resistance will halve the current. This relationship is summarized in Ohm's Law: $V = I \times R$, where V = pd (in volts), I = current (in amperes) and R = resistance in ohms (symbol Ω), a practical unit representing a conductor in which an applied pd of 1 V produces a current of 1 A.

Figure 3.4 shows this in diagrammatic form: a source V gives rise to a current I flowing through the load resistance R. The symbol for V shown in the figure is that for a single-cell battery, a device which uses electrochemical action to maintain a steady pd between its positive (red) and negative (black) terminals. Other sources of unidirectional (direct current) voltage use electromagnetic or petrol-motor generators.

It should be mentioned that all sources of voltage possess a certain internal resistance r. This has been been omitted in the figure but will, of course, reduce the available voltage in practical circuits in proportion to $(I \times r)$. Similarly, the transducers, amplifiers and signal processors to be discussed later possess internal resistance which appears across their output or input terminals as 'output impedance' or 'input impedance', respectively.

The power or rate of doing work in a DC circuit is measured in watts and equals the product of volts and amperes: i.e.

$$W = V \times I \text{ or, from Ohm's Law, } W = I^2 \times R \text{ or } W = V^2/R$$

Two pieces of electronic equipment are said to be correctly 'power matched' when the transfer of power from one (the generator) to the other (the load) is a maximum. The condition for this is that the internal resistance or output impedance of the generator is equal to the input impedance of the load. On the other hand, the principal requirement in microphone circuits is for 'voltage matching' or the transfer of maximum voltage to the load, and this is best met by making the load impedance several times higher than the microphone impedance.

3.4.1 Combinations of resistors

More than one resistor may be present in an electrical circuit, the two basic forms of interconnection being series and parallel, as illustrated in Figure 3.5.

Resistors connected in series have the same current flowing through them. Therefore, from Ohm's Law, the individual voltages across R_1 and R_2 in Figure 3.5(a) are given by $V_1 = IR_1$, and $V_2 = IR_2$. In Figure 3.5(b) R_1 and R_2 are replaced by the equivalent resistance R, i.e. the voltage V produces the same current I and so $V = IR$.

Now, since $V = V_1 + V_2$, we can say that:

$$IR = IR_1 + IR_2 = I\,(R_1 + R_2)$$

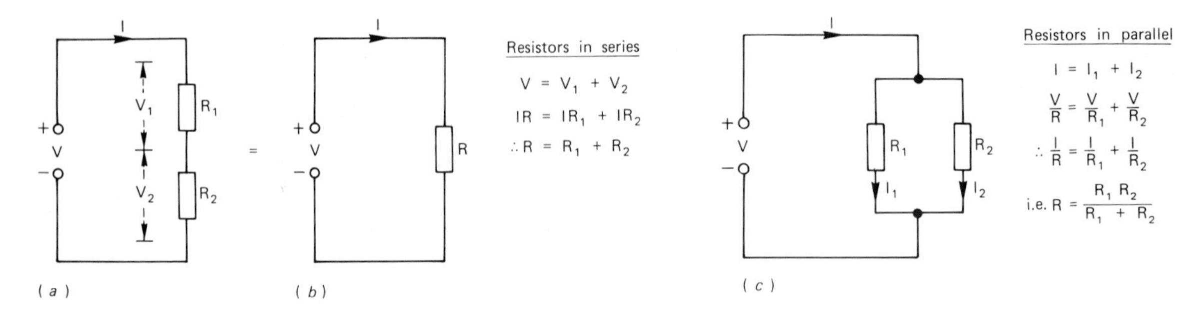

Figure 3.5 Combinations of resistors. (a) Resistors in series can be replaced by (b) a single resistor R, where $R = R_1 + R_2$; (c) resistors in parallel

Therefore $R = R_1 + R_2$ or 'resistances in series are equivalent to a resistance which equals the sum of the individual resistances'.

In the case of resistances in parallel (Figure 3.5(c)) it is the voltage V which is common to both resistors and different currents will flow through them as follows:

$$I_1 = \frac{V}{R_1} \text{and } I_2 = \frac{V}{R_2}$$

Again replacing R_1 and R_2 by the equivalent single resistance R in the expression $I = V/R$, and noting that $I = I_1 + I_2$, we can say:

$$\frac{V}{R} = \frac{V}{R_1} + \frac{V}{R_2}$$

$$\text{i.e. } \frac{I}{R} = \frac{I}{R_1} + \frac{I}{R_2}$$

or 'resistances in parallel are equivalent to a single resistance whose reciprocal is equal to the sum of the reciprocals of the individual resistances'. It follows that when resistances are connected in parallel, the resultant resistance is smaller than any of the individual ones.

A special case of resistances in series or in parallel is met when we consider conductors such as cables of various lengths or thicknesses. Intuitively we can guess that a long cable will introduce more resistance than a short one of the same material (resistance is proportional to wire length) and that a thin cable will offer more resistance than a thick one (resistance is inversely proportional to wire cross-sectional area). Measurements indeed confirm these two proportionality statements, so that the resistances of microphone cables, for example, are quoted in 'ohms per metre' and the resistance of a two-strand cable is only half that of the single-strand version.

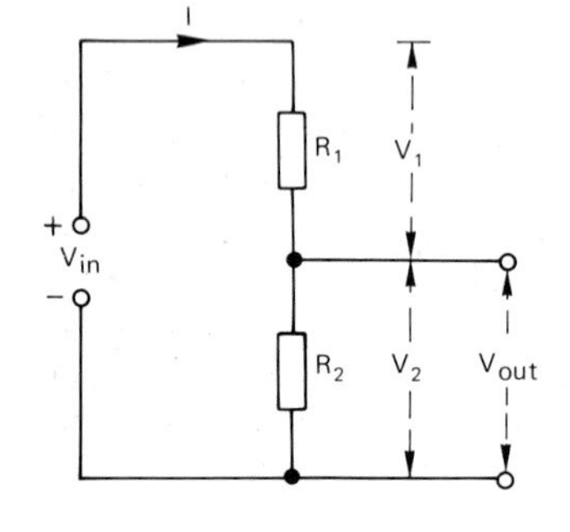

Potential divider

Attenuation ratio $\frac{V_{out}}{V_{in}}$

$= \frac{IR_2}{I(R_1 + R_2)} = \frac{R_2}{(R_1 + R_2)}$

e.g. cct $R_1 = 99\Omega$, $R_2 = 1\Omega$

Attenuation ratio $= \frac{1}{100} = 0.01$

$= -40\text{dB}$

Figure 3.6 The potential divider

3.4.2 The potential divider

In Figure 3.6 two resistors R_1 and R_2 are connected in series, that is, $V_{IN} = I \times (R_1 + R_2)$. This forms a potential divider in which $V_{IN} = V_1 + V_2$ and it is possible to construct an attenuator or loss pad by taking an output from across either R_1 or R_2. Taking V across R_2, for example, attenuates the voltage in the ratio:

$$\frac{V_{OUT}}{V_{IN}} \text{ i.e. } \frac{I \times R_2}{I \times (R_1 + R_2)} \text{ or } \frac{R_2}{(R_1 + R_2)}$$

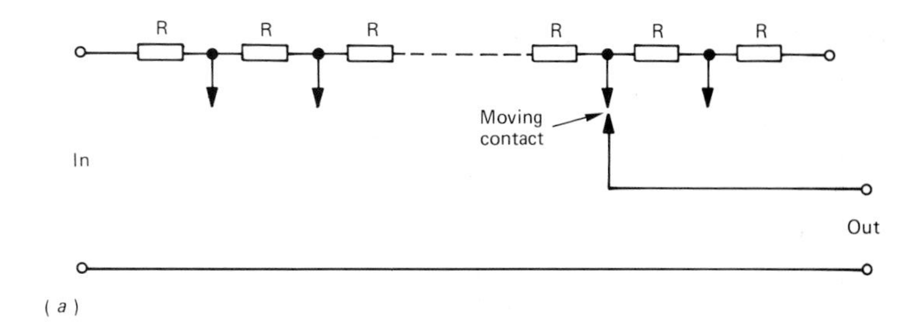

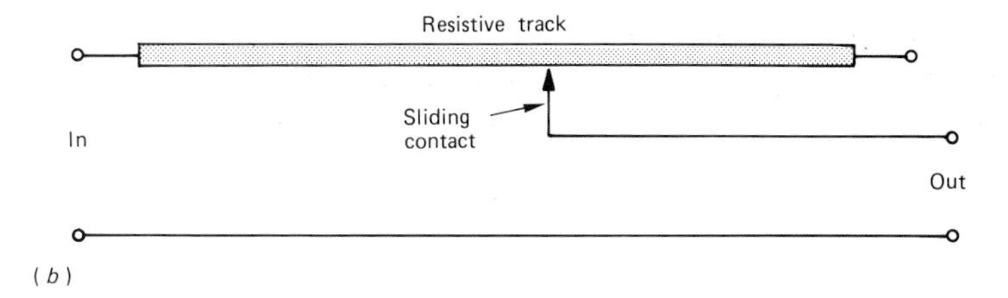

Figure 3.7 The variable attenuator (potentiometer) can comprise (a) a series of fixed resistors; (b) a continuous track of resistive material

e.g. if $R_1 = 99\Omega$ and $R_2 = 1\Omega$, the voltage will be attenuated in the ratio:

$$\frac{1}{100} = 0.01$$

or, in decibels, attenuation equals 20 log 0.01 = −40 dB.

A variable attenuator, fader or potentiometer (pot) can be produced either by connecting a number of resistors in series and moving a spring-loaded slider connector across the junctions or a sliding contact along a continuous track of resistive material (see Figure 3.7).

3.5 Capacitance

If two conductors are separated by an insulator they form a capacitor (old name, condenser) which can be used to store or hold an electrical charge. Figure 3.8 represents a capacitor, perhaps consisting of two metal plates X and Y with an air space between them. When the switch is moved to A, electrons will flow from the negative terminal of the battery and, since they cannot cross the gap, accumulate on plate X to give it a negative charge. This will repel electrons from plate Y, causing them to move towards the positive terminal of the battery, leaving plate Y with a positive charge.

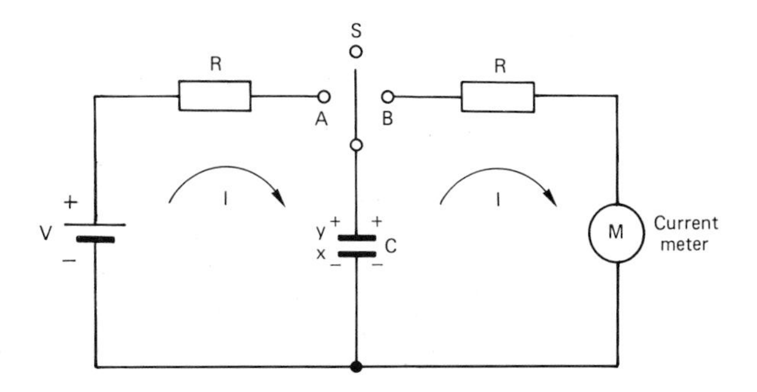

Figure 3.8 Capacitance. Moving switch S to position A will cause a brief flow of current and charge up the capacitor as shown; then moving the switch to B will produce a discharge current which will be indicated on the meter M

Moving the switch back to the open position will leave this state of charge with a static electrical field between the plates. This storage ability is called *capacitance* (symbol C), a practical unit being the microfarad (μF). The standard unit the farad (= 1 million μF) is the capacitance of a capacitor in which the application of 1 V produces a charge Q of 1 coulomb: i.e. $C = Q/V$.

The capacitance of a given pair of fixed plates depends on their area, the distance between them and the type of material or dielectric filling the space. This is summed up in the expression:

$$C = \frac{AK}{2\pi d}$$

where C = capacitance in picofarads (10^{-12} farad), A = area of each plate (cm^2), d = spacing distance (cm), and K = the dielectric constant of the material between the plates (for air $K = 1$; for mica $K = 4$–7).

If the switch is now moved to position B, the stored charge of electrons will flow round the right-hand circuit and give an indication on the current meter M. Two identical resistors R have been included in the diagram, and it should be noted that the charging and discharging operations do not take place instantaneously but over a finite period which depends on the values of both C and R. Typical curves for the voltage and current in the circuit during the charge and discharge periods are shown in Figure 3.9. These curves are of

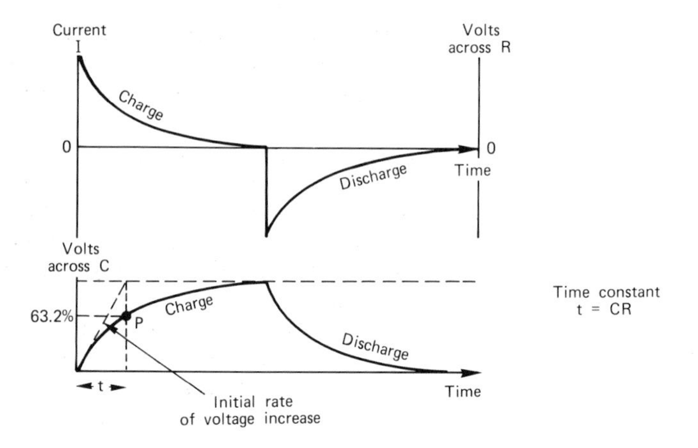

Figure 3.9 Typical curves plotted against time for the current and voltage measured across C during the charge and discharge cycle described in Figure 3.8

the type known as exponential, i.e. their slope is proportional to the value of the varying quantity at any instant. Thus when the current has fallen to half the initial value, its rate of decrease has also fallen to half the initial rate, and so on. Theoretically, therefore, the fully charged state is never reached. In practice, the time that it would take to charge the capacitor at the initial rate (if this were maintained) is used as a reference value. It can be shown mathematically that in this time the charge on the capacitor has in fact reached 63.2% of the final value (see point P on the figure). This time is called the *time constant*. It is independent of the applied voltage V but equals the product of C and R: i.e.

Time constant $t = CR$

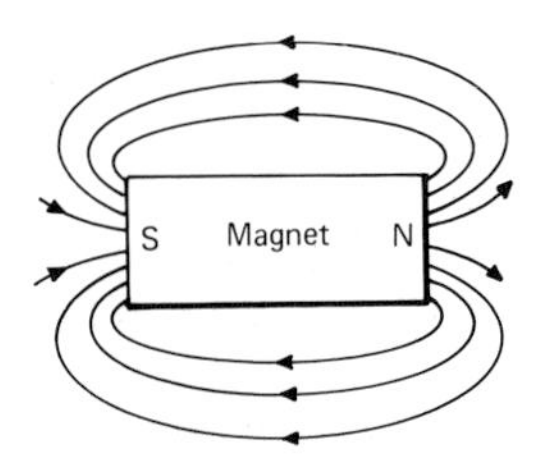

(a)

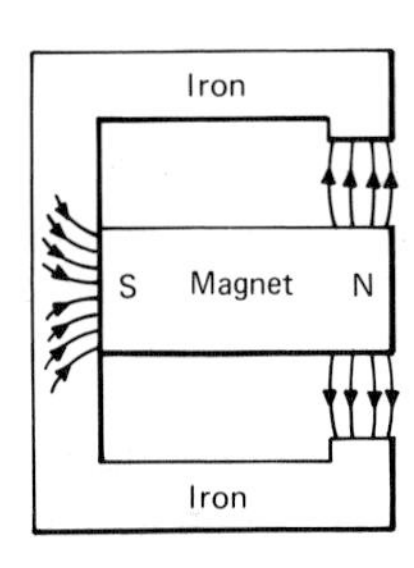

(b)

Figure 3.10 Magnetic fields. (a) Field surrounding a bar magnet; (b) adding iron polepieces will concentrate the field across the gaps

where t = time constant (μs), C = capacitance (μF), and R = resistance (Ω), e.g. in a circuit consisting of a 1 μF capacitor in series with a 1 MΩ resistor, the time constant is 1 s.

3.6 Inductance

The picture of a charged capacitance as storing static electrical energy between its positive and negative plates has an obvious similarity with a bar magnet which stores magnetic energy between its North and South poles. Magnetizable substances such as soft iron or steel comprise groups of molecules or domains which behave like tiny bar magnets. In their normal state, these domains are randomly orientated but they can be brought into a state of North/South alignment by the application of an external field. The material is then said to be magnetized and exhibits a magnetic field (Figure 3.10(a)) which is concentrated around two regions called poles and denoted North and South, respectively, as are the North- and South-seeking poles of a compass needle. The domains of soft iron are comparatively easily turned whereas hard steel is more difficult to magnetize but is more 'permanent' in its storage ability – i.e. it is less easily demagnetized. Figure 3.10(b) shows how the field of a bar magnet can be persuaded to flow round a soft-iron path to develop a concentrated field across a specially shaped air gap. This principle is used, for example, in tape heads and moving-coil microphones and loudspeakers.

The close affinity or interdependence between electrical and magnetic energy is illustrated by the fact that any flow of electric current is accompanied by the setting up of a proportionate magnetic field. Current flowing in a straight wire, for example, sets up a circular field whose direction is defined conventionally by the Right-hand Rule (see Figure 3.11), in which the thumb shows the direction of current flow and the fingers indicate the polarity of the field.

Placing a current-carrying conductor near another separate conductor (Figure 3.12) so that its magnetic field cuts through the second wire has the effect of inducing an electromotive force in the second circuit and hence a flow of secondary current. This, of course, is the way in which electrical interference can occur in a microphone cable, for example, as stray magnetic fields from power cables or radio waves induce currents in the cable.

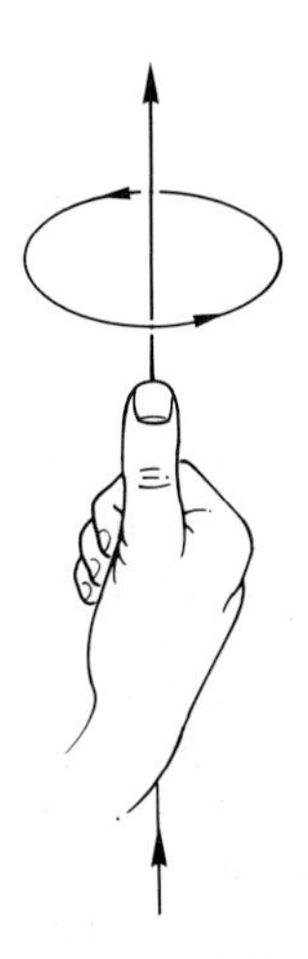

Figure 3.11 Right-hand rule. With the thumb pointing in the direction of current flow, the field polarity is as indicated by the fingers

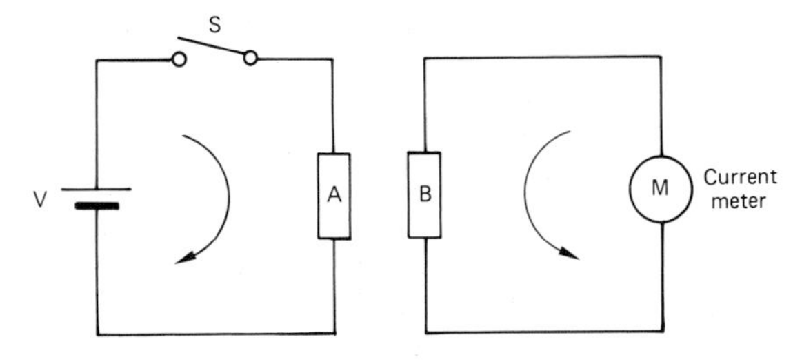

Figure 3.12 Magnetic induction. Closing the switch S causes current to flow through conductor A. The magnetic field built up links with conductor B and induces a secondary emf and current flow

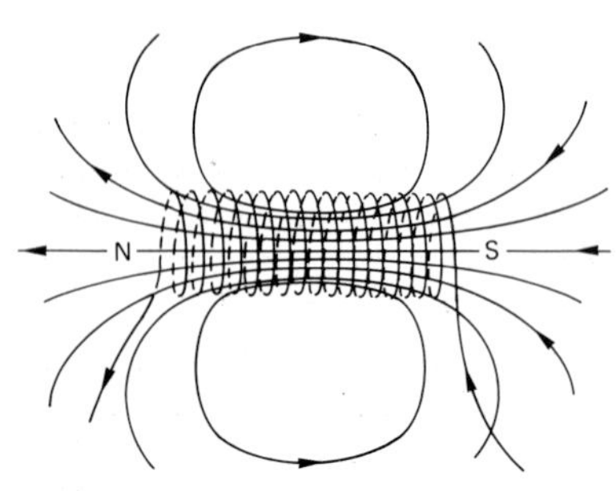

Figure 3.13 Passing current through a coil of wire produces an external magnetic field resembling that of the bar magnet shown in Figure 3.10(a)

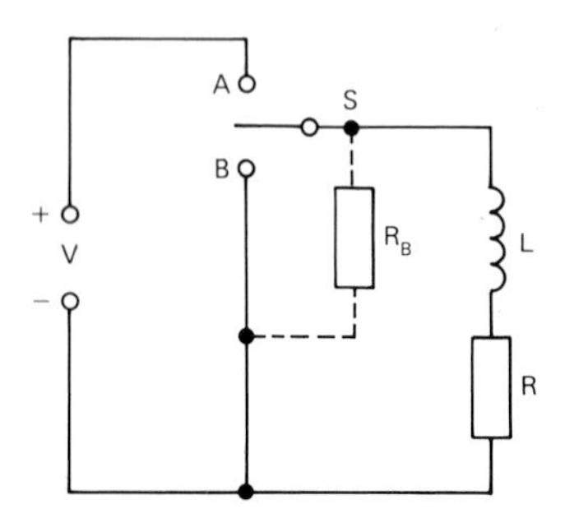

Figure 3.14 Inductance. Moving switch S to position A will cause current to flow through the inductor L and series resistor R and set up a magnetic field around L. Then moving the switch to B will produce a discharge current through R as the field collapses

Note that there must be relative movement between the two conductors, or growth or decay in the magnetic field, so that there is an actual change in the linkage of magnetic lines of force (flux) for this mutual induction to occur. For example, the action of switching on and off the current in the primary circuit will cause the meter in the secondary circuit to kick first in one direction and then the other as the magnetic field first builds up and then decays. When steady DC current is flowing, no induction takes place. However, if there is relative motion between a conductor and a fixed magnetic field, there will again be a change in the flux linkage causing an emf to be induced. The value of the emf is proportional to the rate of change of flux linkage, regardless of whether this is brought about by a change in primary current or by relative motion between field and conductor.

The field surrounding a straight wire is naturally weak and dispersed along the whole length, but winding a wire into a coil produces a concentrated field very like that of a bar magnet (Figure 3.13). The ability of such a coil to store magnetic energy is called its *inductance* (symbol L) and is analogous to the capacitance of a capacitor. The unit of inductance is the henry (symbol H), defined as the inductance of a circuit in which a current change of 1 ampere per second induces a *back emf* (opposing the change) of 1 V.

Figure 3.14 corresponds to the circuit which was shown in Figure 3.8 to illustrate the charging and discharging of a capacitor. Here setting the switch to A will produce a flow of current and set up a magnetic field around the inductor L. Throwing the switch to B will cause the field to collapse, sending a short-term current round the circuit through R. (In practice it will be necessary to bypass the switch with a suitable resistor R_B, otherwise at the instant when the switch becomes 'open circuit' and the current is suddenly reduced to zero a large back emf will be created.)

Naturally, as for the capacitor, the build-up and decay of the magnetic field do not occur instantaneously, since each is opposed by a back emf, but over a finite period which depends on the values of both L and R (see Figure 3.15). Again the curves are exponential and the time constant of the circuit is L/R s. This is the time that the current would take to reach its final value if it continued growing at the initial rate, and is actually the time taken for the current to reach 63.2% of its full value.

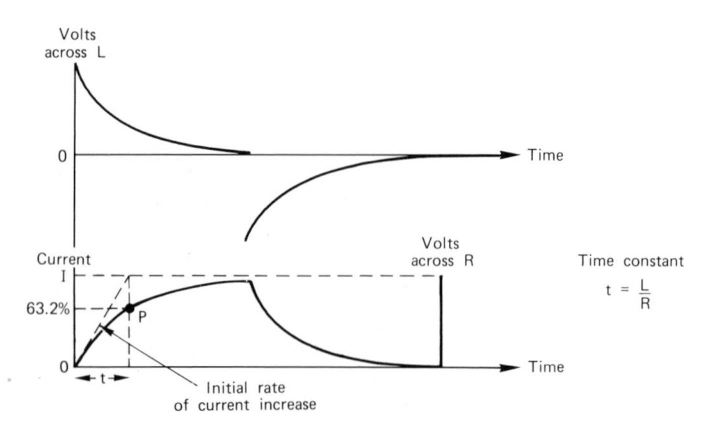

Figure 3.15 Typical curves plotted against time for the current and voltage across L during the charge and discharge cycle described in Figure 3.14

3.6.1 Electromagnetism

In the same way as introducing a dielectric material between the plates of a capacitor increases the capacitance, a core of soft iron may be used to increase the inductance of a coil of wire. This produces an electromagnet behaving very much like an ordinary bar magnet and capable, for example, of attracting iron objects. The relay in Figure 3.16 consists of an electromagnet and a hinged iron lever acting as a switch to provide remote-controlled operation of the cue light. Closing the key produces a flow of current, thus energizing the electromagnet so that it attracts the lever, completes the right-hand circuit and operates the cue light. Other simple examples of remote-controlled systems include buzzers, bells, doors and tape machines.

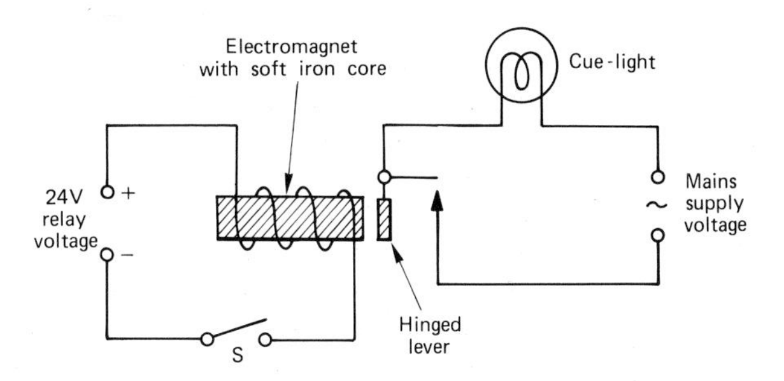

Figure 3.16 Electromagnetism. A simple relay uses an electromagnet and a hinged lever to provide remote-controlled switching of electrical circuits

At another level of sophistication are the various electromagnetic transducers used to convert mechanical energy into electrical energy and vice versa – notably moving-coil microphones and loudspeakers. These handle audio frequencies but are otherwise using the principles of the electric generator (dynamo) and motor, respectively. In a generator, the conductor is driven by an external force making it move within a magnetic field. As it cuts through the lines of magnetic force it has a current induced in it. Conversely, in a motor, current is fed to a conductor set in the field of a permanent magnet and this produces a force causing the conductor to move. The direction of the force is always perpendicular to the plane containing the current and the external field, and its value is proportional to the strength (flux density) of the field, the current and the effective length of the conductor within the field: i.e.

F is proportional to $B \times I \times l$

where F is the force (Newtons (N)), B is the flux density (webers (Wb)), I is the current (A), and l is the length of the conductor (m).

Figure 3.17 illustrates the moving-coil microphone case, where a coil of wire is attached to a thin diaphragm. The diaphragm is set into vibration by the arrival of a sound wave in which the air pressure swings above and below the normal atmospheric pressure (see Section 2.5). The coil is therefore driven to and fro within the field concentrated in the annular (circular) gap of a specially shaped magnet similar to the one shown in Figure 3.10(b). The result is an induced emf causing a flow of current through the external circuit which will reverse its direction during each half cycle of the sound waveform.

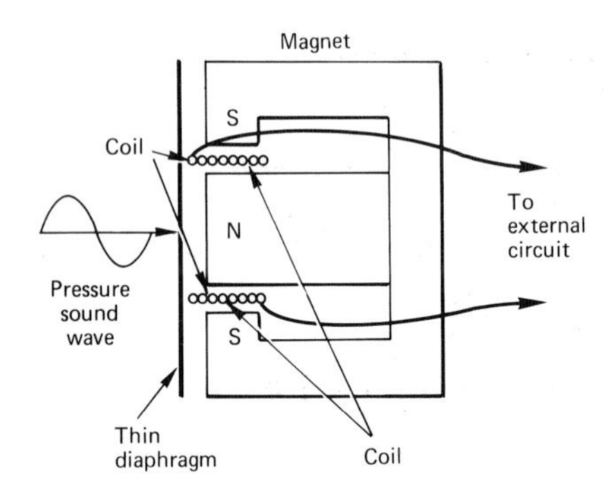

Figure 3.17 The moving-coil microphone. The pressure sound wave sets the diaphragm into vibration and motion of the attached coil in the magnetic field acts as a generator (dynamo) of electric current

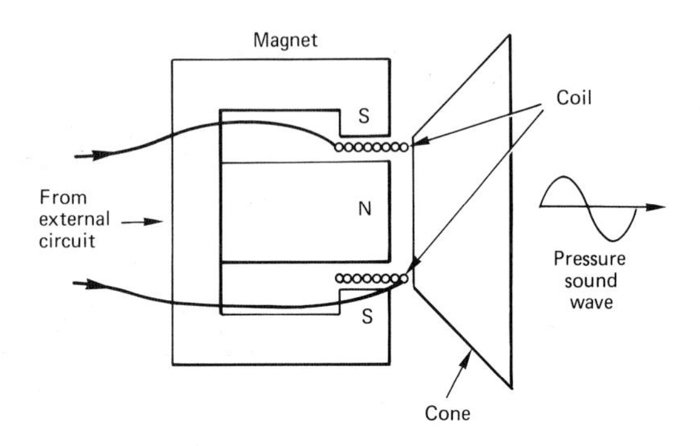

Figure 3.18 The moving-coil loudspeaker. Alternating current from an external circuit flows through the coil. Interaction between the coil field and that of the magnet acts as an electric motor driving the lightweight cone into vibrations which produce a pressure sound wave

Figure 3.18 shows the converse situation in a moving-coil loudspeaker which again consists of a coil immersed in an annular magnetic field but with the coil attached at the apex of a lightweight cone designed for efficient transfer of energy to the air. When an external emf causes current to flow through the coil the latter will experience a force causing it to move forwards or backwards, depending on the relationship between the direction of current flow and the fixed polarity of the field. A continuously reversing emf, of the kind just envisaged as being generated by sound waves arriving at the diaphragm of a moving-coil microphone, will set the coil and cone into to and fro vibration, causing radiation of equivalent sound waves into the air.

3.7 Alternating current

The voltage sources and electric current considered at the beginning of this chapter were of the undirectional DC type supplied by batteries. However, the signal currents met in telephony, sound recording and broadcasting are continually reversing their direction of flow. We have just seen this in the above description of moving-coil microphone and loudspeaker operation, where alternating emf and current were associated with the alternating air pressure in sound waves, i.e. nominally in the audio frequency range from 20 Hz to 20 kHz. Alternating currents (AC) are also the basis for the mains electricity supply in most countries, when the frequency is 50 Hz or 60 Hz, and for radio and television transmission and reception, using the much higher frequencies required for efficient radiation of electromagnetic waves over long distances.

The principal characteristics of an AC voltage source are its amplitude (peak voltage), its fundamental frequency (in hertz), its waveform (a sinewave when only a single frequency is present, but otherwise more complex) and its envelope – indicating the rate of initial growth (attack) and decay. These characteristics are illustrated in Figure 3.19.

Figure 3.19 Principal characteristics of an AC signal. (a) Amplitude; (b) frequency; (c) waveform; (d) envelope

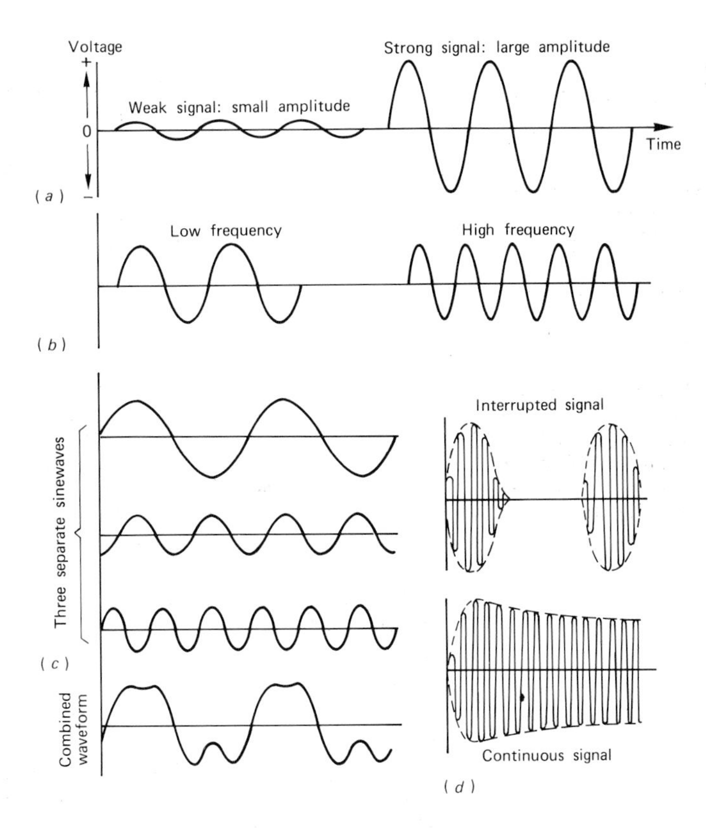

As was established for sound sources in Chapter 2, the simplest AC voltage generator performs simple harmonic motion to produce the familiar single-frequency sinewave. Figure 3.20(a) shows a 100 V peak AC voltage generator connected across a 500Ω resistive load, and Figure 3.20(b) shows the resultant voltage waveform, the plus and minus signs indicating, for example, clockwise and anticlockwise current flow. The variation in current with time can be calculated from Ohm's Law. Thus, at the peaks where $V = 100$ V, $I = 100 \div 500 = 0.2$ A; at half-volts $I = 0.1$ A and so on, enabling the broken curve for current to be drawn. It is seen to have the same shape and phase as the voltage waveform though with a different vertical scale calibrated in amperes instead of volts.

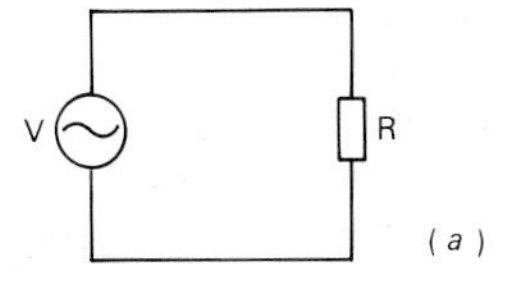

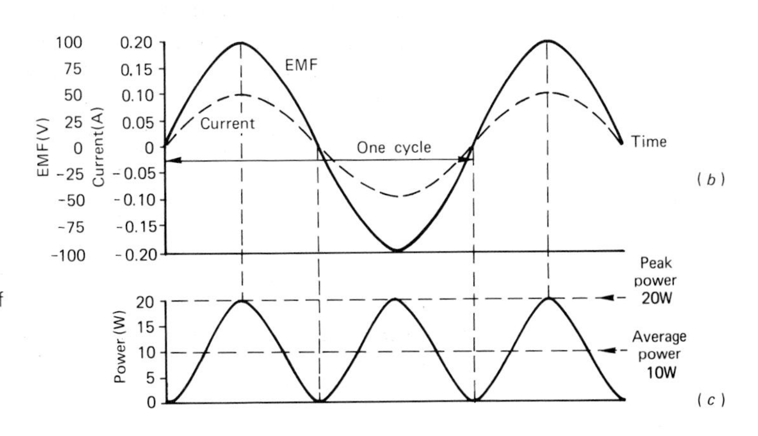

Figure 3.20 AC voltage, current and power. (a) Applying a 100 V peak emf across a resistive load of 500 Ω; (b) the curves of voltage and current plotted against time; (c) plot of instantaneous power

Power rating in DC circuits was specified as simply the product of voltage and current, i.e. W (W) = V (V) × I (A). Things are obviously more complicated in AC circuits since V and I are continually changing and indeed oscillating between positive and negative values. The power rating in AC circuits is therefore specified by comparison with the DC rating which would produce the same heating or lighting wattage. This will be the case if the average power taken from both types of electrical supply is the same.

In Figure 3.20(c), instantaneous values of power ($W = V \times I$) at various points in the AC cycle have been calculated by multiplying the instantaneous values of V and I in the 500Ω load. The resulting power curve has twice the frequency of the V and I curves since multiplying two negative numbers produces a positive answer (confirmed in practice by the fact that the electrical power supplied to a load by an AC source peaks twice in each voltage cycle).

Examining the power waveform, it is seen to have an average value of 10 W, i.e. half the 20 W peak value. This power rating has been arrived at by applying an AC voltage with a peak value of 100 V to a 500Ω load. It is equivalent to a DC situation where some DC voltage V is applied, and V_1 can be calculated from the expression

$$W = V_1^2/R$$

$$\text{i.e. } V_1^2 = W \times R = 10 \times 500 = 5000$$

$$\text{Therefore } V_1 = \sqrt{5000} = 70.7 \text{ V}$$

In other words, an AC voltage produces the same dissipation of power in a given resistive load as a DC voltage of 0.707 times the peak voltage. This is called the root mean square (rms) value since it corresponds to squaring the voltage at every instant throughout the cycle, taking the mean or average of all these and finally taking the square root of the result. It may be noted that $0.707 = 1/\sqrt{2}$, so that

$$V_{rms} = \frac{1}{\sqrt{2}} V_{peak}$$

Figure 3.21 shows the peak, rms and average values (ignoring sign) of a sinewave.

In effect, we have shown that applying an AC voltage to a resistance is, in general, similar to applying DC. The resulting rms current is proportional to the rms voltage as laid down by Ohm's

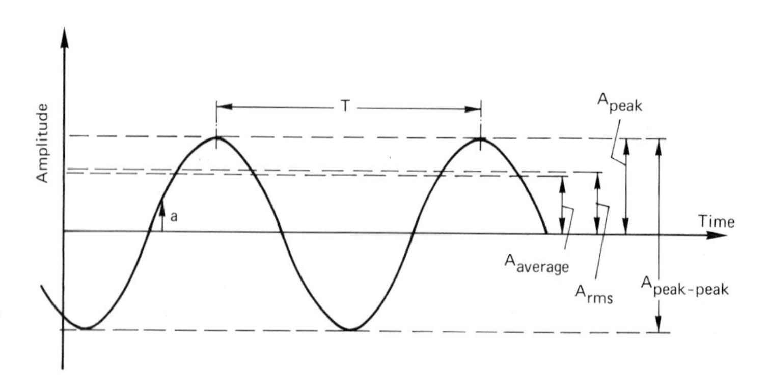

Figure 3.21 Comparison of peak, rms and average values of a sinewave (ignoring sign)

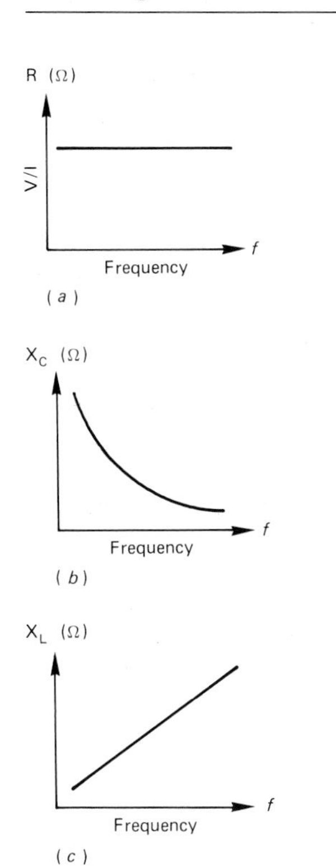

Figure 3.22 Graphs of resistance and reactance against frequency. (a) Resistance; (b) capacitive reactance; (c) inductive reactance

Law, i.e. $R = V/I$, and is the same at all frequencies. This is summed up in Figure 3.22(a) where the graph of R against frequency is a straight horizontal line.

3.7.1 AC and capacitance

Applying AC to a capacitance (Figure 3.23(a)) cannot, of course, produce a current through C since the air or dielectric in the gap between the capacitor plates acts as a block to current flow. However, charge and discharge currents do flow in the circuit on either side of the capacitor and the continual reversal of the applied voltage during each half cycle of the AC may be imagined as producing current flow into and out of the capacitor as the charge on the plates builds up to a maximum and then decays to zero, first in one direction or polarity, and then in the other.

In fact as a kind of bonus, when the applied voltage reverses its direction the discharge belonging to the previous half cycle forms the new charging current. Interestingly, the AC voltage and current are not in phase, as for resistance; current leads voltage by 90°. This is illustrated in Figure 3.23(b) and explained by the fact that, when V is a maximum, the capacitor is fully charged and the current is therefore zero. Similarly, when V starts to fall, the charge also falls at the same rate and so the current (which equals rate of flow of charge) increases and reaches a maximum when the V slope is maximum, i.e. at $V = 0$.

The opposition to AC current flow of a capacitor is measured in ohms and called its *reactance* (symbol X_c) to differentiate it from resistance. Since the current is proportional to both the charge-storing capacity C and the rate of charging and discharging (which in turn is proportional to frequency) we deduce from $X_c = V/I$ that:

$$X_c \text{ is proportional to } \frac{1}{fC}$$

The graph of X_c against frequency therefore takes the form shown in Figure 3.22(b), approaching V/I at DC ($f = 0$) and zero at very high frequencies. It can be shown mathematically that

$$X_c = \frac{1}{2\pi fC}$$

where X = reactance (Ω),
f = frequency (Hz), and
C = capacitance (F)

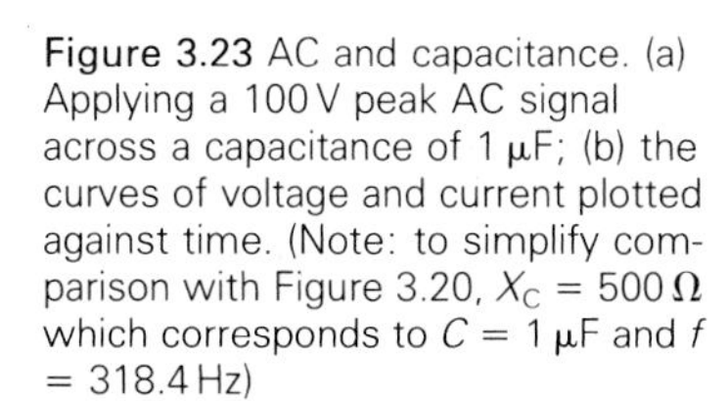

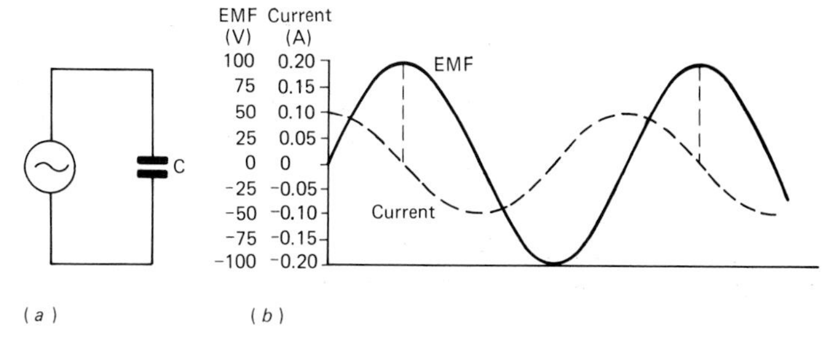

Figure 3.23 AC and capacitance. (a) Applying a 100 V peak AC signal across a capacitance of 1 μF; (b) the curves of voltage and current plotted against time. (Note: to simplify comparison with Figure 3.20, $X_C = 500\ \Omega$ which corresponds to $C = 1\ \mu F$ and $f = 318.4$ Hz)

e.g. a 1 μF capacitor will have a reactance of 3180 Ω at 50 Hz, 318.0 Ω at 500 Hz and 31.80 Ω at 5000 Hz.

The opposition to current flow in a circuit containing both resistance and reactance is called *impedance* (symbol Z) and can be calculated from the expression:

$$Z = \sqrt{(R^2 + X^2)} \text{ (i.e. taking rms values)}$$

where Z = impedance (Ω),
R = resistance (Ω), and
X = reactance (Ω).

3.7.2 AC and inductance

Similar reasoning can be used to examine the way that an inductance reacts to an applied AC voltage (Figure 3.24). In this case it is a magnetic field which is built up, rather than a static electrical field, first in one direction (polarity) and then the other during each half cycle of the applied AC. Here we find that voltage leads current by

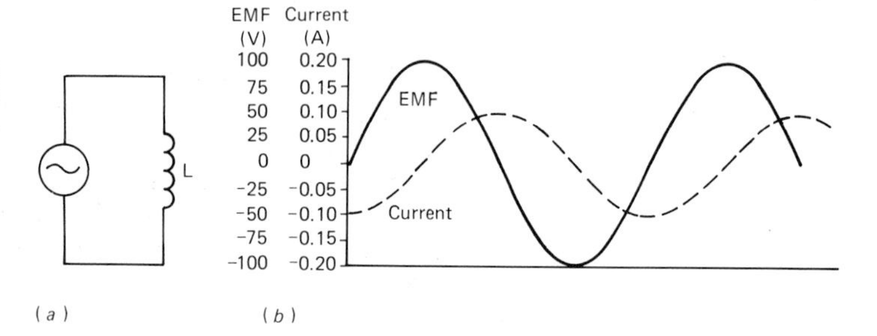

Figure 3.24 AC and inductance. (a) Applying a 100 V peak AC signal across an inductance of 0.2 H; (b) the curves of voltage and current plotted against time. (Note: to simplify comparison with Figures 3.20 and 3.23, X_L = 500 Ω, which corresponds to L = 0.2 H and f = 398.1 Hz)

90°, since voltage is a maximum when the rate of current increase is a maximum, i.e. current is zero, and the inductive reactance (X_L) is proportional to both frequency and inductance. In fact $X_L = 2\pi fL$, producing a graph of X_L against frequency which takes the form shown in Figure 3.22(c), approaching V/I at high frequencies and zero at DC ($f = 0$) since an inductor will always possess some DC resistance.

3.7.3 Transformers

The process of mutual inductance, whereby changes in the linkage between the magnetic field surrounding a current-carrying conductor and a second conductor induces an emf in the latter, was illustrated in Figure 3.12. Deliberate harnessing of this effect provides the basis of a *transformer* (Figure 3.25(a)) in which two coils of wire, insulated from each other, are wound on the same central former or soft-iron core in such a way as to ensure maximum flux linkage efficiency (minimum losses). Application of an AC voltage V_p in the primary circuit will cause an alternating current to flow through the primary winding. Assuming no losses, this will induce a voltage V_s in the secondary winding in exact proportion to the ratio between the number of primary and secondary turns, since V_p will induce the

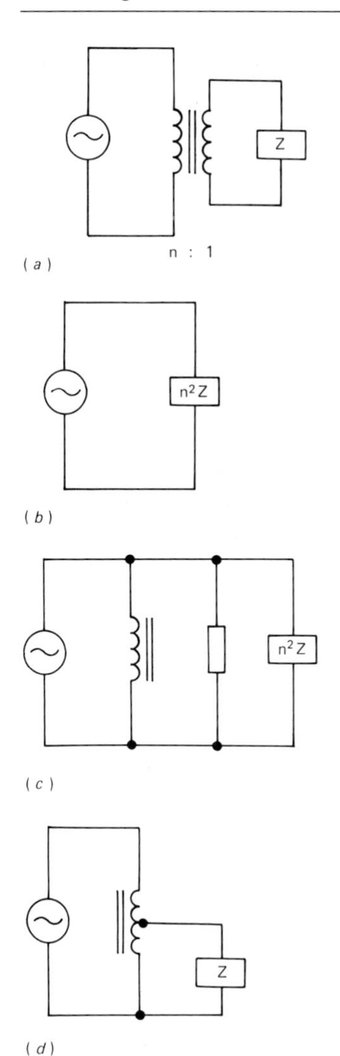

Figure 3.25 The transformer. (a) Transformer with turns ratio n:1; (b) load is transformed to $n^2 Z$; (c) transformer losses drawn as parallel inductance and resistance; (d) auto-transformer having a single-tapped winding

same voltage in every secondary turn. If the turns ratio is n (there are n times as many turns in the primary as in the secondary) it follows that the primary voltage is also n times the secondary voltage, i.e. $V_p = nV_s$.

This explains a transformer's ability to step up voltages (n is less than 1) or step down voltages (n is greater than 1). Obviously, a 1:1 transformer produces no voltage change but it provides a convenient method of isolating one circuit from another as far as DC or some kinds of interference are concerned.

When there are no losses, the power being taken from the transformer equals the input power, i.e. $V_p I_p = V_s I_s$. Therefore any step-up in voltage is accompanied by a proportionate reduction in current and vice versa. In other words, since $V_p = nV_s$, the currents are in inverse ratio: i.e.

$$I_p = I_s/n$$

A transformer is also useful for impedance transformation or *matching*. Thus the impedance or load 'seen' by the generator in Figure 3.25(a) is:

$$Z_p = \frac{V_p}{I_p} = \frac{nV_s}{(1/n)I_s} = n^2 \frac{V_s}{I_s} = n^2 Z_s$$

So the effect of the transformer is to present the source with a load which equals the load in the secondary multiplied by the turns ratio squared (Figure 3.25(b)). For example, using a 1:3 microphone transformer it would be necessary to provide a 9000Ω load if the microphone design called for 1000Ω loading. Note that in practical transformers there will be some inevitable losses due to wire resistance and magnetizing currents in the core. These can be indicated for more precise calculations by a parallel resistance and inductance, as shown in Figure 3.25(c). For some applications a secondary winding is not used, the step-down turns ratio being obtained by simply taking an output from a tapped connection on the single winding. This is called an auto-transformer (see Figure 3.25(d)).

3.7.4 Filters

One application of the variable reactance with frequency of capacitors and inductors is to construct frequency-selective filters to attenuate low frequencies (high-pass filter) or high frequencies (low-pass filter). This is illustrated in Figure 3.26(a), where a voltage V_{IN} is applied to R and C in series, forming a potential divider on the lines of the purely resistive attenuator shown in Figure 3.6. Taking the output across R will attenuate the voltage in the ratio V_{OUT}/V_{IN}, i.e. the attenuation ratio is:

$$\frac{IR}{I\sqrt{(R^2 + X_c^2)}} = \frac{R}{\sqrt{(R^2 + X_c^2)}}$$

Note that at the frequency for which $X_c = R$, the attenuation ratio is $R/\sqrt{2R^2} = 1/\sqrt{2} = 0.707$, which corresponds to 3 dB.

This −3 dB point on the filter response curve is called the *cut-off frequency* and its value can be calculated from:

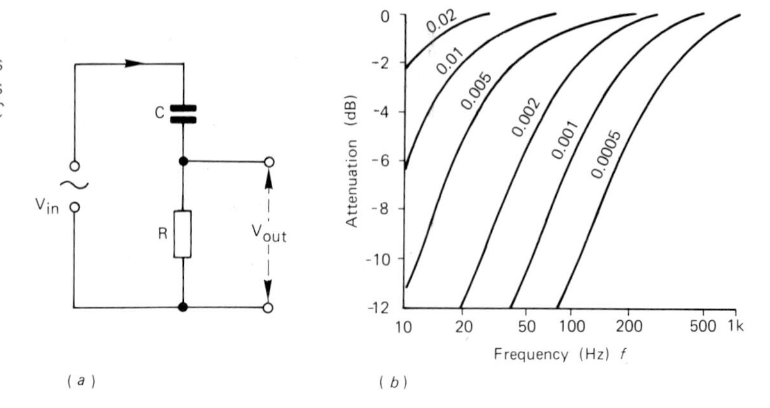

Figure 3.26 High-pass RC filter: applying AC input voltage across R and C in series, taking output across R; (b) the resulting attenuation curves for R = 1 MΩ and various values of C in microfarads as shown

$$R = X_c = \frac{1}{2\pi fC} \quad \text{i.e. } f = \frac{1}{2\pi RC}$$

The appearance of the time constant RC in this expression (as discussed in Section 3.5) should be noted, and indeed a filter is often specified in terms of its time constant rather than its cut-off frequency. For example, taking $R = 1\,\text{M}\Omega$ and $C = 0.001\,\mu\text{F}$ gives:

$$f = \frac{10^6}{2\pi \times 10^6 \times 10^{-3}} = \frac{10^3}{6.28} = 159\,\text{Hz}$$

and the time constant $RC = 10^6 \times 10^{-9} = 1\,\text{ms}$. It may also be noted in passing that rearranging the expression produces:

$$f \times RC = \frac{1}{2\pi} = 0.159$$

This is a useful rule-of-thumb number for quick calculation of the cut-off frequency for a given time constant and vice versa.

Figure 3.26(b) shows the attenuation curves for $R = 1\ \text{M}\Omega$ and various values of C. It will be seen that the filter slope straightens out at 6 dB per octave (per halving of frequency) below about the −10 dB level and is the same for all values of cut-off frequency.

Figure 3.27 illustrates the operation of a low-pass filter again using R and C in series but taking the output across C. This has the same

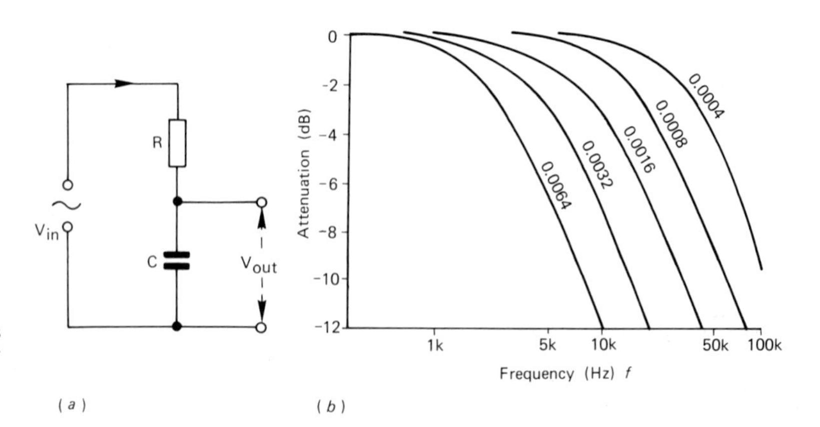

Figure 3.27 Low-pass RC filter. (a) Applying AC input voltage across R and C in series, taking output across C; (b) resulting attenuation curves for R = 10 000 Ω and various values of C in microfarads as shown

cut-off frequency (−3 dB point) and time constant as before and again the attenuation slope approaches 6 dB per octave. Similar filters can be constructed using inductors in place of capacitors. (Technical readers will appreciate that a number of assumptions have been made in the above basic explanations in the interests of keeping things simple. For instance, the AC generator is assumed to have zero internal impedance and the load across the output is assumed to be infinite, or at least high enough for its impedance in parallel with R or C to have very little effect on the result.)

3.7.5 Tuned circuits

It was shown in Figure 3.22 that the reactance of a capacitor decreases with frequency while that of an inductor increases. In any circuit possessing both types of reactance it therefore follows that some frequency exists at which these reactances are the same. This is called the resonant frequency, and the circuit thus formed is called a tuned circuit. When the reactances are equal, $X_C = X_L$, or

$$\frac{1}{2\pi fC} = 2\pi fL$$

Solving this for f we arrive at the expression for the resonant frequency:

$$f_R = \frac{1}{2\pi\sqrt{(LC)}}$$

Figure 3.28 shows a circuit comprising C, L and R in series. At the resonant frequency f_R the reactances are equal, as we have said, and opposite in phase. This means that they effectively cancel each other, and the total impedance of the circuit is given by $Z = \sqrt{[R^2 + (X_L^2 - X_C^2)]}$. Therefore at f_R the only component opposing current flow is R and so current rises to a maximum. This is shown in Figure 3.28(b), which plots current against frequency for two values of R. Clearly, when R is small, the circuit is more sharply tuned and rises

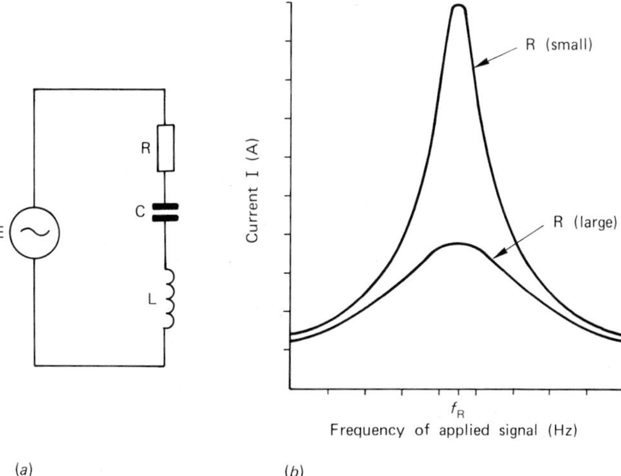

Figure 3.28 Series-tuned circuit. (a) Applying AC input voltage across R, C and L in series; (b) the current rises to a maximum at the resonant frequency for which $X_C = X_L$

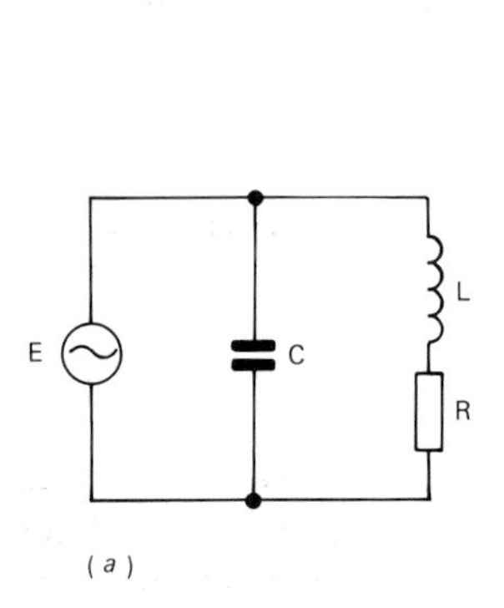

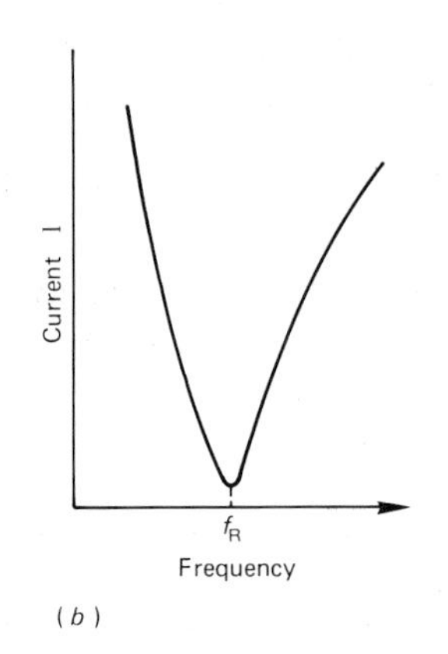

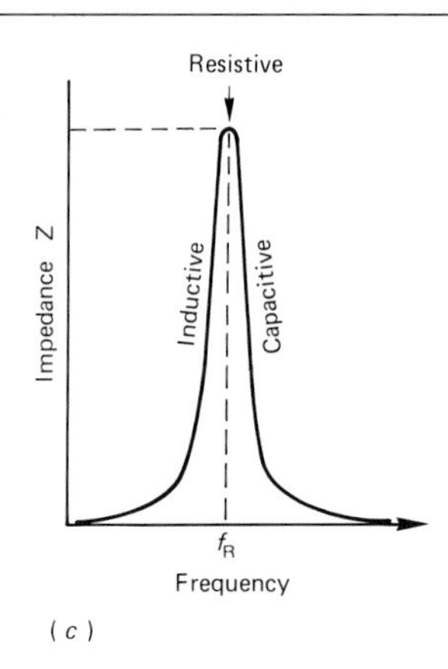

Figure 3.29 Parallel-tuned circuit. (a) Applying AC input voltage across C in parallel with L and R in series; (b) current drawn from the generator falls to a minimum at the resonant frequency; (c) impedance of the tuned circuit rises to a peak at the resonant frequency and is inductive or capacitive at frequencies below or above resonance respectively

to a narrow peak at resonance. Increasing R (or damping) gives broader tuning and a direct analogy is seen to exist with the resonance effects in mechanical and acoustical systems performing simple harmonic motion as illustrated in Figure 2.17 on page 29. In fact these analogies are much used by designers of microphones and other electro-acoustic devices, mass being equivalent to inductance, compliance (the reciprocal of stiffness) equivalent to capacitance and friction (or damping) equivalent to resistance.

Figure 3.29(a) shows a parallel tuned circuit with R included in the L arm. Again resonance occurs at the frequency for which the two reactances X_C and X_L are equal. However, in this case, though a high value of current circulates in the right-hand closed circuit, the net value of the impedance presented to the source rises to a maximum at resonance and so the current drawn from the generator falls to a minimum, limited only by R. This is indicated in Figure 3.29(b), which plots current supplied by the generator against frequency, showing a downwards dip at f_R. The graph of impedance against frequency shown in Figure 3.29(c) is seen to be purely resistive at f_R, being predominantly inductive at lower frequencies and capacitive at higher ones. The analogy with acoustical/mechanical resonance is again obvious.

3.8 Watts, volts and decibels

The logarithmic unit the decibel was introduced in Section 2.7 as a convenient means of expressing the difference between acoustic powers, intensities and sound-pressure levels. The same benefits apply when sound signals are converted into electrical form. Thus the difference between two electrical powers W_1 and W_2 in decibels is given by:

$$\text{dB} = 10 \log \frac{W_1}{W_2}$$

In microphone circuits it is common to express power levels with respect to a reference (zero) power W_{ref} of 1 mW and to indicate this by adding m as a suffix to give dBm. For example, a microphone might give an electrical power output for a given input SPL of 1 mW. This could be expressed as a power level in dBm by calculating:

$$10 \log \frac{10^{-6}}{10^{-3}} = 10 \log 10^{-3} = -30\,\text{dBm}$$

Again, since $W = V^2/R$, the ratio between two voltages V_1 and V_2 can be expressed in decibels as:

$$\text{dB} = 10 \log \frac{V_1^2}{V_2^2} = 20 \log \frac{V_1}{V_2}$$

By convention, a commonly used reference voltage V_{ref} is that corresponding to 1 mW in 600 Ω, at one time a standard impedance in many studio installations. This voltage may be calculated from the formula:

$$W_{ref} = \frac{V^2_{ref}}{R}; \text{ i.e. } V_{ref} = \sqrt{\frac{R}{W_{ref}}} = \sqrt{\frac{600}{10^{-3}}} = \sqrt{0.6} = 0.775\,\text{V}$$

Strictly speaking, this reference voltage should be used only when the circuit impedance is 600 Ω, but this is often ignored and the voltage is measured regardless of impedance, and quoted in decibels referred to 0.775 V (1 mW in 600 Ω) and the symbol dBu (or dBv in the USA) used to indicate this. Another reference voltage sometimes used (for example in quoting microphone sensitivity) is 1 V, when the symbol used is dBV.

This agreed use of decibels allows the output voltage (sensitivity) of microphones to be quoted simply, as well as the gain of amplifiers (positive dB) and the loss in attenuators or filters (negative dB).

3.9 Active devices

So far, we have limited the discussion to passive components arranged into circuits or networks allowing voltages to be stepped up or down (by transformers), attenuated (by resistive potential dividers), filtered (by frequency-dependent networks, including capacitors or inductors) or tuned to specific frequencies (by resonant combinations of C and L). The majority of audio circuits into which microphones must work also include active devices in various forms. These make more versatile signal processing possible, notably rectification, in which AC is changed into DC, and amplification, in which signal levels can be boosted, the extra power being drawn from batteries or the mains supply. Two basic types of active device will be outlined here: the thermionic valve (vacuum tube), in which electrons pass through a vacuum or gas, and the transistor, in which the electrons pass through solid semiconductors.

Historically, the valve came first (around 1904) and comprises an evacuated glass or metal tube into which are inserted electrical contact points called electrodes. The names of the different valve types indicate the number of electrodes: thus a diode has two electrodes, a triode has three electrodes and so on (see Figure 3.30). The electron stream enters the valve from the electrode shown at the bottom of the figure. This is called the *cathode*, and usually consists of a nickel pipe which gives off a steady stream of electrons when its

Figure 3.30 Thermionic valves or tubes. (a) Diode; (b) triode

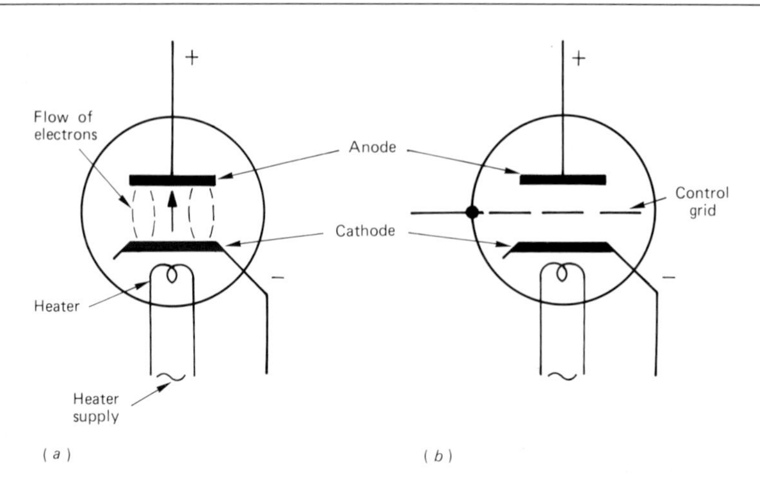

temperature is raised due to current passing through a 'heater' filament threaded inside it. The second electrode in a diode is called the *anode*, and it will be readily appreciated that a diode is a one-way device. Applying an alternative voltage between anode and cathode will result in pulses of current flowing during the half-cycles when the anode is positive (and attracts the electrons) with no current flow during the negative half-cycles. This process of *rectification* is illustrated in Figure 3.31 along with the smoothing effect introduced by adding the capacitor C. Further smoothing circuitry is added in practical AC/DC supply units and is essential in the polarizing voltage supplies and head amplifiers used in condenser microphones, for example, since any residual AC ripple would appear in the microphone audio output.

Figure 3.30(b) shows a triode valve formed by adding a third electrode known as the *control grid*. This mesh-type electrode, devised by De Forrest in 1906, gives sensitive control of the flow of electrons passing through it to the anode. Because of its closer proximity to the cathode, quite small changes in the grid voltage (when an AC signal is applied) will produce relatively large voltage swings in the anode load. This process of voltage boosting or

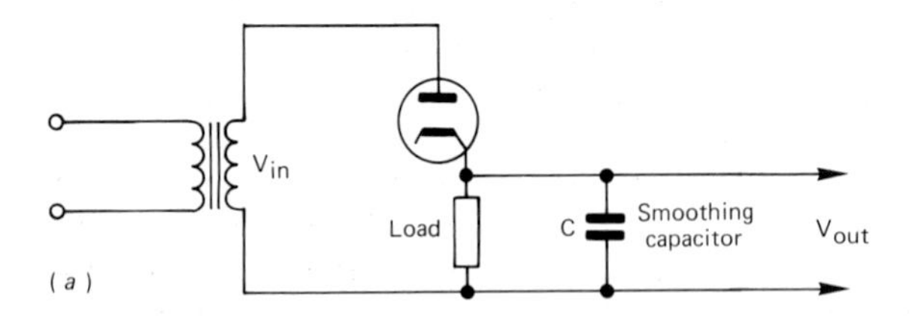

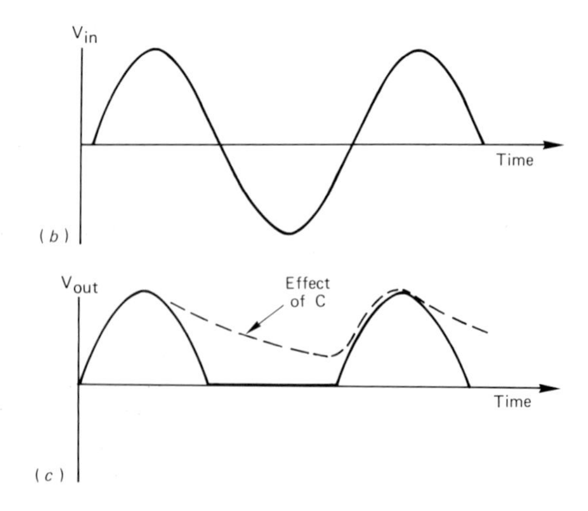

Figure 3.31 The diode rectifier. (a) Basic circuit of half-wave rectifier; (b) the applied sinewave; (c) the half-wave rectified output voltage showing the smoothing effect of the capacitor

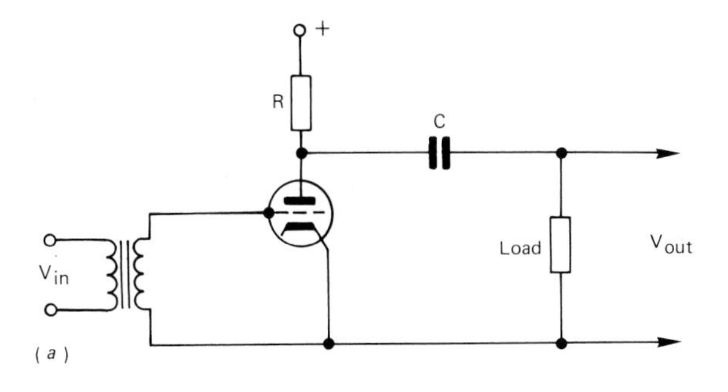

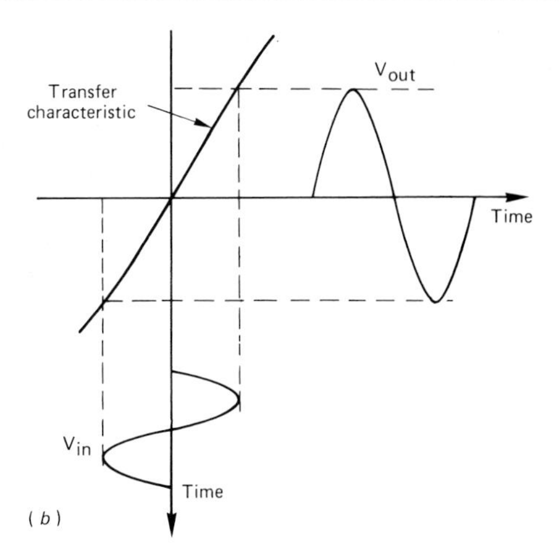

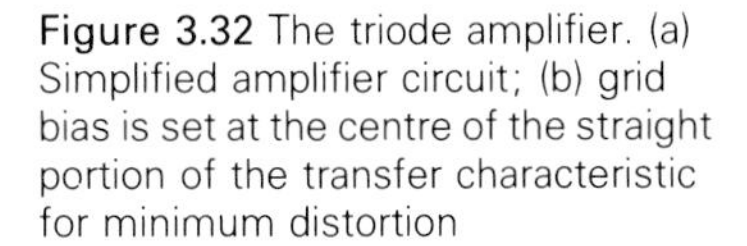
Figure 3.32 The triode amplifier. (a) Simplified amplifier circuit; (b) grid bias is set at the centre of the straight portion of the transfer characteristic for minimum distortion

amplification is illustrated in Figure 3.32. The gain of the amplifier circuit thus formed is given by the ratio V_{OUT}/V_{IN} and is normally expressed in decibels:

$$\text{Gain} = 20 \log \frac{V_{OUT}}{V_{IN}}$$

e.g. if a signal input of 1 mV produces a voltage of 1 V across the load, the gain is 1000:1 or 60 dB.

Valves suffer from a number of practical disadvantages and since about 1948 they have been progressively superseded by transistors in most audio equipment, including the elementary rectifier and amplifier applications just described.

The emergence of transistors followed the identification of 'semiconductor' crystalline elements such as germanium and silicon whose atoms have so-called valence electrons in their outer orbits. These are easily removed and capable of current carrying, while the positively charged atom with one or more electrons removed is called a 'hole' and is also capable of current carrying (in the opposite direction or polarity). The degree of conductivity of silicon (and, to a lesser extent, germanium) is much less than that of metals and so these elements are referred to as semiconductors. Introducing small amounts of selected impurity elements can make a semiconductor material either predominantly a negative current carrier (*n-type*) or a positive current carrier (*p-type*).

The first practical use of this discovery was to form a *p–n* junction diode, which behaves very like the valve diode already discussed, by doping the two sections of a semiconductor with appropriate amounts of *p* and *n* impurities. Figure 3.33 illustrates how both valve and *p–n* junction diodes produce rectifier action, allowing current to flow in one direction but not the other.

Following the emergence of the semiconductor diode, intensive research was applied to developing a solid state equivalent of the thermionic valve triode and a solution was found around 1948. Basically, the idea is to form the *n–p–n* transistor shown in Figure 3.34(a), in which a *p*-type layer is placed between two *n*-type sections. The *p* layer is given a small positive bias with respect to the

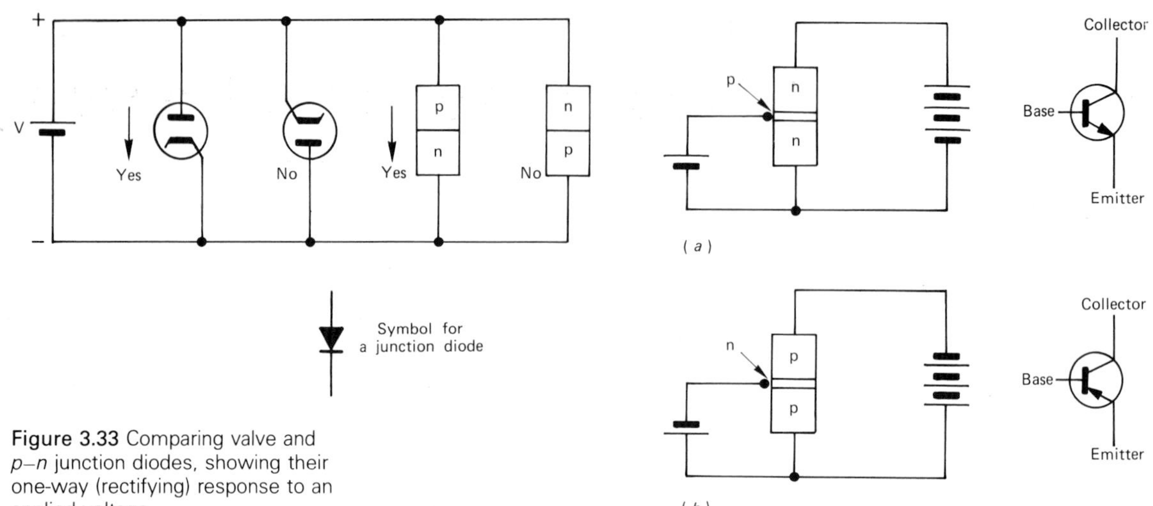

Figure 3.33 Comparing valve and *p–n* junction diodes, showing their one-way (rectifying) response to an applied voltage

Figure 3.34 Illustrating the connections to (a) *n–p–n* and (b) *p–n–p* transistors with their circuit symbols

(more heavily doped) lower *n* section. In effect, this lower *n* region acts as a supplier of electrons. It therefore corresponds to the cathode in a valve and is called the *emitter*. The upper *n* region is called the *collector* and resembles the anode in a valve. The central *p* layer is called the base and controls the rate of electron flow in much the same way as the grid in a valve. As well as the *n–p–n* type of transistor, the *p–n–p* type also exists, as shown in Figure 3.34(b).

This is not the place to describe semiconductor action in detail or the numerous types of transistor circuit in use. It will be enough to list some of the principal advantages of transistors over valves:

1. No fragile vacuum construction;
2. No heater supply needed, with the risk of induced AC hum signals; transistors operate at normal temperatures;
3. Low operating voltages; transistor circuits are safer and ideally suited to battery operation in portable equipment;
4. Low power consumption;
5. Extremely small and inexpensive; complex circuits can be housed in very little space and integrated circuits, microprocessors or chips measuring only a few millimetres can replace scores or even hundreds of discrete transistors;
6. Extreme reliability and almost indefinite life.

4 Microphone technology

4.1 Design requirements

The desirable features of any microphone depend to some extent on the particular applications for which it is intended. However, a study of the principal characteristics which distinguish one microphone from another will prove helpful both in identifying which physical and technological principles have an important bearing on microphone performance and as a guide to using microphones creatively. Manufacturers' specifications for some widely used microphones are listed later in this chapter and in Chapter 5. The following notes are intended to help in understanding and comparing these specifications.

4.1.1 Frequency response on-axis

The microphone should be even-handed in the way it responds to sounds over the whole frequency range of interest. In high-fidelity terms this means that the graph of signal output voltage plotted against frequency for a constant acoustic level input at all frequencies over the range 20–20 000 Hz (normally taken to encompass the full range of human hearing) should be a straight horizontal line. While this requirement seemed unattainable in the early days, modern microphones can come very close to it, if we limit the measurements to the axial response, i.e. when the source is placed on the main reference axis of the microphone in an anechoic test-room – where no reflections take place from the room boundaries (see Figure 4.1).

International standards and manufacturers' published response curves generally refer to this 'free-field' response obtained by using a calibrated sound source at a specified distance in an anechoic test-room or duct. In practice, the distance between the microphone and the source can have a number of important effects on frequency response (quite apart from the obvious change in the direct-to-reverberant sound ratio, to be discussed later). For this reason, some manufacturers also specify the diffuse-field frequency response. In the best designs of directional microphone, the 0 axial free-field and

Figure 4.1 Microphone testing with a calibrated source in an anechoic room. (Courtesy Bruel & Kjaer)

the diffuse-field response curves are very nearly parallel. Omnidirectional (pressure-operated) microphones, however, tend to exhibit a treble peak in the free-field response when the diffuse-field response is designed to be flat (see Figure 4.2).

The full 20 Hz to 20 kHz spectrum may not be necessary or even desirable in some applications. For example, a narrower range may be specified for microphones to be used in vehicles, aircraft or hearing aids in order to optimize speech intelligibility in noisy surroundings. Sometimes the ideally flat response may be abandoned in favour of one which emphasizes or suppresses certain frequency bands to provide improved communication or a special effect. A vocalist will often choose a particular microphone because of its tendency to enhance some desired vocal quality, typically rolling off below 100 Hz and peaking broadly around 5 kHz. Lavalier or clip-on microphones need an equalized response to correct for diffraction effects, and so on.

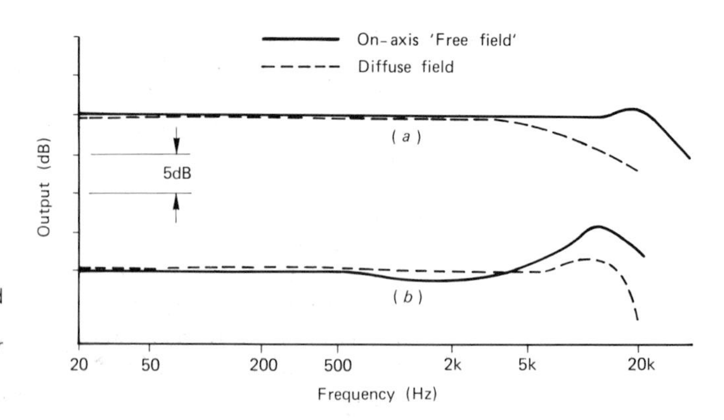

Figure 4.2 Free-field and diffuse-field response. Frequency plots for omnidirectional microphones designed for a flat response in (a) free field; (b) diffuse field

4.1.2 Directivity

In practical situations, of course, sounds do not arrive at a microphone solely along its axis. They may originate from sources located in any direction, and in addition there will be numerous reflected sound waves from walls and obstacles, all contributing in large or small measure to the microphone's total output signal. As will become clear in Sections 4.2 and 4.3 below, microphones can be designed either to respond equally to sounds from all angles or to discriminate against those arriving from specific directions.

The directional behaviour of microphones is most easily illustrated by plotting on circular or polar graph paper their response to sounds at a fixed sound-pressure level for all angles in a particular plane. The most common polar response shapes are illustrated in the idealized diagrams of Figure 4.3. They include the circle (denoting a

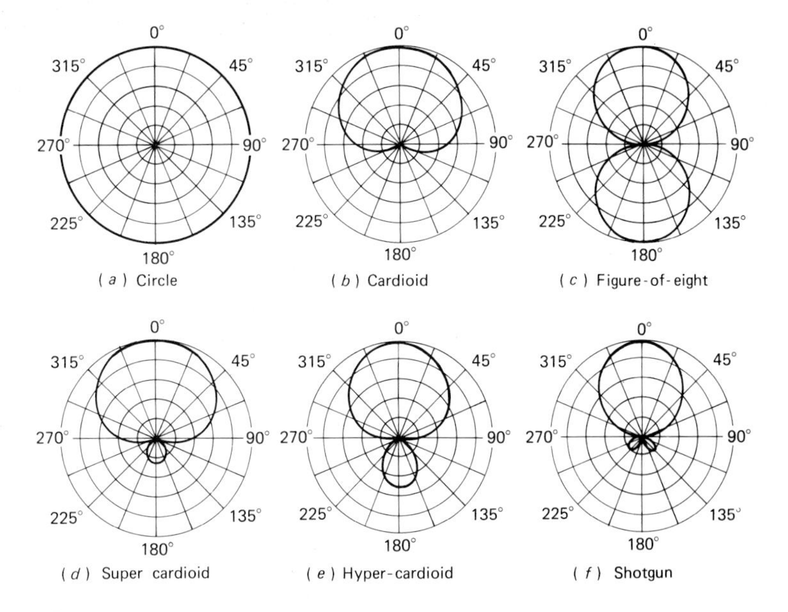

Figure 4.3 The principal microphone directivity patterns

non-directional or 'omnidirectional' microphone which responds equally at all angles), the figure-of-eight (a bidirectional microphone which responds equally at front and back but discriminates against sounds arriving at the sides) and the cardioid or heart-shape (a unidirectional microphone which responds over a wide frontal angle but discriminates against sounds arriving at the back). Strictly speaking, these two-dimensional directivity patterns should be rotated about the microphone's axis to show the solid shape representing the directivity in three-dimensional space and Figure 4.4 illustrates an artist's impression of these solid figures for five of the basic patterns.

Not surprisingly, microphone directivity is often a principal reason for choosing between different models for particular applications. As well as fixed omni, bidirectional or cardioid microphones (plus a few other types such as hypercardioid and supercardioid), there are models offering switched or continuously variable directivity. These

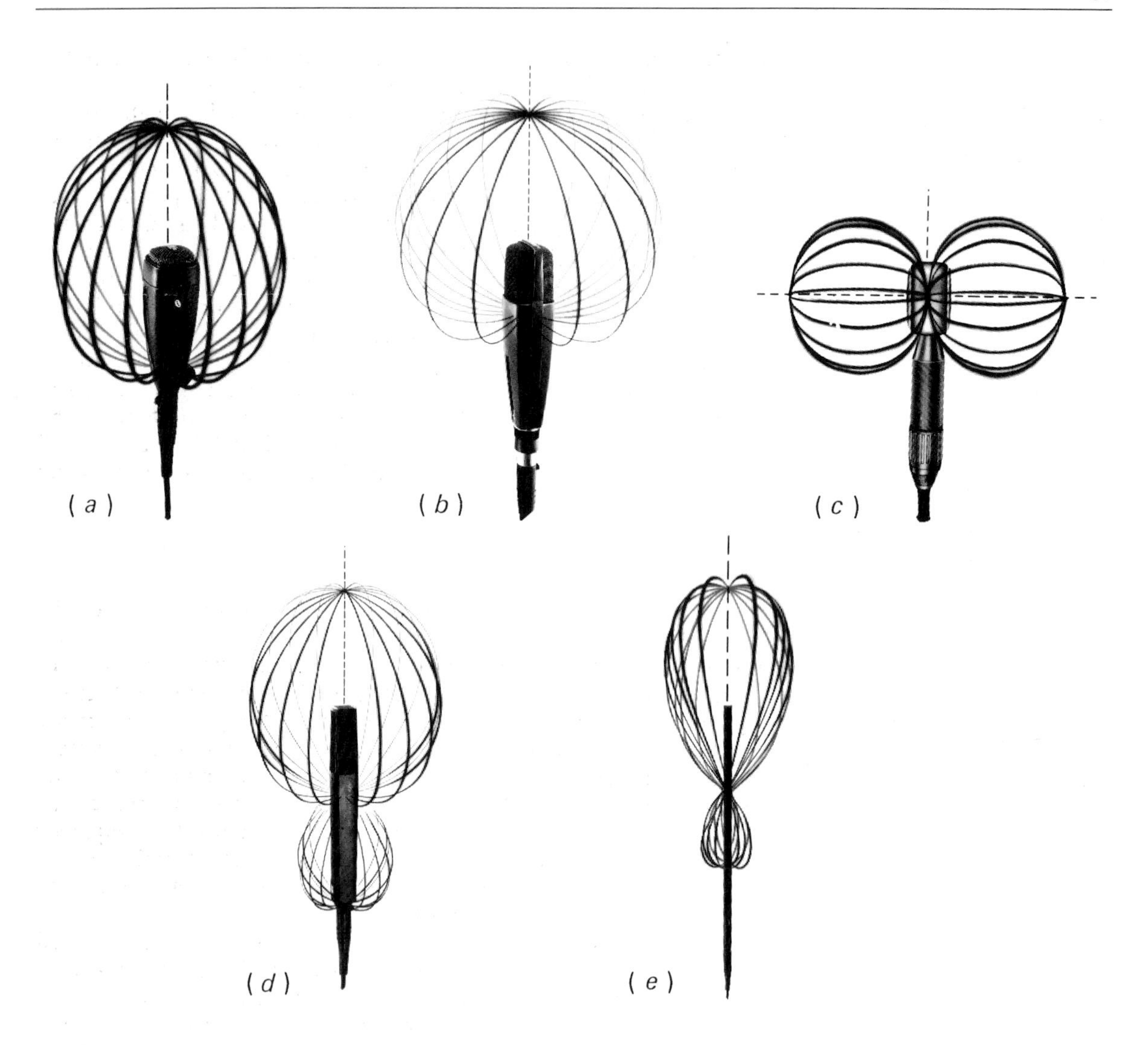

Figure 4.4 Artist's impression of typical three-dimensional directivity patterns. (Courtesy Sennheiser)

provide a useful degree of versatility, though usually at increased cost and complexity in design. There are also microphones issued as a series in which a common body unit can provide a choice of directivity patterns by simply swapping over the screw-on head capsule (see, for example, Figure 4.32).

4.1.3 Frequency response off-axis

Whatever the nominal directivity pattern of the microphone, it should ideally maintain the same frequency response at all angles; i.e. there is a need for polar pattern uniformity. If this is not maintained, the total output comprising axial and random off-axis sounds will fail to provide the even-handed frequency coverage specified for the axial response. Older designs seldom meet this

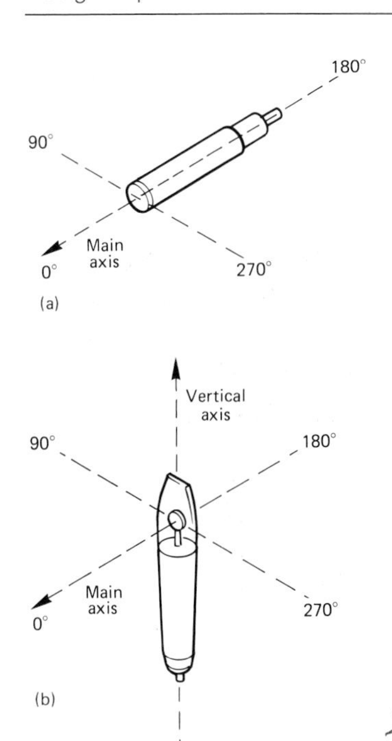

Figure 4.5 Microphone shapes and diffraction. (a) The cylindrical end-fire type; (b) the assymmetrical side-fire type

requirement, the most common fault being a falling off in treble response from about 5 kHz upwards at oblique angles.

A principal reason for directional response variability is the extent to which the microphone itself acts as a physical obstacle to sound waves. Diffraction effects are naturally related to frequency (i.e. wavelength) and can lead to peaks and dips in response which may vary with angle of incidence. Reducing the microphone dimensions has the effect of pushing the diffraction problem further up the frequency scale. As a general rule, therefore, smaller microphones have potentially better axial and directional uniformity than larger ones. Also the thin cylindrical microphones whose diaphragm is set at right angles to the main axis have a better chance of achieving a uniform response at high frequencies than the four-square designs in which the diaphragm is in the main plane of the microphone body. The differences are illustrated in Figures 4.5 and 4.6. Figure 4.7 illustrates the variety of microphone shapes and sizes produced by one well-known manufacturer.

4.1.4 Sensitivity

The sensitivity or conversion efficiency of a microphone, i.e. the output voltage produced for a stated incident sound-pressure level, should ideally be as high as possible. This helps the microphone to discriminate against system noise and electrical interference along the signal path. In practice, this requirement is at odds with those already mentioned. For example, high sensitivity is easier to achieve with a large diaphragm which is driven by a greater amount of acoustic energy, whereas, as we have seen, a smooth frequency response calls for small physical dimensions. It also happens that a flat frequency response can be designed into some microphones only by the incorporation of a fair amount of resistive damping, which has the effect of reducing overall sensitivity.

It is unfortunate that microphone manufacturers have not agreed on a single method for specifying sensitivity, and this often makes it difficult to compare the printed microphone specifications directly. The most common method is that recommended by the IEC and British Standards. This specifies the open-circuit (unloaded) voltage output for an input SPL of 1 pascal, i.e. 10 microbars (10 μbar). As was shown in Figure 2.16, this corresponds to a fairly loud sound since 1 Pa = 94 dB above the threshold of hearing on the SPL scale. Commercial microphones vary considerably in sensitivity, but typical ratings using this IEC method range from about 2 to 20 mV/Pa. A slightly different rating sometimes used quotes the output voltage for 0.1 Pa (1 μbar) which is a widely accepted reference SPL corresponding to the level of normal speech at 20 cm (74 dB SPL). In the USA the sensitivity is often stated not in terms of the actual voltage but the voltage level in decibels with reference to 1 V (written dBV), or sometimes 0.775 V (written dBu). As a rule of thumb, when published specifications have to be compared, 1 mV/Pa = − 60 dBV, 10 mV/Pa = −40 dBV and 20 mV/Pa = −34 dBV. Power output ratings are also found in some microphone manufacturers' literature.

An overload or maximum SPL value is sometimes given at which the total harmonic distortion will not exceed a stated value, say 0.5% or 1.0%.

(a)

(b)

Figure 4.6 Typical end- and side-fire microphone designs. (a) AKG C460B; (b) Neumann U47 (introduced in 1960 and still popular as a vocal microphone)

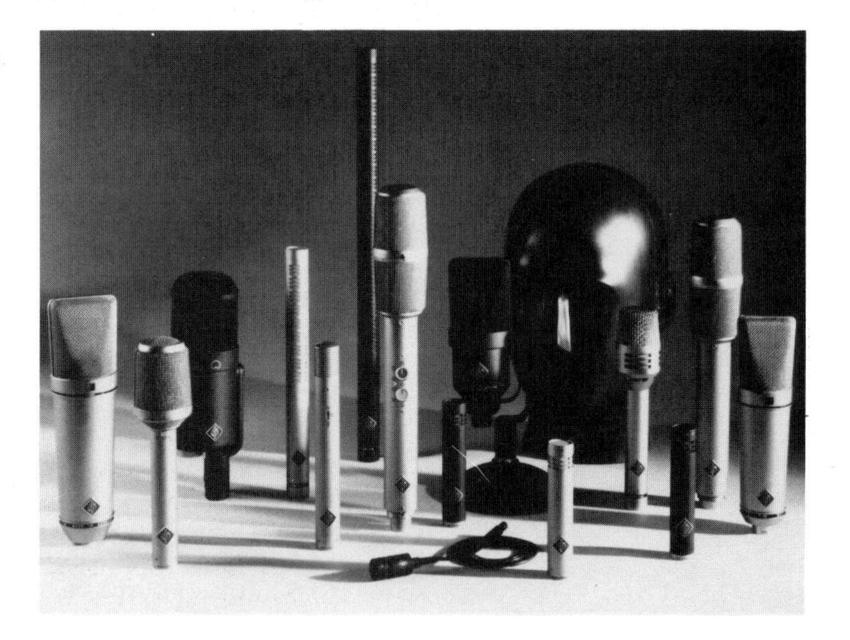

Figure 4.7 Microphone shapes, showing some of the products from just one manufacturer. (Courtesy Neumann)

4.1.5 Self-noise

The inherent noise level of a microphone, and more specifically that of any associated built-in amplifier, should be as low as possible. This need to preserve optimum signal-to-noise ratio has been further highlighted by the advent of digital recording and consumer media such as the compact disc, where residual noise is very low indeed.

(Typical S/N ratios and usable dynamic ranges of compact disc players are quoted at around 96 dB.)

Microphone inherent self-noise is usually specified as the equivalent SPL level which would give the same output voltage. This can be measured on test equipment having a flat frequency response (unweighted) or more usually via an IEC 'A' weighting filter which allows for the reduced sensitivity of the ear at high and low frequencies and therefore more accurately indicates the audibility or nuisance value of the noise. Typical values are down around 15–20 dBA SPL, which is about the ambient noise level in the best sound studios, and corresponds to a thermal noise value of about −129 dBm.

As an alternative, some manufacturers quote a 'signal-to-noise ratio'. This relates the A-weighted equivalent noise level to 1 Pa (94 dB SPL). For example, a microphone having S/N ratio 74 dBA has an equivalent noise level of 94 − 74 = 20 dBA SPL.

4.1.6 Distortion

The waveform of the electrical output signal from a microphone should be as faithful a replica of the input acoustic waveform as possible. In other words, non-linear distortion should be kept as low as possible. In practice, non-linear distortion is mainly associated with high signal levels which may produce overload or saturation effects in the microphone transducer mechanism or in the built-in amplifier (in the case of condenser microphones).

The distortion rating of a microphone is often quoted as the maximum SPL which can be handled for a given value of total harmonic distortion (THD). A typical value might be 130 dB for 0.5% THD. Combining this maximum SPL with a noise rating of, say, 20 dBA would produce a microphone able to cover a dynamic range of 110 dB. Used properly, and with correct setting of gain levels throughout the recording equipment chain, this would be enough to reproduce practically all musical signals without introducing audible distortion on the loudest signals or audible noise during the musical silences.

4.1.7 External influences

A well-designed microphone will avoid the picking up of stray mains frequency 'hum' fields or radio frequency interference, even in worst-case situations near television monitors, cameras, lighting regulators or transmitting stations. Similarly, microphones for outdoor work or close vocals require a low response to wind noise, otherwise there may be audible rushing noises or 'popping' on plosives like 'p' and 'b'. Handheld microphones must be rugged and insulated against mechanical shocks and vibrations. Both these latter requirements must be met for television and film boom operation.

4.2 Transducer types

With the exception of a few exotic types based on a heated wire or a cloud of ions, all practical microphones make the conversion of acoustical energy to electrical energy via the mechanical vibrations in response to sound waves of a thin, light diaphragm. Generally, this diaphragm will be circular in shape and clamped at its periphery, though other shapes occur, including the familiar thin ribbon stretched between clamps at each end. It appears therefore that the energy conversion takes place in two stages – though, of course, they happen simultaneously: acoustical-to-mechanical and mechanical-to-electrical.

The second stage may use any one of half-a-dozen electrical generator principles and the microphones tend to be categorized accordingly, e.g. moving-coil, condenser, etc. However, before describing each of these transducer categories in detail it will be logical to distinguish between the two main ways in which microphones of any transducer type first extract energy from the sound wave. These are respectively called pressure operation and pressure-gradient (or velocity) operation.

4.3 Pressure operation

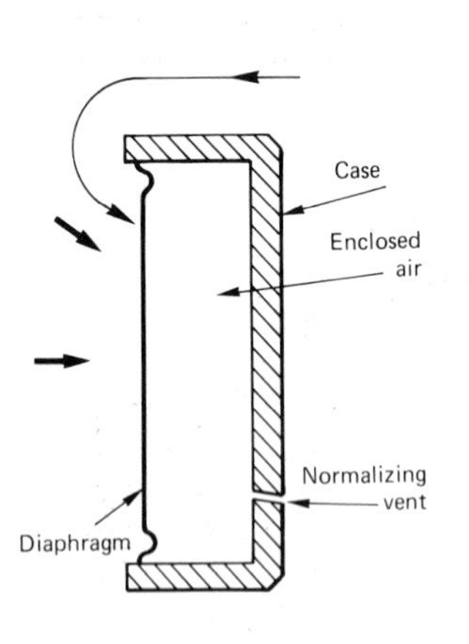

Figure 4.8 Pressure-operated microphone, showing enclosed back with small vent hole

The distinguishing feature of pressure-operated microphones is that the rear surface of the diaphragm is enclosed so that the actuating force is simply that of the instantaneous air pressure at the front (Figure 4.8). A small vent hole is cut through the casing to equalize the long-term external and internal air pressures. Then the swings of external pressure in the incident sound wave above and below normal atmospheric pressure will cause the diaphragm to move inwards and outwards accordingly. The force on the diaphragm is equal to the product of the sound pressure (per unit area) and the area of the diaphragm, and is essentially independent of frequency.

At least at low frequencies (long wavelengths) the effective force will be the same for all angles of incidence. Therefore a plot of the microphone's polar response, measuring the output voltage for a given SPL arriving at various angles in the stated plane, will show a perfect circle (as seen in Figure 4.3(a)). By symmetry, with the usual circular casing construction, the same circular plot will apply in all planes about the microphone axis and so its three-dimensional directivity pattern is a sphere, as outlined in Figure 4.4(a). A purely pressure-operated microphone is therefore non-directional and is referred to as *omnidirectional*.

This simple description begins to fail at higher frequencies as the sound wavelength in air approaches the dimensions of the microphone casing. The microphone then acts as an obstacle to the sound wave and two interference factors come into play, both tending to reduce the electrical output for sounds arriving at oblique angles compared with those arriving on-axis.

First, when the wavelength is about equal to the microphone diameter or less, reflection and diffraction take place. This means that axial sound waves are reflected back along their original path and a standing wave is established having twice the original amplitude (as discussed in Chapter 2, page 35). The result is called pressure doubling and can boost the high-frequency axial response by up to 6 dB (see Figure 4.9). The effect is sometimes aggravated by the existence of a cavity resonance in front of the diaphragm. This treble

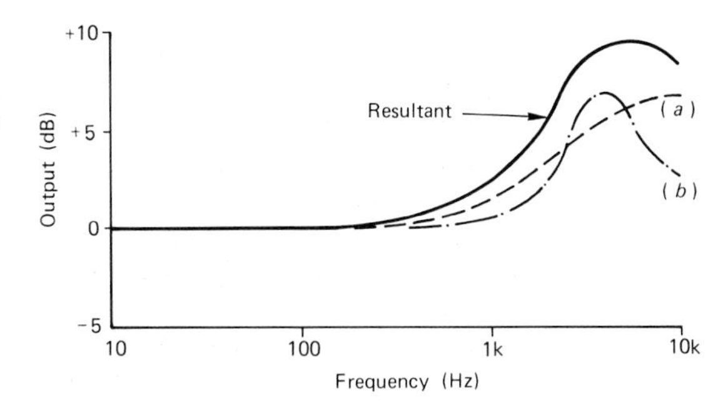

Figure 4.9 Pressure-operated microphone, showing typical peak in high frequencies on-axis due to (a) diffraction (pressure doubling) and (b) cavity resonance

emphasis is usually undesirable, and explains why it is often better to speak or sing across the microphone diaphragm rather than straight at it.

Second, oblique incidence of high-frequency waves means that the phase is no longer uniform across the face of the diaphragm. This leads to partial cancellation and reduced electrical output. These effects are progressively more pronounced at shorter wavelengths (see Figure 4.10) and so pressure-operated microphones generally become more narrowly unidirectional with increasing frequency (see

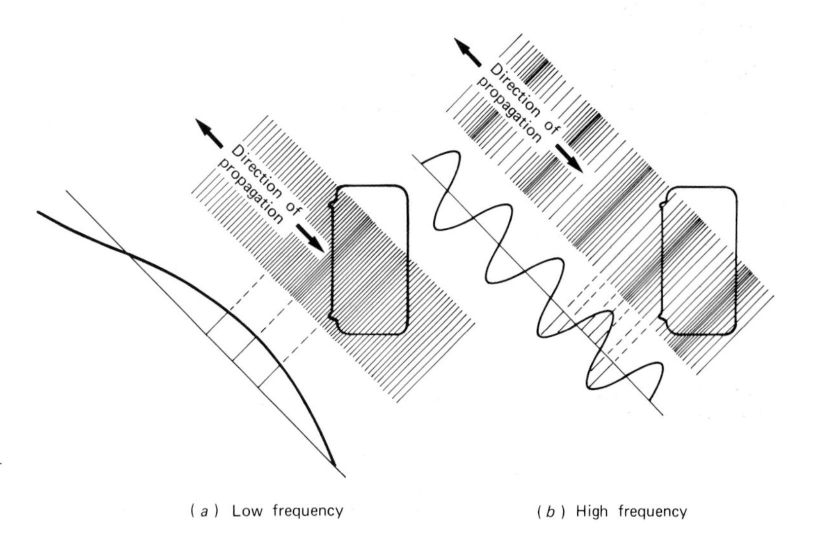

Figure 4.10 Off-axis response is reduced at high frequencies (short wavelengths) due to phase cancellations across the face of the diaphragm

Figure 4.11(a)). Another way to represent this progressive loss of treble at greater angles of incidence is shown in Figure 4.11(b), where the frequency response is separately plotted on rectangular co-ordinates for specific angles of incidence.

The effect is purely a physical one, regardless of the type of transducer, and related strictly to the ratio D/λ, where D is the microphone diameter and λ is the wavelength. Thus halving the front diameter of the microphone will double the frequency at which a given narrowing of the directivity pattern occurs. This increased directivity at high frequencies has several practical consequences. For example, when a less bright tonal balance on the direct sound is wanted, the user can tilt or rotate the microphone angle to give a

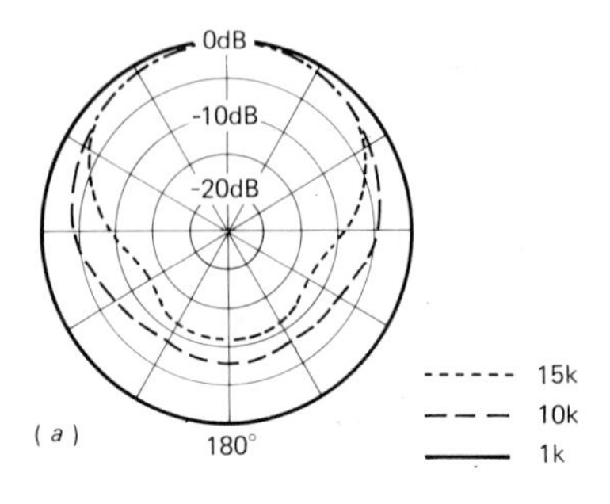

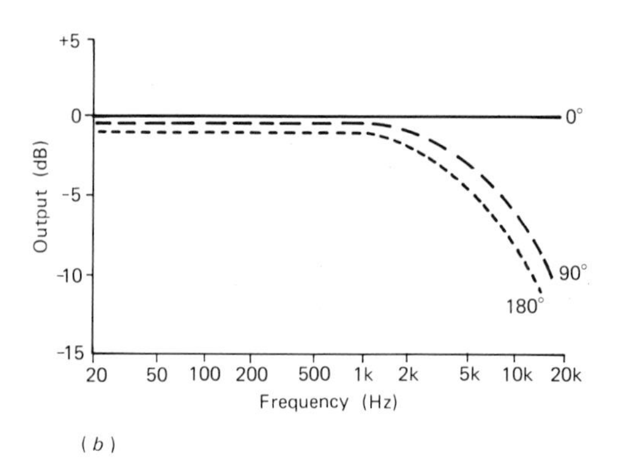

Figure 4.11 A pressure-operated microphone is essentially omnidirectional, but becomes more unidirectional at high frequencies, as shown here in (a) polar graph and (b) rectangular graph forms

controlled amount of treble attenuation. On the other hand, random reflected sound and the pick-up from other voices or instruments off-axis may appear bass-heavy, since they arrive at angles for which the treble frequencies are attenuated.

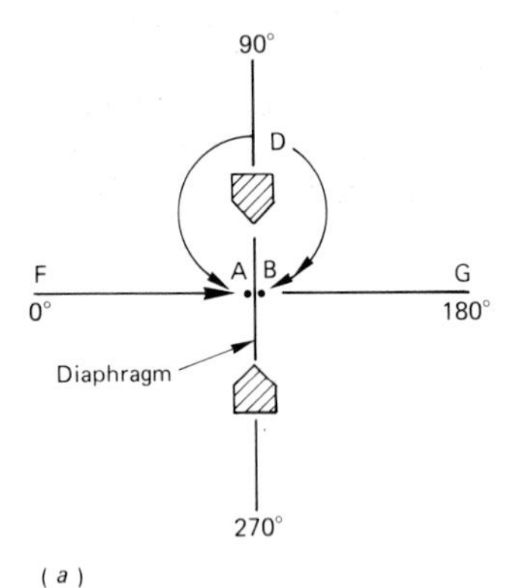

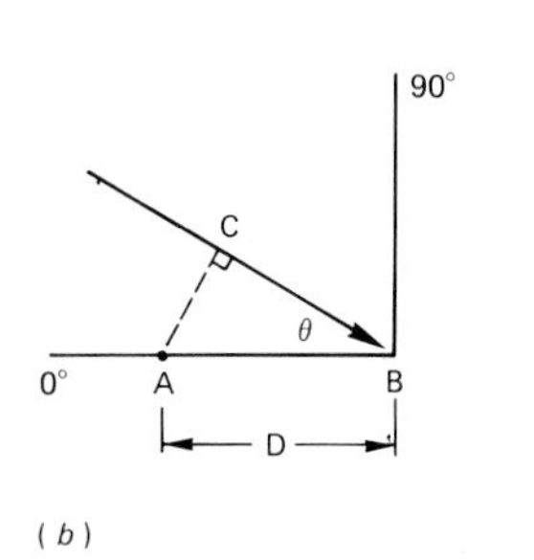

Figure 4.12 Pressure gradient microphone. (a) For on-axis sounds the path length difference D is a maximum; (b) at oblique angles D reduces in proportion to cos θ

4.4 Pressure-gradient operation

Certain microphones, notably the standard ribbon to be discussed later, are built with both faces of the diaphragm equally open to the air. The effective force on the diaphragm at any instant is then not simply due to the sound pressure at the front but to the difference in pressure, or pressure gradient (PG), between front and back. This has an important bearing on the system's directivity, as can be seen in Figure 4.12(a).

At frequencies low enough for diffraction effects to be ignored, the wave arriving on-axis (0°) will produce alternating pressures on the front and back faces A and B of the diaphragm which do not coincide exactly in time. Their time of arrival or phase will be separated by an interval which is directly proportional to the extra distance AB travelled to reach the more remote face. This extra distance D will be a maximum for axial sounds arriving at either 0° or 180° but will steadily diminish and eventually fall to zero as the angle of incidence increases from 0° to 90° or 270°. The resultant pressure difference and therefore force on the diaphragm may be taken to be proportional to this effective distance. As is shown in Figure 4.12(b), this shorter distance is CB and, since the ratio of this to D is $CB/AB = \cos\theta$, the microphone output at any angle θ is given by the expression $Y = X \cos\theta$, where Y = sensitivity (response) at θ and X = the maximum sensitivity on-axis ($\theta = 0°$). The value of cos θ for any angle can be obtained from cosine tables, as was described in Section 2.1.

Plotting cos θ on a polar graph produces the familiar bidirectional figure-of-eight diagram (see Figure 4.13(a)). The significance of the + and − signs is that there is a phase reversal of the force acting on the diaphragm for sounds arriving earlier at the back than the front. This plot uses a linear scale for output level and shows, for example, that the output falls to half at $\theta = 60°$. However, it is generally more helpful as a description of relative sensitivities with angle to use a logarithmic (dB) scale, as shown in Figure 4.13(b). From this we see,

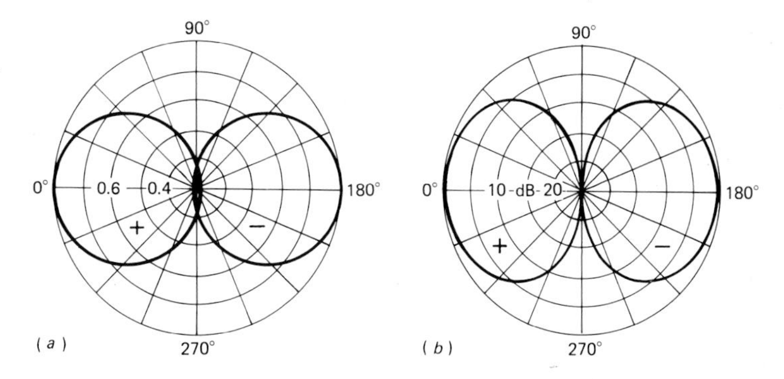

Figure 4.13 A pressure-gradient microphone has a figure-of-eight polar pattern, shown here on (a) linear and (b) the more usual logarithmic scale

for example, that defining an arbitrary frontal 'acceptance angle' as the arc within which the response remains within 3 dB limits gives 90° for the basic bidirectional microphone.

As with pressure operation, the above simple description of PG operation assumes that the wavelength λ is large compared with D. In fact the pressure gradient or pressure difference (and therefore the force on the diaphragm) is a function of the ratio D/λ (i.e. increases with frequency at a rate of 6 dB per octave), reaching maximum at a critical frequency f_c where $D/\lambda = 0.5$. This is illustrated for various ratios of D/λ in Figure 4.14. At frequencies higher than f_c, the PG output falls steeply but the designer usually plans the geometry so that a change to pressure operation sets in at

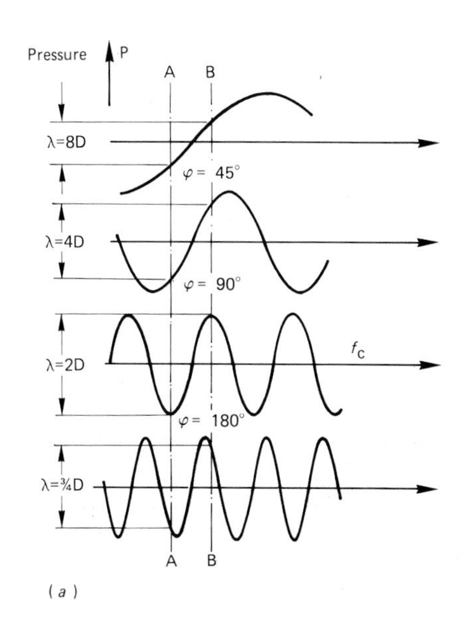

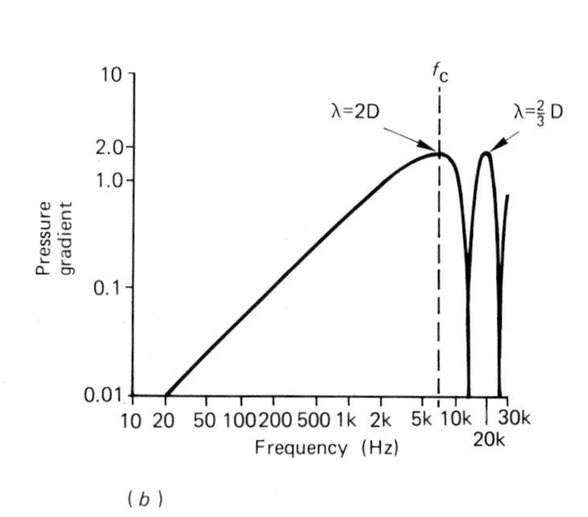

Figure 4.14 Variation of pressure gradient with frequency. (a) Pressure difference between points A and B increases with frequency until f_c, where $\lambda = 2D$; (b) plotting PG against frequency for $D = 25$ mm

about this point. The designer must opt for mass-controlled operation, i.e. with a low resonant frequency, so that the falling velocity/frequency graph exactly matches the PG rising graph.

The availability of bidirectional PG microphones since the late 1930s in addition to the simple omnidirectional types has allowed a greater flexibility in layout and balance methods, as will be discussed in later chapters.

4.4.1 Proximity effect

PG-operated microphones are characterized by a pronounced boost in low-frequency output when placed close to the sound source. This is called the 'proximity effect', and is caused by the additional pressure difference between points A and B introduced by the inverse square law increase in intensity at closer distances. This has more of an influence at low frequencies and therefore progressively tips up the bass end response. The effect is most marked in pure PG

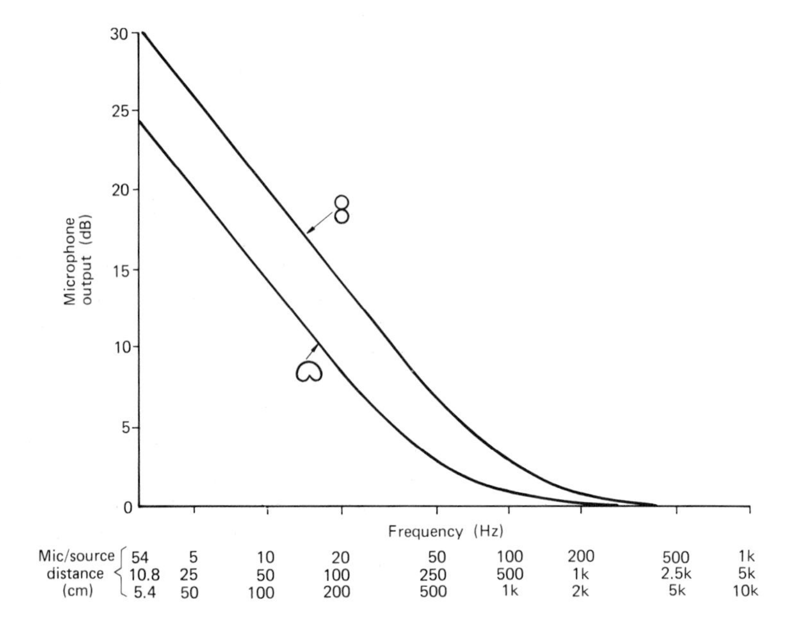

Figure 4.15 Proximity effect, showing bass tip-up with close-miking at distances of 54, 10.8 and 5.4 cm for both figure-of-eight and cardioid microphones

(figure-of-eight) microphones, as shown in Figure 4.15. In combination pressure-plus-PG microphones as discussed in the next section, the bass tip-up is still present but is less steep (see curve for cardioid).

4.5 Combination microphones

A whole family of directivity patterns became possible in the 1940s when microphones were introduced combining the characteristics of pressure and PG operation. Early models simply built separate pressure (omni) and PG (figure-of-eight) microphone capsules into a single casing. Making the axial sensitivities of the two units equal, say A, and adding their electrical outputs together produced a combined polar pattern as shown in Figure 4.16. At the front (left-hand side of the figure) the outputs of the two elements OA and OB are in phase and will add to give a combined output OC, which reaches a maximum value of 2A at 0°. At 90° the output of the PG element OB has fallen to zero and the combined output is reduced to A (−6 dB). At 180° (the back) the PG signal is in reverse phase and will cancel the pressure element's contribution to reduce the combined output to zero. The result is the heart-shaped (cardioid) pattern which has become very popular in many applications.

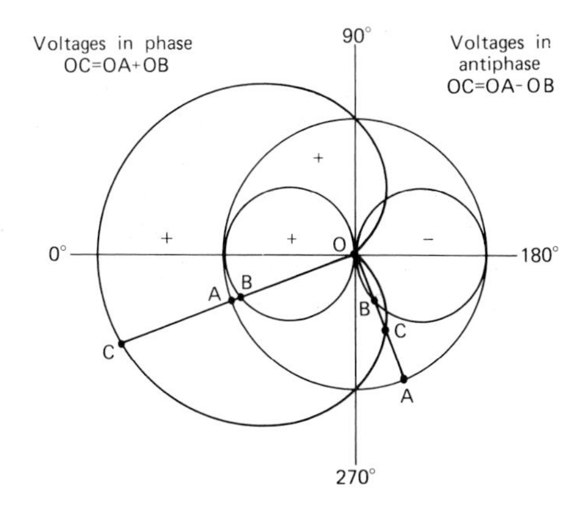

Figure 4.16 Derivation of cardioid directivity pattern by combining pressure and pressure gradient elements of equal sensitivity

If some means is introduced of adjusting the relative axial sensitivities of the PG and pressure elements, various intermediate polar patterns can be obtained of the kind already illustrated in Figures 4.3 and 4.4. The general expression for the change of response with angle for any microphone combining pressure and PG operation is:

$$X = A + B \cos \theta$$

1. *Putting B = 0* leaves the omni pattern, a circle of radius A.
2. *Putting A = 0* leaves the figure-of-eight pattern with a maximum value at 0° and 180° of B.
3. *Putting A = B* produces the cardioid pattern illustrated in Figure 4.16.
4. *Making B progressively larger than A* shifts the polar pattern to shapes somewhere between cardioid and figure-of-eight. For example, when B = 2A approximately, a small rear lobe appears and the pattern is called supercardioid. When B = 3A the rear lobe is larger and the pattern is called hypercardioid.

All these basic directivity patterns have characteristics which make them suitable in particular microphone applications. Some of these characteristics are listed in Figure 4.17 and their importance may be summarized as follows.

Acceptance angle (i.e. frontal arc over which output does not fall by more than 3 dB rel. 0°): shows that both the hypercardioid and supercardioid have a wider acceptance angle than the figure-of-eight but the cardioid (and, of course, the omni) have the widest pick-up of all.

Output at 90° rel. 0° (dB): shows that the supercardioid has better side source rejection than the cardioid, the hypercardioid is better still, and the figure-of-eight is best of all.

Output at 180° rel. 0°(dB): shows that the supercardioid has better rear sound rejection than the hypercardioid but the cardioid is best of all.

Angle at which output is zero: shows the angle(s) at which best rejection of unwanted sound sources takes place.

Random energy efficiency (REE): shows the fraction of all incident sound reproduced compared with the omni, for which REE = 1. Both the figure-of-eight and cardioid respond to only one-third of all

Figure 4.17 Comparing the operational characteristics of the basic polar patterns

CHARACTER-ISTIC	OMNI-DIRECTIONAL	CARDIOID	SUPER-CARDIOID	HYPER-CARDIOID	BIDIRECT-IONAL
Polar response pattern	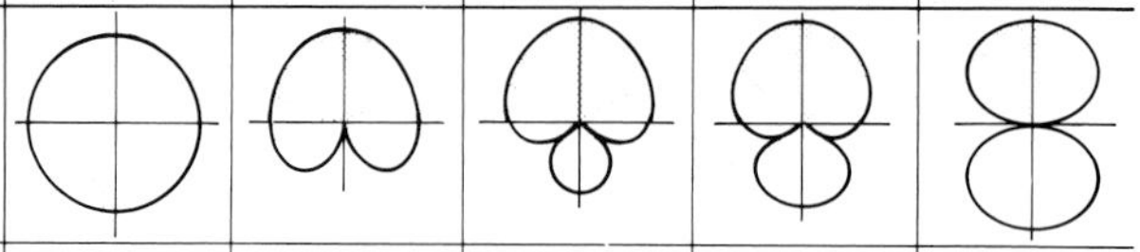				
Polar formula	1	0.5 + 0.5 cos θ	0.375 + 0.625 cos θ	0.25 + 0.75 cos θ	cos θ
Acceptance angle	—	131°	115°	105°	90°
Output at 90° (rel. 0°)	0	-6dB	-8.6dB	-12dB	-∞
Output at 180° (rel. 0°)	0	-∞	-11.7dB	-6dB	0
Angle at which output = 0	—	180°	126°	110°	90°
Random energy efficiency (REE)	1	0.333	0.268	0.250	0.333
Distance factor (DF)	1	1.7	1.9	2	1.7

NOTE:
1 = Drawn shaded on polar pattern
2 = Maximum front-to-total random energy efficiency for a first order cardioid
3 = Minimum random energy efficiency for a first order cardioid

random sound, the hypercardioid only one-quarter and the supercardioid slightly more than one-quarter.

Distance factor (DF): this is a corollary of the above and shows the increased distance from the source at which the various directional microphones can be placed for a given ratio of axial (direct) to random sound pick-up compared with the omni, for which DF = 1. It is seen that the figure-of-eight and cardioid can be placed 1.7 times, the hypercardioid 2.0 times and the supercardioid 1.9 times further than an omni for the same perceived amount of ambience. This effect is illustrated in Figure 4.18.

Instead of physically combining different microphone elements in a single case, which meant that early cardioids were bulky and unreliable as to directivity in the vertical plane, later designs utilize either dual-diaphragm or acoustic delay (phase shift) techniques in a single capsule.

The dual-diaphragm principle was introduced by Braunmühl and Weber of Neumann and comprises identical diaphragms on each side

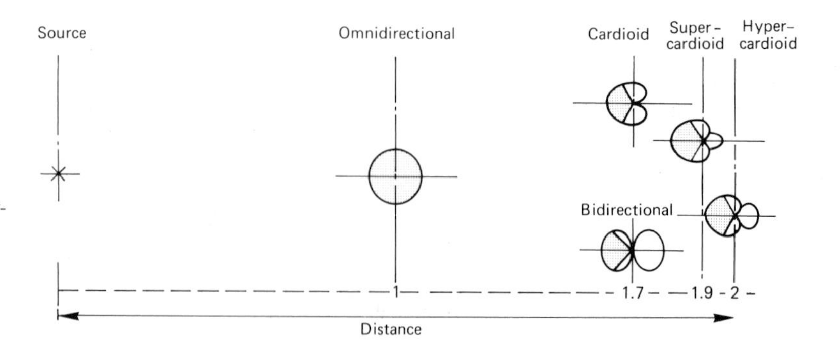

Figure 4.18 Distance factor. Comparing the relative source/microphone distances at which the basic polar patterns pick up the same ratio of direct to random sound. (Courtesy Bruel & Kjaer)

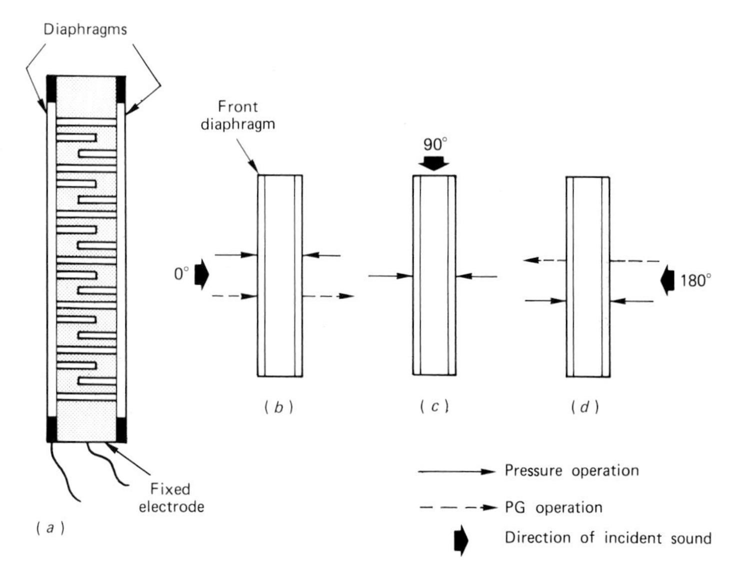

Figure 4.19 Dual-diaphragm cardioid condenser microphone, showing (a) the pattern of holes and cavities in the central fixed plate and how the pressure and PG forces combine for sounds arriving at (b) 0°, (c) 90° and 180°

of a central block having some holes bored all the way through to provide for PG operation and some cut only part of the way through to act as 'acoustic stiffness' chambers (Figure 4.19). In a simplified explanation of the original design in which the rear diaphragm was not electrically connected, the output at 0° is due to a combination of pressure operation (which tends to move both diaphragms inwards and outwards on alternate half-cycles) and PG operation (in which only the front diaphragm moves while the rear diaphragm remains stationary due to equal and opposite pressure and PG forces). At 90° the PG forces on both diaphragms fall to zero, reducing the output to half, and at 180° the front diaphragm experiences equal antiphase forces and the output falls to zero.

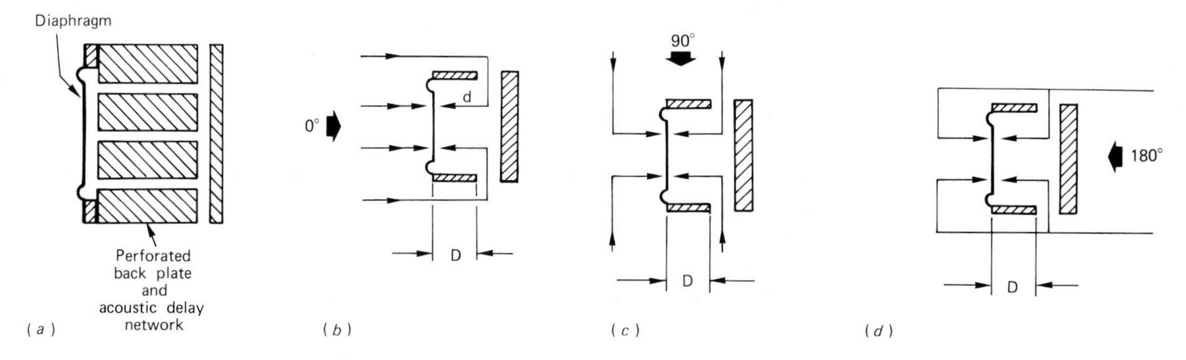

Figure 4.20 Single-diaphragm cardioid microphone, showing the acoustic delay network built into the perforated back plate and how the front and back forces on the diaphragm vary for sounds arriving at (b) 0°, (c) 90° and (d) 180°

Today single-diaphragm cardioid microphones are also common in which acoustic delay networks produce the required phase relationships. This is illustrated in Figure 4.20. The effective path length difference for 0° incidence is $D + d$, where D is the external distance between front and back and d is the internal distance from the outside to the rear of the diaphragm. Careful design of the acoustic phase-shift network can make D and d effectively equal, at least at low frequencies, so once again the pattern obtained is a cardioid with the resultant path length, respectively, $2D$, D and 0 for incident sounds at 0°, 90° and 180°.

In a later version of the dual-diaphragm configuration both diaphragms are electrically connected and, with suitable switching of polarity and relative sensitivity of the back-to-back cardioids thus formed, a range of directivity patterns is obtained. Using the expression derived earlier for the directivity of a combined pressure and PG-operated microphone, the formula for two cardioids back-to-back becomes:

$$X = [\mathrm{A} + \mathrm{B} \cos \lambda] + [\mathrm{A} + \mathrm{B} \cos (\lambda + 180°)]$$

If the rear diaphragm element is switched off, the microphone is a simple forward-facing cardioid. If the two elements are combined in phase or 180° out-of-phase, the result is omni or figure-of-eight, respectively. This system will be discussed further in Section 4.6.3.

4.6 The microphone as a current generator

Having examined the various ways in which acoustic forces drive the diaphragm into vibrations which are analogous to the incident sound waveform, it is now appropriate to describe how this oscillatory mechanical energy can be used to generate a corresponding electrical signal. In fact every imaginable form of mechanical/electrical conversion has been tried, though only the most common will be described here in approximate order of importance.

4.6.1 The moving-coil (dynamic) microphone

This category of microphone relies on the principle outlined in Chapter 3 (page 55) that motion of a conductor in a magnetic field generates an emf causing a flow of current in the conductor (as in a dynamo). In the usual configuration the diaphragm is shaped as shown in Figure 4.21, and carries a circular insulated coil of copper or aluminium wire centred in the annular air gap between N and S polepieces of a permanent magnet. The magnitude of the induced emf is given by

$$E = Blu$$

where B is the flux density, l is the effective length of the coil in the gap and u is the diaphragm velocity.

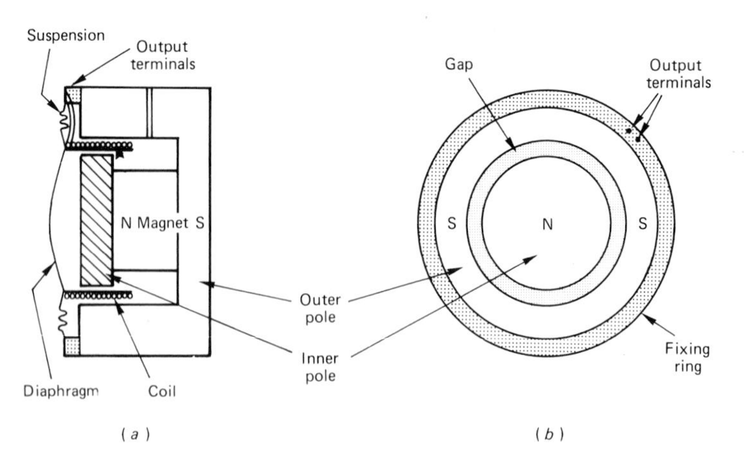

Figure 4.21 Moving-coil (dynamic) microphone. Basic construction in (a) side and (b) plan view

It will be seen that designing for high sensitivity is a matter of making B and l as large as possible. The proportionality between E and u means that the moving-coil microphone is a constant velocity device (see Figure 2.19 on page 31). It is therefore best suited to pressure operation, where the force is constant at all frequencies, and resistance control. The mechanical resonance is set at some mid-frequency, say 800 Hz, and smoothed as far as possible by introducing acoustic damping (resistance). To avoid the serious reduction in sensitivity that too much damping would cause, acoustic reactances as well as resistances are often built in to arrive at the desired flat frequency response.

The moving-coil microphone is basically a low-impedance source. Older designs had a source impedance of around 5 – 20 Ω and needed a matching step-up transformer located at the input to the microphone amplifier or (better for cable matching) built into the microphone casing. However, some modern designs use many turns of very thin-gauge wire having a DC resistance of about 200 Ω. This permits direct matching to standard microphone amplifier inputs (nominally around 1000 Ω) and the use of long connecting cables.

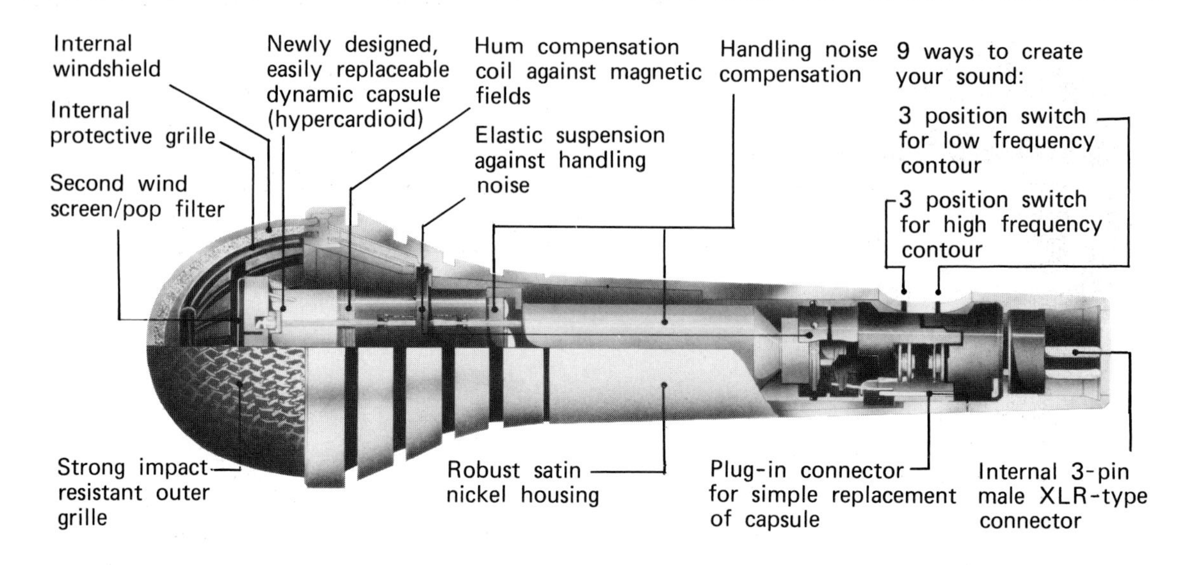

Figure 4.22 The AKG D330BT moving-coil microphone

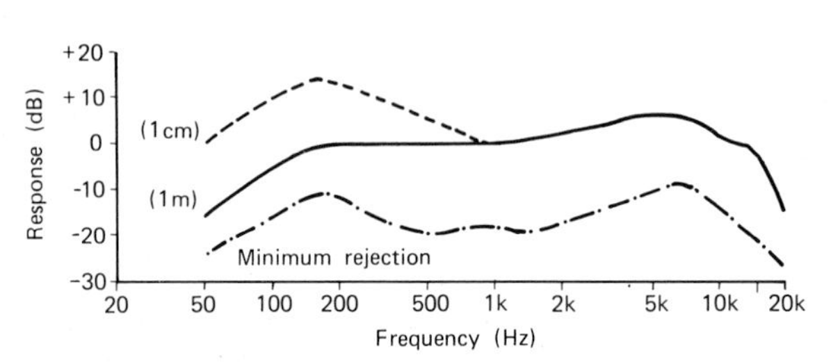

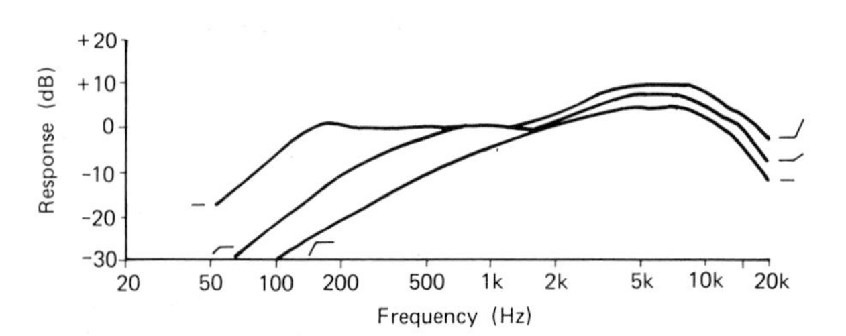

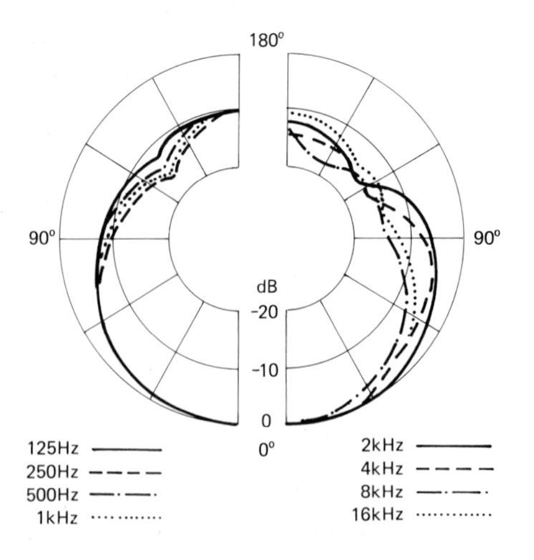

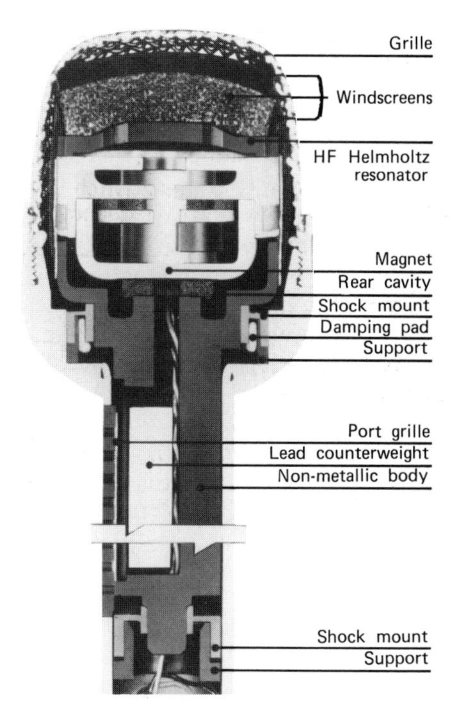

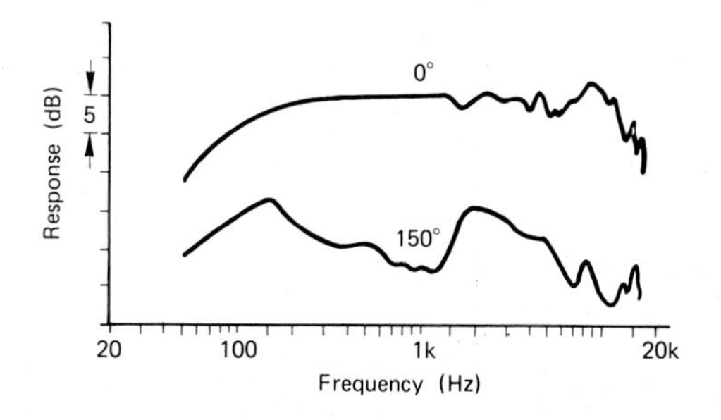

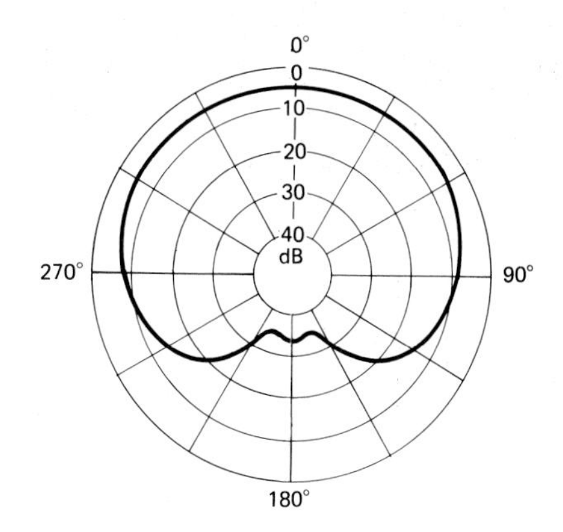

Figure 4.23 The Electro-Voice RE18 moving-coil microphone

Key
1. Memraflex grille screen
2. High-density Acoustifoam™ windscreen
3. Low-density Acoustifoam™ windscreen
4. Cloth side port windscreen
5. High-frequency-extending Helmholtz resonator
6. High-flux magnet structure
7. Fully isolated rear cavity
8. Butyl rubber front shock mount
9. Silicon oil-filled impact damping pad
10. Non-metallic mount support
11. Variable-D® port grille screen
12. Resonance-lowering lead counterweight
13. Non-metallic transducer body
14. Butyl rubber rear shock mount
15. Non-metallic mount support

Examples of moving-coil microphone design

1. AKG D330 (Figure 4.22, opposite page)
Transducer type: moving-coil pressure gradient
Polar pattern: hypercardioid
Frequency range: 50–20 000 Hz
Sensitivity: 1.2 mV/Pa (−58 dBV)
Max.SPL for 0.5% THD: 128 dB (50 Pa)
Impedance: 370 Ω
Features: robust double grille, die-cast housing, elastic suspension, double foam/fabric wind/pop filter, hum-reducing coil, three-position bass (100 Hz) and presence (4 kHz) switches
Main applications: handheld or stand-mounted vocal microphone

2. Electro-Voice RE18 (Figure 4.23, this page)
Transducer type: moving-coil pressure gradient
Polar pattern: supercardioid
Frequency range: 80–15 000 Hz
Sensitivity: −56 dB ref. 1 mW/10 Pa
Impedance: 150 Ω
Features: robust grille, steel case, butyl rubber plus viscous damping fluid shock-isolating system, standard and high-density foam wind/pop filter, hum-reducing coil, Variable-D design (using slotted rear ports) to reduce proximity effect
Main applications: handheld or stand-mounted vocal or commentator's microphone

4.6.2 The ribbon microphone

A special adaptation of the electromagnetic or dynamic principle is used in the ribbon microphone. Again there is a permanent magnet

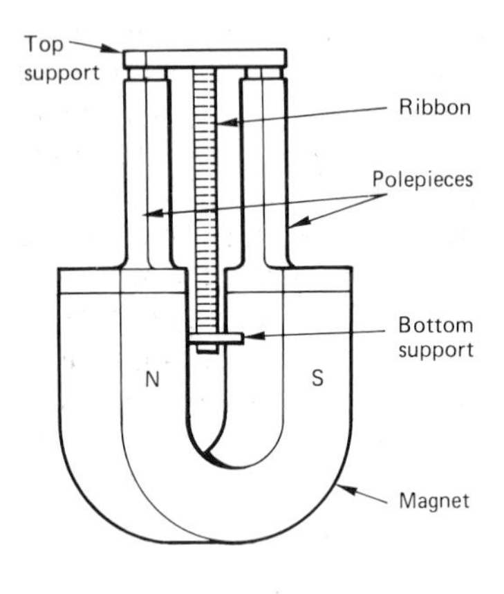

Figure 4.24 Ribbon microphone: basic construction

with specially shaped polepieces producing a concentrated magnetic field across an extended gap in which is suspended a thin metal ribbon (Figure 4.24). This ribbon acts both as diaphragm and conductor (i.e. coil). Typically, the ribbon is made of aluminium foil about 0.1 mm thick and 2–4 mm wide. It is mildly corrugated to prevent curling at the edges and allow accurate tensioning to set a design resonance frequency of the order of 2–4 Hz. Needless to say, the DC resistance of the ribbon is very low and a built-in step-up transformer is essential to provide a standard matching impedance. The low mass, only about 0.2 mg, gives the ribbon excellent transient response. Against this it must be said that sensitivity to wind noise or mechanical shocks is high. Ribbon microphones are therefore unsuitable for most outdoor applications or boom operation in television and film.

Again, as for the moving-coil microphone, the induced emf is given by $E = Blu$, making this a constant-velocity system. There is the difference, however, that the driving force in a PG-operated device is proportional to frequency (see Figure 4.14). Therefore the ribbon microphone must be mass controlled (Figure 2.19), which means placing its main resonance well below the required frequency range at around 2 Hz, as previously mentioned.

While a few back-enclosed pressure-operated ribbon microphones have been designed, the most common construction has both ribbon surfaces equally exposed to the air to give PG operation and the familiar figure-of-eight directivity. The earliest BBC Type A ribbon microphone appeared in 1936, and was an immediate success, for example, with drama producers. Actors could address each other from opposite sides of the microphone and make entrances or exits with predictable changes in apparent perspective simply by stepping sideways into the less sensitive side angles. This proved so useful that the BBC had special round carpets made with 100° and 80° segments marked 'Live' and 'Dead', respectively, and arcs at 3, 4 and 5 ft radius.

By enclosing all or part of the rear surface of the ribbon with an acoustically absorbing or phase-shift network it is possible to produce a purely pressure-operated omni microphone or a combination cardioid.

4.6.3 The condenser (capacitor or electrostatic) microphone

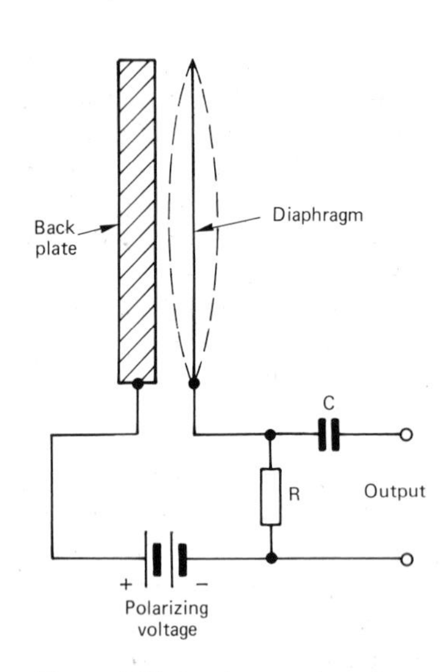

Figure 4.25 Condenser microphone: basic principle

This category of microphone is so called because the transducer element is a simple condenser (or capacitor in modern parlance) whose two electrodes or plates are formed by the thin conductive diaphragm and a fixed backplate separated by a narrow air gap (Figure 4.25). A polarizing DC voltage is applied across the two electrodes via a very high resistance R. This establishes a quasi-constant charge Q on the capacitor. The capacitance of the assembly is given by

$$C = kA/x$$

where A is the effective diaphragm area, x the electrode spacing and k the dielectric constant of air.

Since the voltage across the capacitor is $E = Q/C$, we can write $E = (Q/kA)x$, which shows that E is proportional to x. Therefore, when the diaphragm vibrates in response to the incident sound-pressure wave the applied DC polarizing voltage will be effectively modulated by an AC component in linear relationship to the diaphragm vibrations. This AC component is taken through the DC-blocking coupling capacitor C to provide the microphone's output signal.

The use of a low-loss air dielectric and the basic simplicity of the generator mechanism make the capacitor microphone a potentially quiet and linear transducer. However, the very narrow gap needed in the interests of adequate sensitivity, and the very high impedance represented by the microphone capacitance of only about 25–50 pF, calls for high precision in manufacture and the provision of a high-impedance 'head' amplifier located as close to the electrodes as possible (see Figure 4.6). Both these factors increase the cost and complexity of the design.

The linear relationship between E and x means that the capacitor microphone is a constant-amplitude device. In the case of a pressure-operated capacitor microphone this automatically produces a flat output response. It is therefore appropriate to use stiffness control (see Figure 2.19) and place the mechanical resonance well above the working frequency range – a relatively easy matter in view of the low mass of the diaphragm. For PG-operated capacitor microphones, having a perforated backplate to allow sounds to reach both sides of the diaphragm as for the standard ribbon, the force on the diaphragm is proportional to frequency (at least at low frequencies) and the displacement must be given a 6 dB/octave falling response. Therefore resistance control is used, choosing a mid-frequency resonance and adding resistance damping.

Head amplifiers have been improved in design over the years as more compact, and quieter, solid-state components have become available. Ironically, however, something of a cult has developed for employing the seemingly obsolete valve (tube) amplifier microphones for their subjectively perceived 'naturalness' of tone and smooth onset of non-linear distortion. At least one company (AKG) has even recommenced manufacture of an early valve microphone and renamed it 'The Tube'.

As a rule, however, the spread of digital recording has set a premium on dynamic range and minimum inherent noise as primary microphone design criteria. Since the amplifier is at least as potent a limiting factor as the capacitor capsule itself, exotic solid-state amplifiers are now the rule with previously unattainable figures for intrinsic noise at one end of the range and low non-linear distortion at the other. The field-effect transistor (FET) selected for a microphone head amplifier must have very low mid-band self-noise, low- and high-frequency noise being more a function of the FET gate resistor and capsule acoustic friction, respectively.

High-level performance is quoted as the maximum SPL for which total harmonic distortion remains less than 1% (as also mentioned in Section 4.1.6). The West German Hi-Fi standard calls for this overload SPL to be no less than 114 dB (10 Pa). Studio microphones achieve much better figures than this (say, 120–124 dB SPL), and it can be raised further (e.g. for use as a close drum microphone) by incorporating a switchable 10 dB or 20 dB pre-attenuator (overload protection switch). FETs have sometimes been replaced by operational amplifiers (Op-Amps) giving up to 144 dB SPL capability. This

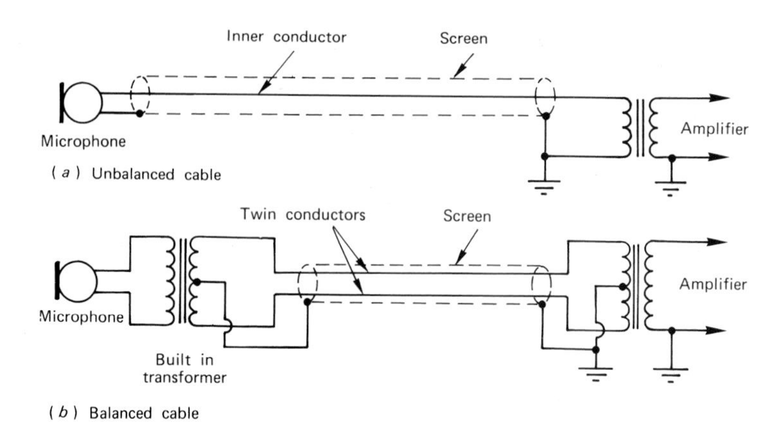

Figure 4.26 Microphone cables: (a) unbalanced; (b) balanced

is beyond the pressure level of even the loudest music sources, but the added headroom is claimed to provide audibly improved handling of transient peaks.

Supply cables for condenser microphones have become less complicated and less prone to faults. Early valve-amplifier microphones needed a multi-core cable with separate provision for the heater and anode voltages. Today a two-wire shielded cable is the norm. This can be unbalanced, mainly for amateur or industrial applications, with one conductor carrying the audio signal, the other the DC supply voltage and the cable braiding or shield acting as a common return path. However, studio microphones traditionally employ a balanced cable arrangement in which the audio signals use both conductors to maintain identical potentials relative to the earth or 'ground' potential of the cable shield, microphone casing, plug exterior, etc. (Figure 4.26). This permits long cable runs combined with good rejection of stray hum fields and radio-frequency interference. As the figure shows, any stray induced emf will produce identical currents in both signal conductors and, since these flow in opposite directions round the circuit, they will contribute zero signal across the load.

Two methods of conveying the DC supply along the signal conductors are in common use.

A-B powering (Figure 4.27a) typically employs a well-filtered 12 V DC supply with its positive pole connected to one conductor and the negative pole to the other via identical 180 Ω resistors. This DC is prevented from reaching the load, i.e. the mixer input circuit, by blocking capacitors as shown. The two conductors and the microphone circuit must be 'floating', i.e. electrically isolated from the case and cable shield. However, the junction between the 180 Ω resistors, and therefore one pole of the 12 V supply, is connected to the shield to ensure the necessary screening effect. It is a limitation of A-B powering that connecting other microphones presents problems and dynamic microphones with output transformers will distort or even suffer damage.

Phantom powering (Figure 4.27(b)) overcomes these difficulties and allows the same cable to be interchangeable between capacitor and dynamic microphones without necessarily switching off the supply voltage. Two standards are in use, 48 V + 4 V with maximum

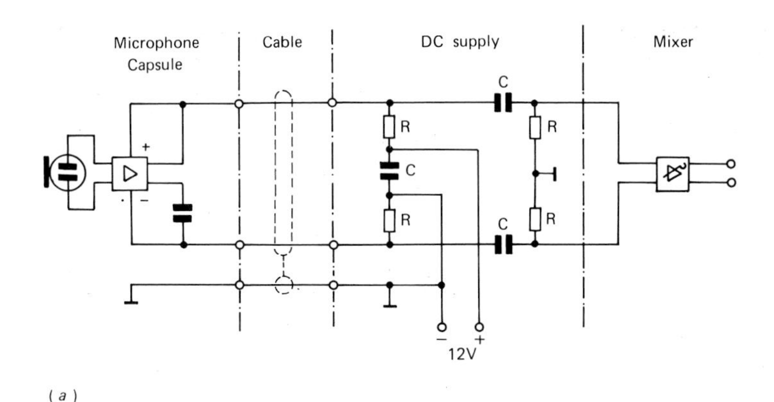

Figure 4.27 Condenser microphone power supply systems. (a) A-B power; (b) phantom power

current 2 mA, and 12 V + 1 V with maximum current 10 mA. The DC is divided so that one half of the current flows to the microphone through each conductor, returning to the source via the cable shield. The 'phantom' splitting of the DC voltage is achieved by taking the positive pole to the junction between two identical resistors (6.8 K for 48 V or 680 Ω for 12 V). The DC is thus decoupled from the audio circuit along with any inherent noise. No DC blocking capacitors are needed provided the following mixer or amplifier has the usual balanced and floating circuit.

The 48 V phantom supply is built into many mixing consoles and is naturally ideal for capacitor microphones designed for 48 V. A 12 V supply is often found in ancillary equipment for other purposes and therefore readily available. In microphones designed for 12 V the higher polarizing voltage needed is generated using a built-in supersonic oscillator and rectifier.

A further possibility is to use an external battery supply unit which has the advantage of giving very low noise operation. A few larger microphones, notably the Neumann U87, have a compartment for the insertion of small batteries giving up to 200 h running. Electret microphones (described in the next section) draw so very little current that some models have no on/off switch for their internal batteries, which need replacing only about once a year.

Variable-directivity capacitor microphones are popular, combining pressure and PG operation as described in Section 4.5. Figure 4.28(a)

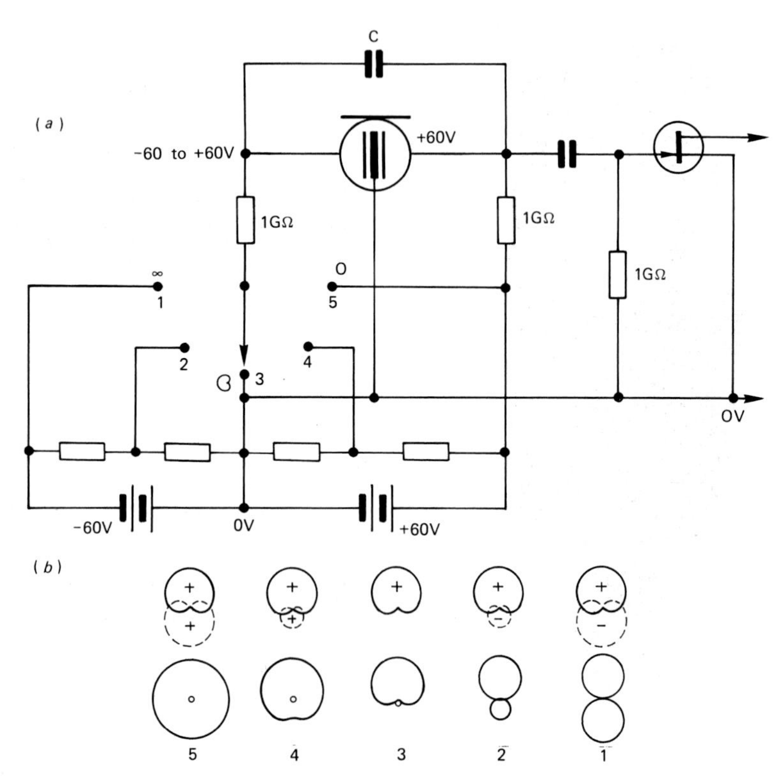

Figure 4.28 Variable-directivity dual-diaphragm condenser microphone. (a) Basic circuit; (b) polar patterns produced for the five switch positions

illustrates a dual-diaphragm type effectively forming two cardioid elements back-to-back, whose signal output is, of course, proportional to the polarizing voltage applied. The right-hand diaphragm in the figure has a fixed +60 V potential with respect to the common backplate, to form a forward-facing cardioid. The voltage to the rear diaphragm is taken via a network allowing the potential to be switched from −60 V through 0 V to +60 V. Since the audio signal outputs of the two elements are connected through capacitor C, the five directivity patterns shown in Figure 4.28(b) are obtained. Remote-controlled switching provides a welcome degree of flexibility and, of course, additional switch positions are possible or a continuously variable potentiometer may be fitted.

Examples of condenser microphone design

1. AKG C414B-ULS (Figure 4.29)
Transducer type: condenser, double 25 mm diaphragm
Polar pattern: cardioid, hypercardioid, omni, figure-of-eight
Frequency range: 20–20000 Hz
Sensitivity: 12.5 mV/Pa (−38 dBV)
Max. SPL for 0.5% THD: 140, 150, 160 dB with pre-attenuation
Impedance: 180 Ω
Equivalent noise level: 25 dB (14 dBA)
Total dynamic range: 126 dB
Features: four switched polar patterns, three-position high-pass filter (12 dB/octave at 75 Hz or 150 Hz), switched attenuator at −10 dB or −20 dB, foam-type windscreen (transformerless version available, C414B-TL)
Main applications: music

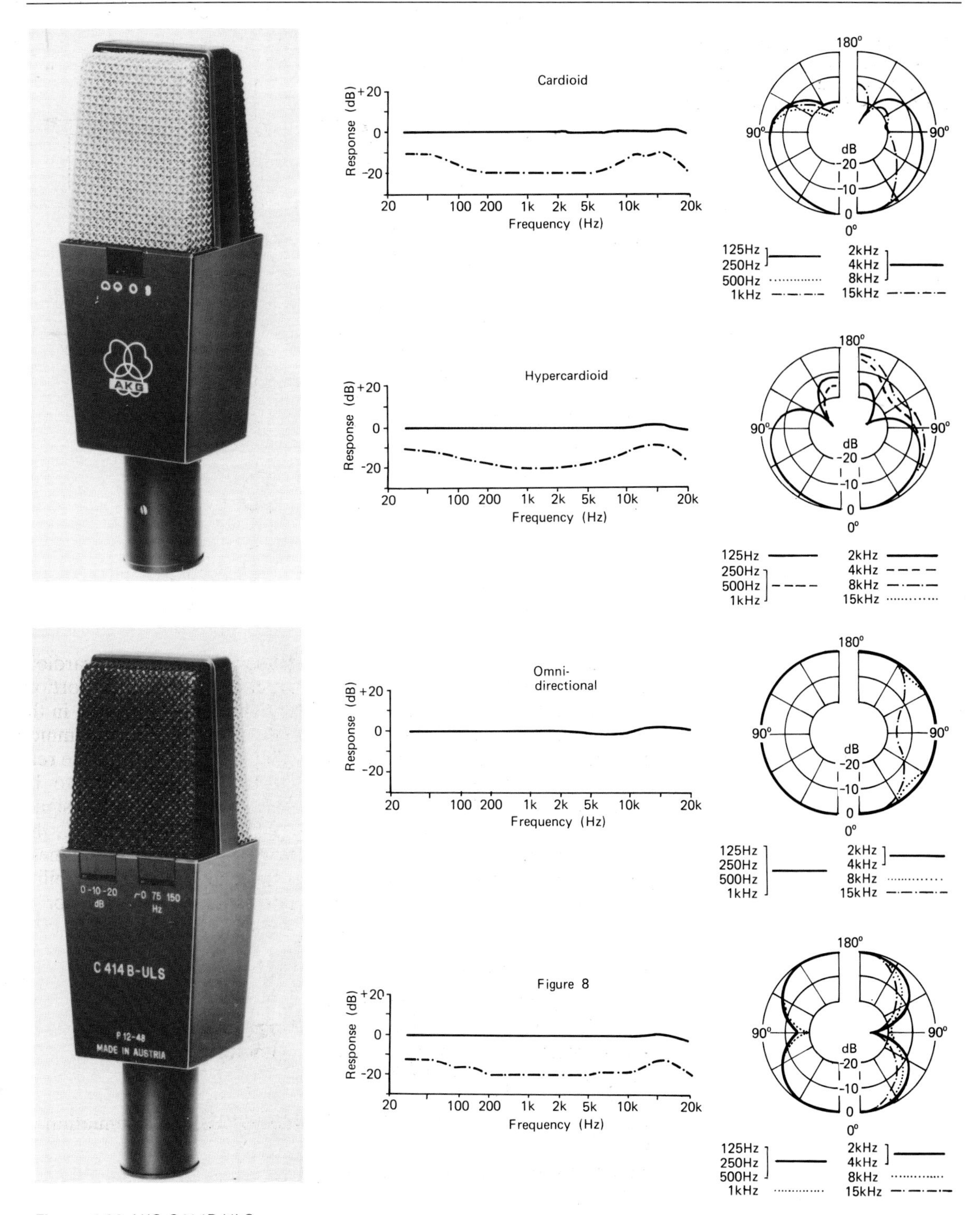

Figure 4.29 AKG C414B-ULS condenser microphone

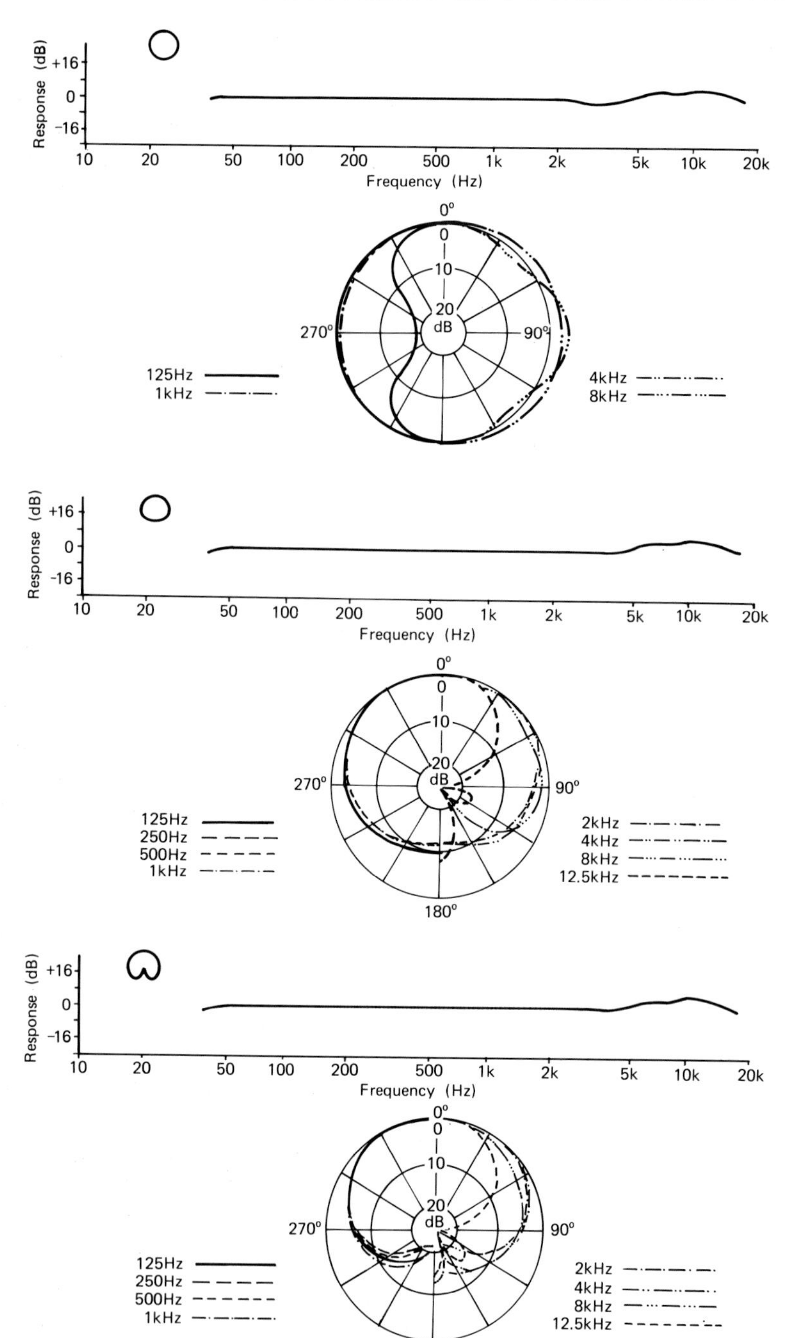

Figure 4.30 Beyer MC 740 condenser microphone

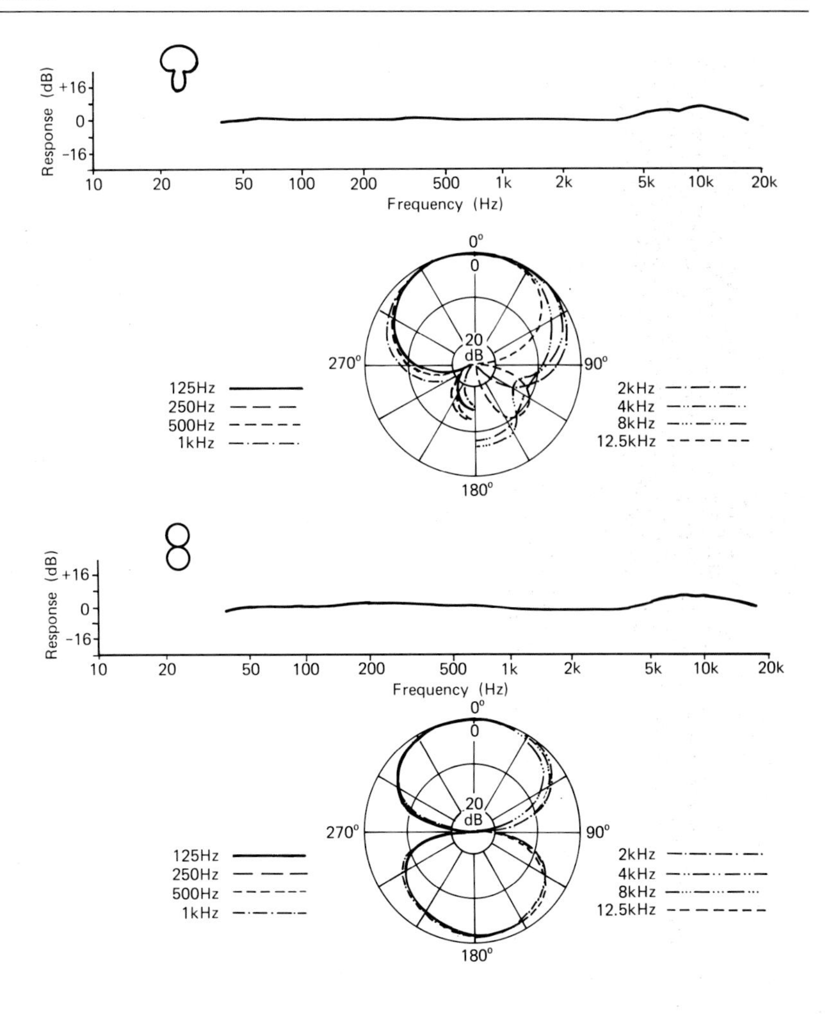

2. Beyer MC 740 (Figure 4.30)
Transducer type: condenser
Polar pattern: cardioid, hypercardioid, wide cardioid, omni, figure-of-eight
Frequency range: 40–20000 Hz
Sensitivity: 10 mV/Pa (−40 dBV)
Max. SPL for 0.5% THD: 134 dB or 144 dB with pre-attenuation
Impedance: 150 Ω
Equivalent noise level: 17 dB
Features: five switched polar patterns, 10 dB attenuator switch
Main applications: music, announcements, discussions

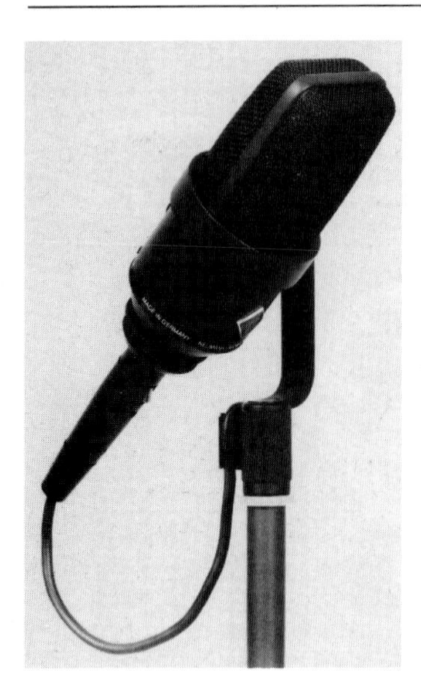

Figure 4.31 Neumann TLM170i condenser microphone

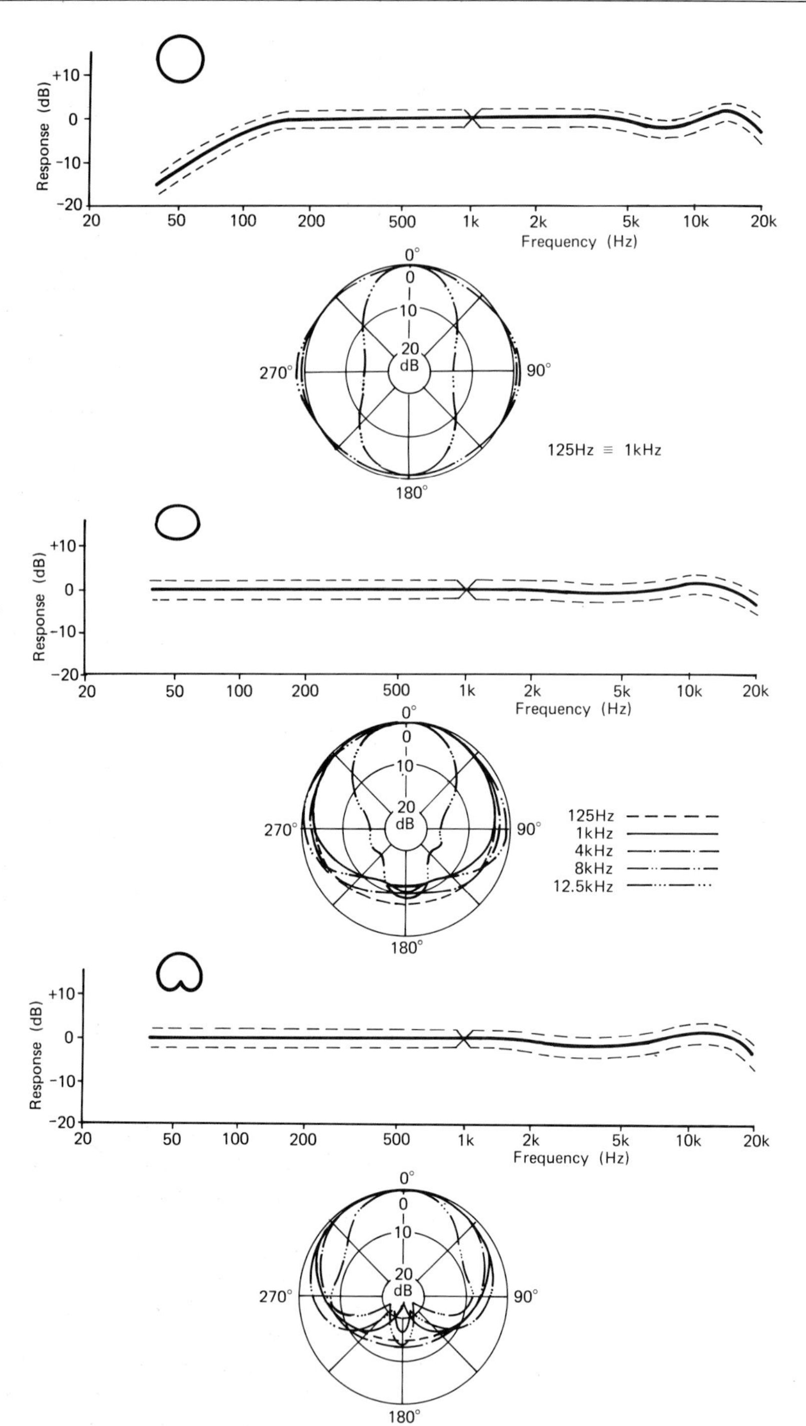

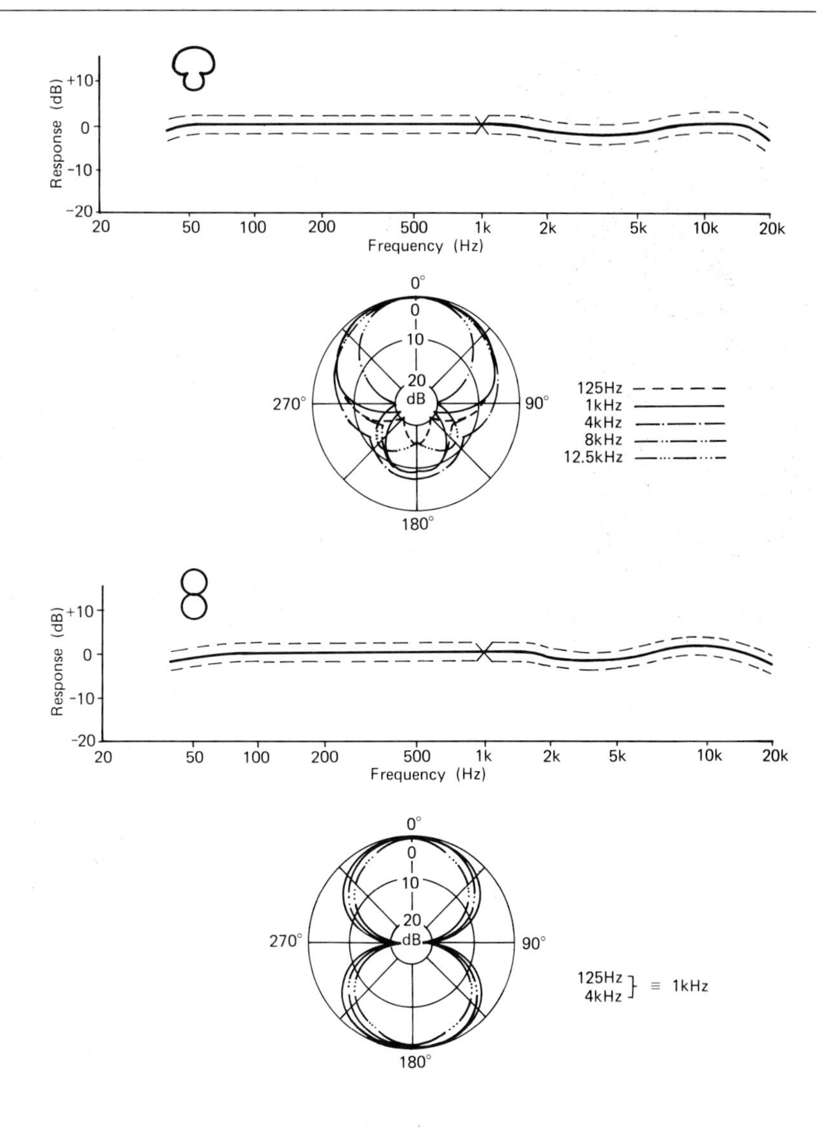

3. Neumann TLM170i (Figure 4.31)

Transducer type: condenser, double 28 mm diaphragm

Polar pattern: cardioid, hypercardioid, wide cardioid, omni, figure-of-eight

Frequency range: 40–18 000 Hz

Sensitivity: 8 mV/Pa

Max. SPL for 0.5% THD: 140 dB or 150 dB with pre-attenuation

Impedance: 100 Ω

Equivalent noise level: 26 dB (22 dB DIN)

Features: five switched polar patterns, transformerless circuit, very low noise, 10 dB attenuator switch, high-pass filter switch, tiltable elastically suspended bracket, 126 dB overall dynamic range

Main applications: music (linear diffuse-field frequency response for all polar patterns)

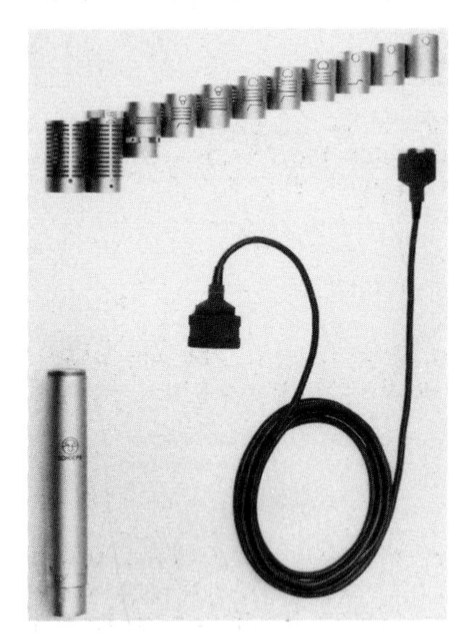

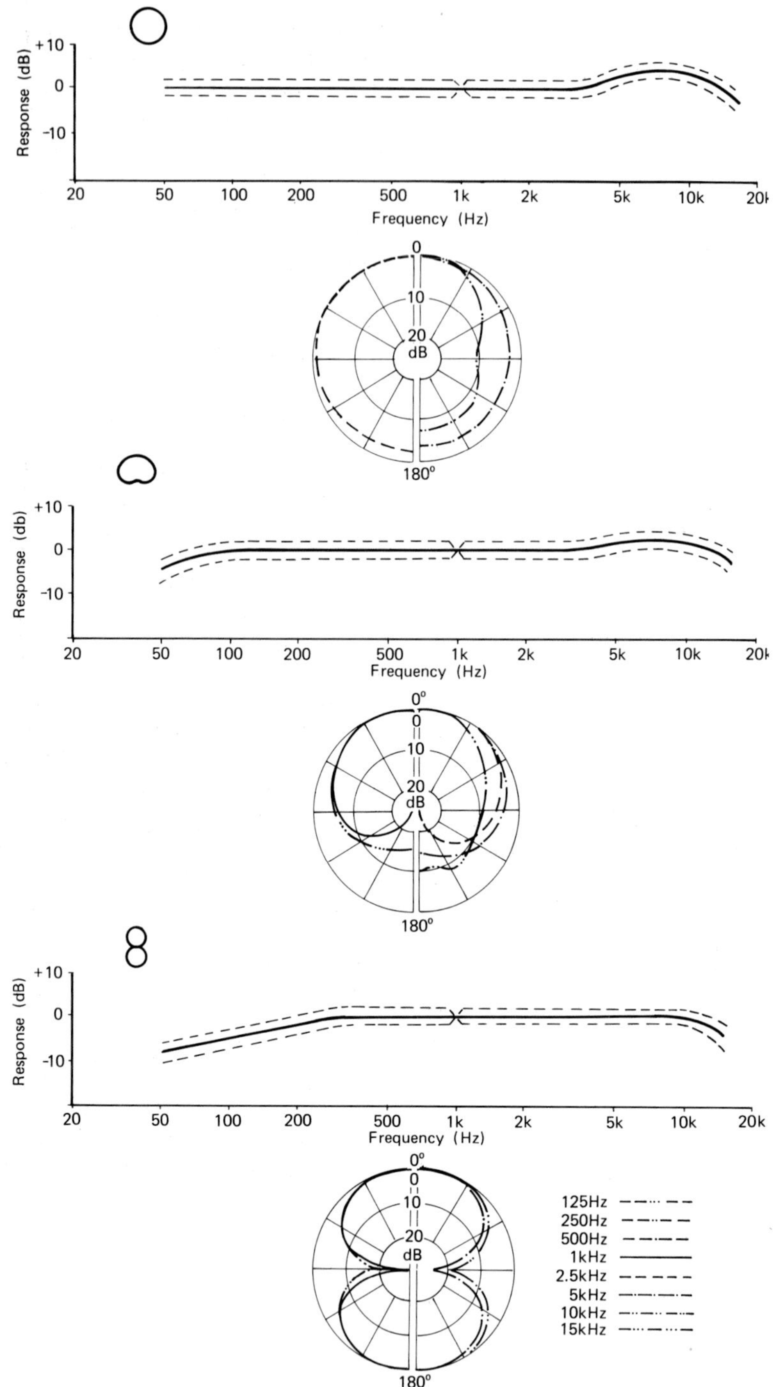

Figure 4.32 Schoeps Colette Series condenser microphones, showing response curves for the MK6 three-pattern capsule

4. Schoeps Colette Series (Figure 4.32)
Transducer type: condenser
Polar pattern: range of eight different capsules covering four single patterns plus two- and three-pattern switchable versions
Frequency range: omni 20–20000 Hz; others 40–20000 Hz
Sensitivity: 9–12 mV/Pa
Max. SPL for 0.5% THD: 130–132 dB
Impedance: recommended load at least 600 Ω
Equivalent noise level: 17–19 dBA
Features: wide range of interchangeable capsules and electrically active connectors and accessories, transformerless output
Main applications: music

4.6.4 The electret microphone

As we have seen, only a low-voltage DC supply is required to power the built-in amplifier of a condenser microphone. Yet there remains the practical disadvantage that (although virtually no current is drawn) a relatively high (60–120 V) DC voltage has also to be provided to polarize the plates. This inconvenience has been eliminated in recent years, with a simultaneous simplification in construction, by the introduction of the electret microphone (Figure 4.33).

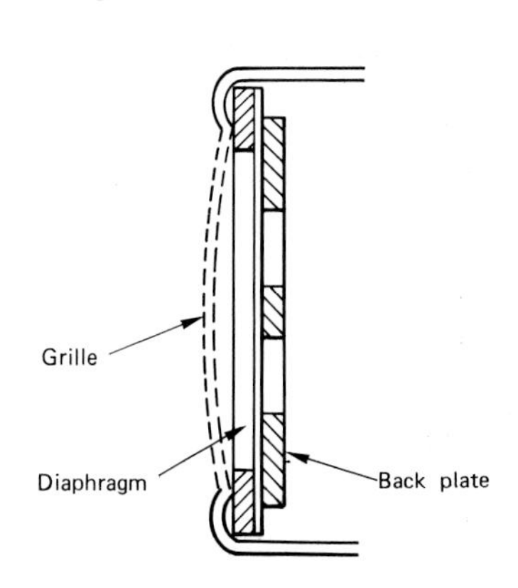

Figure 4.33 Electret microphone: basic construction

This uses a permanently polarized material (usually polytetrafluorethylene) which can be regarded as the electrostatic equivalent of a permanent magnet. The electret-foil material is given its permanent charge, positive or negative, while in a strong electric field and heated either by corona discharge or electron bombardment. Contemporary electret microphones may use either a polarized diaphragm or a neutral diaphragm with the fixed plate coated with electret material ('back polarized').

The latter arrangement avoids the problem that polarizing the diaphragm introduces a degree of compromise, a stable electret requiring a thick diaphragm while mechanical considerations demand a thin one made of selected material. As well as eliminating the need for a high DC voltage supply, electret microphones can be made very small and rugged. It is also possible to construct viable push–pull electret designs capable of reproducing high SPL signals with low distortion. However, variable-directivity designs of the sort described above using switched polarizing voltages are not possible.

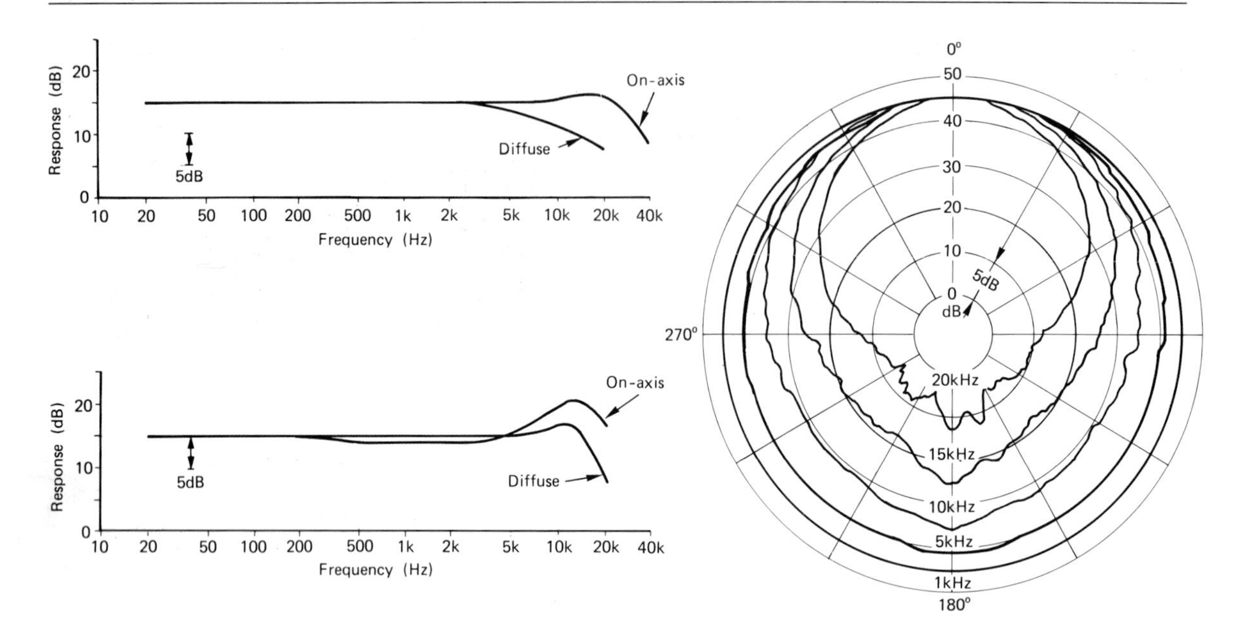

Figure 4.34 Bruel & Kjaer 4003/4006 electret condenser microphone

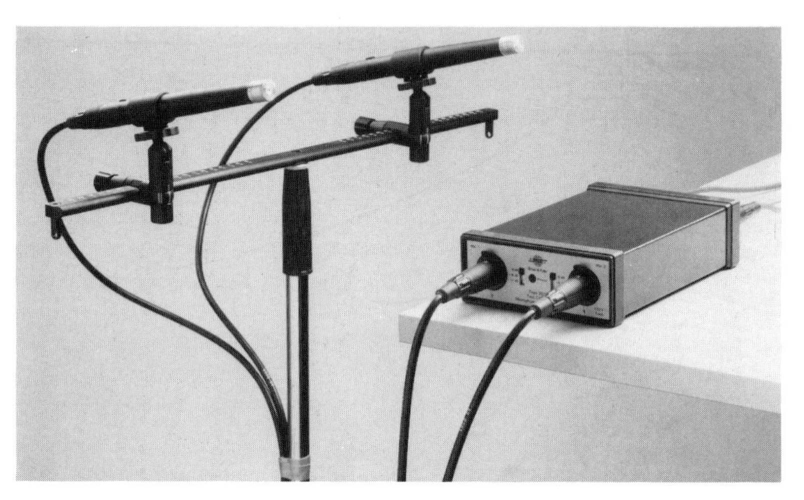

Examples of electret microphone design

1. Bruel & Kjaer 4003/4006 (Figure 4.34)
Transducer type: condenser (electret)
Polar pattern: circle (omni)
Frequency range: 20–20000 Hz ± 2 dB
Sensitivity: 4003 50 mV/Pa (−26 dBV); 4006 12.5 mV/Pa (−38 dB V)
Max. SPL for 1% THD: 135 dB
Impedance: 30 Ω
Equivalent noise level: 15 dB
Features: very low self-noise, choice of protection grid/nose cone gives linear free-field (close-miking) or diffuse-field frequency response, true omni response at high frequencies
Main applications: music

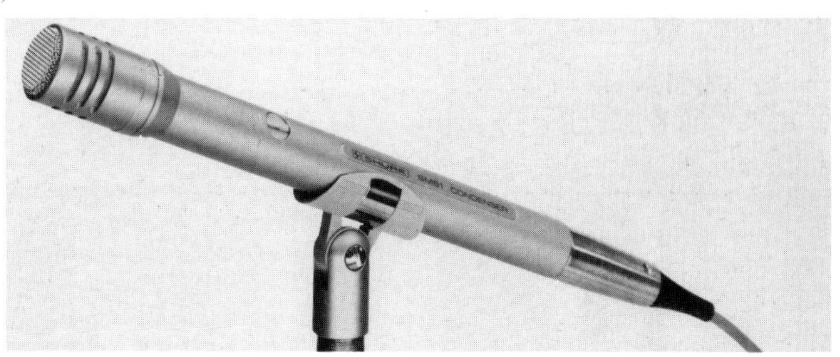

Figure 4.35 Shure SM81 electret condenser microphone

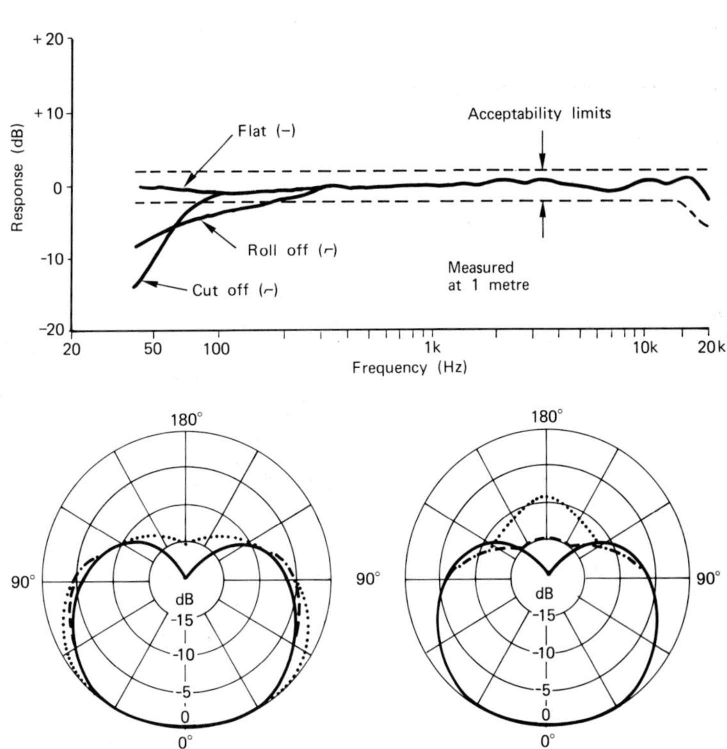

2. Shure SM81 (Figure 4.35)
Transducer type: condenser (electret)
Polar pattern: cardioid (unidirectional)
Frequency range: 20–20000 Hz
Sensitivity: 0.56 mV (−65 dB V)
Max. SPL for 0.5% THD: 131 dB at 250 Hz
Impedance: 150 Ω
Equivalent noise level: 19 dB
Features: robust steel housing, pop filter grille, switched high-pass filter at 6 dB/octave at 100 Hz to compensate for proximity effect, or 18 dB/octave as rumble filter, switched 10 dB attenuator
Main applications: music

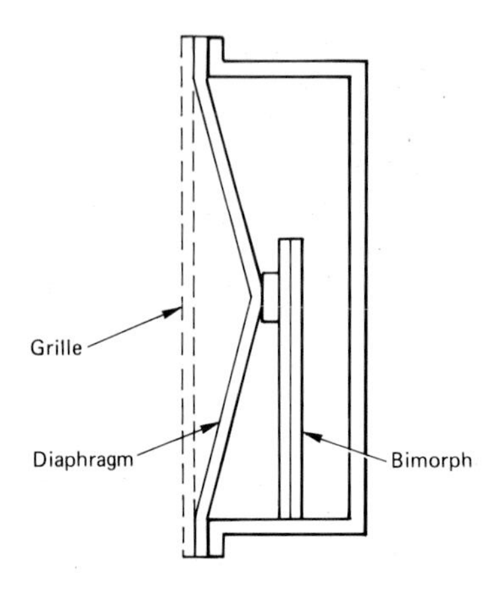

Figure 4.36 Piezoelectric (crystal) microphone: basic construction

4.6.5 The piezoelectric (crystal) microphone

Extremely robust and inexpensive microphones can be produced using crystalline or ceramic materials which possess piezoelectric properties. Suitable wafers or slabs of these materials generate a potential difference between opposite faces when subjected to torsional or bending stresses. The usual piezoelectric microphone capsule comprises a sandwich or 'bimorph' of oppositely polarized slabs joined to form a single unit with metal electrodes attached (Figure 4.36). The vibrations of a conventional diaphragm are transmitted to the bimorph by a connecting rod giving rise to an alternating voltage at the output terminals proportional to the effective displacement.

4.6.6 The carbon (loose-contact) microphone

Venerated because it has been around for more than a hundred years and still found in millions of telephones around the world, the carbon microphone cannot achieve the standards of response and low self-noise demanded for modern studio applications. The principle (Figure 4.37) relies on the variation in DC resistance of a package of small pieces or granules of some conductive material when subjected to variations in external pressure which will cause their areas of contact to increase or decrease.

The modern version uses granules of carbonized hard coal, in a form which goes back to an Edison patent dated 1889. A complex series of operations produces granules of even dimensions (passed through a 60–80 mesh) without flat-shaped or iron-containing particles. Two gold-plated cups enclose the carbon cavity and are attached respectively to the thin diaphragm and a fixed backplate. An energizing source of DC current is applied across the granules' cavity through a load impedance across which the AC audio modulation due to diaphragm vibrations is developed and taken to the output terminals. The load impedance may be a resistor or, more often, the primary winding of a step-up transformer.

The arrangement effectively constitutes an amplifier with about 40–60 dB gain, but it has a number of inherent disadvantages. The capsule generates high self-noise and exhibits non-linear distortion

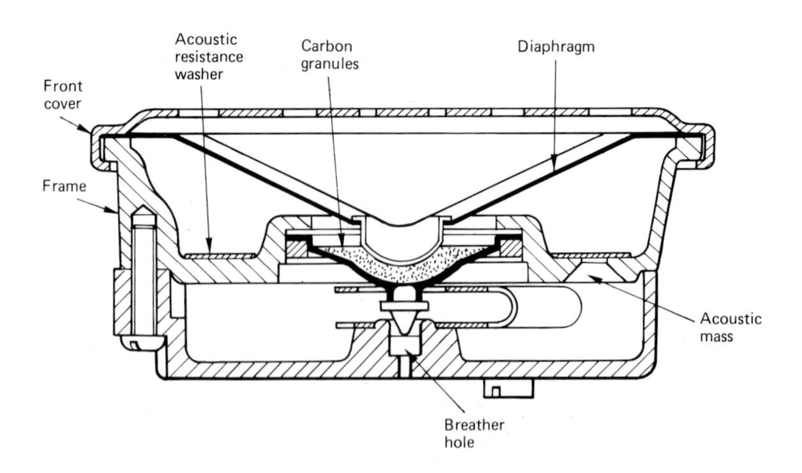

Figure 4.37 Carbon (loose contact) microphone: a modern telephone mouthpiece design

and instability coupled with reduced sensitivity if the granules pack together (requiring a sharp knock to shake them loose again). Improved linearity was obtained in the early days of broadcasting by using a push–pull dual-button carbon transducer. However, this was overtaken during the 1930s by other types of transducer allied to electronic amplification – a fate which seems likely to affect today's carbon microphones as cheap solid-state amplifiers take over the telephone networks.

4.6.7 Other types of microphone

For completeness, other types of microphone transducer constructions and design philosophies should be given a brief mention, even though some of them have not reached general commercial exploitation.

Among variations on the electromagnetic principle, two types of inductor microphone may be cited. In the first, an RCA model once used mainly for broadcast speech, a straight conductor was moulded into a plastic diaphragm and held in a magnetic air gap, much as in the moving-coil microphone. More recently a printed ribbon microphone was described by Tabuchiet in which a spiral of aluminium ribbon is printed onto a flat diaphragm of polyester film. Pressure-gradient operation applies, as for the standard open-backed ribbon, though an omni version with a damped rear cavity has been described.

A genuine pressure ribbon was developed by Olson in the late 1940s, having an acoustic folded pipe labyrinth at the rear and a narrow flared pipe at the front. Its narrow shape made it unobtrusive for television use, but it has been replaced by modern alternatives. Fixed-coil magnetic microphones have also been designed in which the diaphragm carries a drive pin to activate an armature within a magnetic assembly to produce induced current in the coil.

Ingenious vacuum tube, transistor and even integrated circuit microphones have appeared which may yet lead to novel designs for the future. In general, the diaphragm carries a drive pin to oscillate some part of the circuit element. In the vacuum tube microphone (Olson) it was a movable electrode which modulated the flow of electrons. In a transistor microphone described by Sikorski it was a sapphire indenter which applied stress to the emitter region of the transistor. In the integrated circuit microphone developed by Sank there is no diaphragm and the transducer is simply a directly actuated strain-gauge bridge deposited on a silicon chip. Excellent frequency response was reported but the signal-to-noise ratio was low.

The *boundary microphone*, also referred to as a PZM (pressure zone microphone), is not so much a new type of transducer as a novel approach to microphone placement (patented by Long and Wickersham). In essence, it comprises a small capsule microphone, frequently an electret, housed in a flat receptacle in such a way that the diaphragm is located no more than about 1 mm from any flat surface on which the unit is placed (see Figure 4.38). The principal aim is to eliminate the irregular dips in frequency response (comb filter effect, as described in Section 2.11) which can readily arise in normal microphone placement due to interference between the direct wave from the source and that reflected from the floor, etc.

Figure 4.38 PZM (boundary) microphone: basic construction

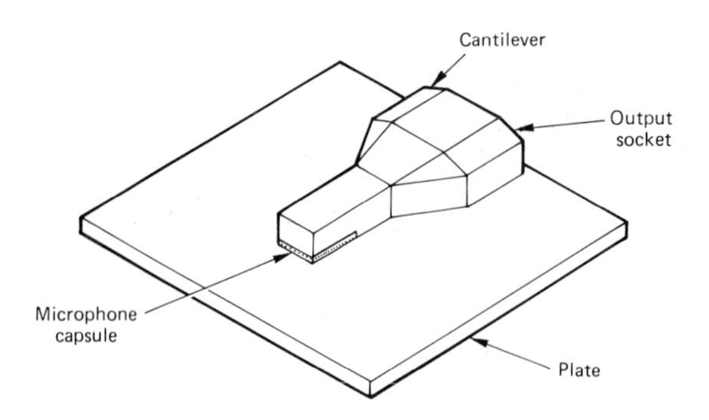

The two situations are contrasted in Figure 4.39. In Figure 4.39(a), the microphone receives two sound waves, one direct and one reflected, so that alternate addition and subtraction will occur at frequencies for which the path length difference results in a phase shift of 0° or 180°, respectively, giving the familiar comb filter effect discussed in Section 2.11 (see Figure 4.39(b)). With a boundary microphone (Figure 4.39(c)) the extreme proximity to the floor produces no significant interference effect at wavelengths in the audio range, and the response is uniform (Figure 4.39(d)).

However, there are are several side-effects. One of these may be taken as beneficial, that the microphone capsule is so enclosed as to avoid the usual boost in axial (free-field) high-frequency response relative to random reverberant (diffuse-field) sound (discussed in Section 2.14). There is also a 6 dB increase in sensitivity due to pressure doubling, with smooth and consistent response at all angles within the hemispherical polar pattern. Against this must be weighed such disadvantages as the pick-up of audience or stage noise.

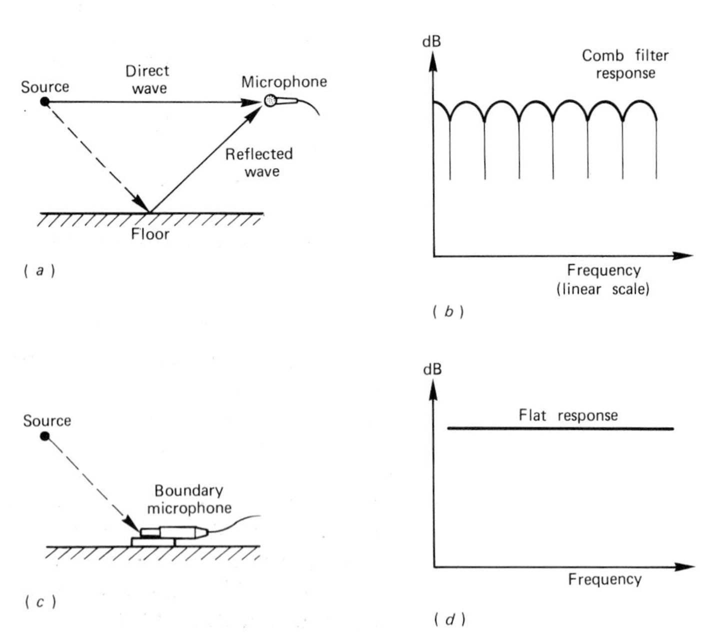

Figure 4.39 How the boundary microphone avoids comb filter distortion. (a) Floor reflection produces a delayed wave; (b) comb filter cancellations result; (c) boundary microphone receives only a direct wave; (d) a flat response results

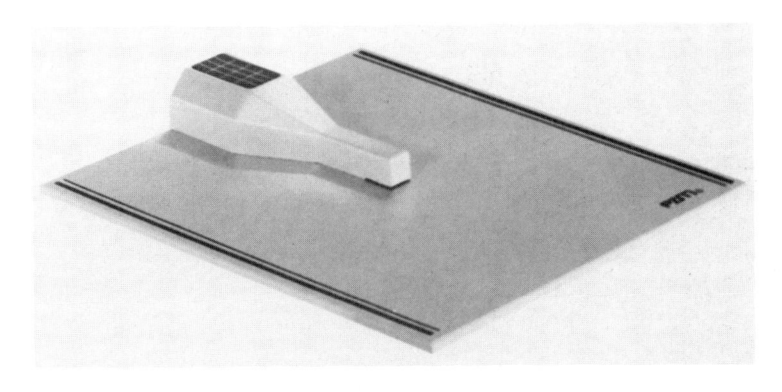

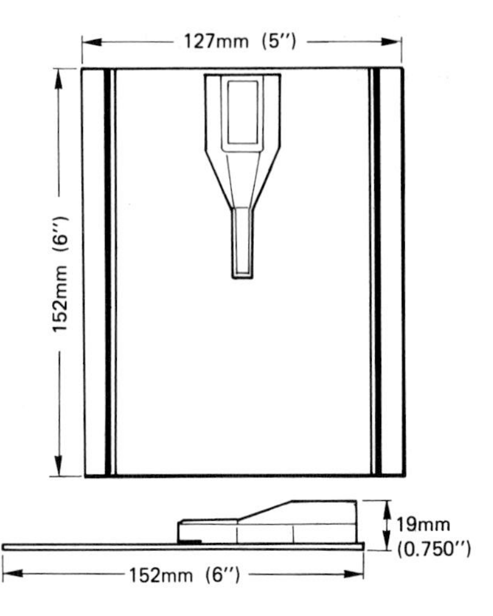

Figure 4.40 Crown PZM-6FS boundary microphone

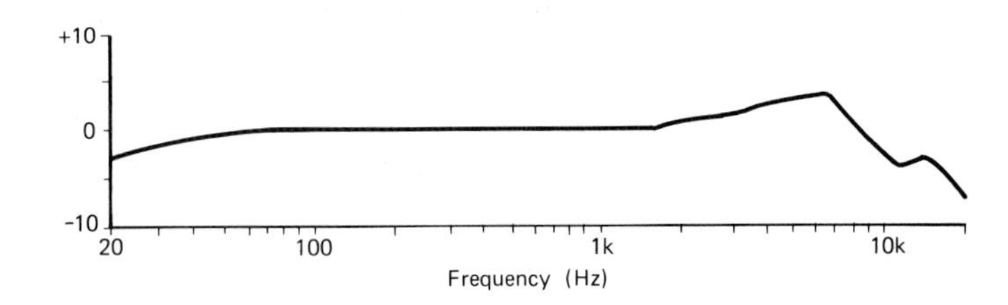

Example of boundary microphone design

Crown PZM-30FS (Figure 4.40)
Transducer type: condenser (electret), boundary type
Polar pattern: hemispherical
Frequency range: 20–15000 Hz
Sensitivity: −67 dB V (−47 dB re 1 mW)
Max. SPL for 3% THD: 150 dB
Impedance: 240 Ω
Equivalent noise level: 22 dB
Features: cantilever housing for capsule and electronics is mounted on a 127 × 152 mm plate
Main applications: placed on a table, podium, wall or floor for all-round pick-up for music or sound reinforcement

Highly directional microphones are in demand for such applications as picking up bat-on-ball or birdsong sounds at a distance. They can also be adapted for close-talking in noisy surroundings, covering audience or dance-routine sounds from outside the camera angle in television, etc. The principle is to pick up axial sounds efficiently but to discriminate against sounds arriving from all other directions. In other words, the object is to secure better random energy efficiency than the 0.25 listed in Figure 4.17 for the hypercardioid – which has the highest directivity obtainable from a first-order pressure gradient microphone. Two basic design approaches have been used, the parabolic reflector and the line, rifle or shotgun microphone.

The idea of using a *parabolic reflector* to collect sound energy over a wide area of the incident (plane) wavefront and reflect it back onto a microphone element located at the focal point was mentioned briefly in Section 2.11 and illustrated in Figure 2.24(b). The parabola shape ensures that all sounds arriving in line with the axis are bounced towards the microphone, giving a considerable gain in axial signal level compared with that of the unassisted microphone. At the same time, waves arriving off-axis are scattered and do not contribute much to the microphone output.

Inevitably, the reflecting action – and the discrimination against unwanted off-axis sounds – is confined to middle and high frequencies where the reflector diameter is large compared with the wavelength. This is illustrated in Figure 4.41 and explains why a bass-cut filter is often included in the microphone circuit. Precise aiming is important if the optimum pick-up of distant sound sources is required and so a gunsight is often fitted and the microphone is usually located slightly away from the focal point to avoid too fine tuning of the high frequencies.

Two or more pressure-gradient microphone elements can be combined to give second-order, third-order, etc. cosine law response with progressively enhanced directivity. However, this approach has practical bandwidth limitations and the *shotgun* or *line microphone* has proved more viable. In its simplest form this consists of an

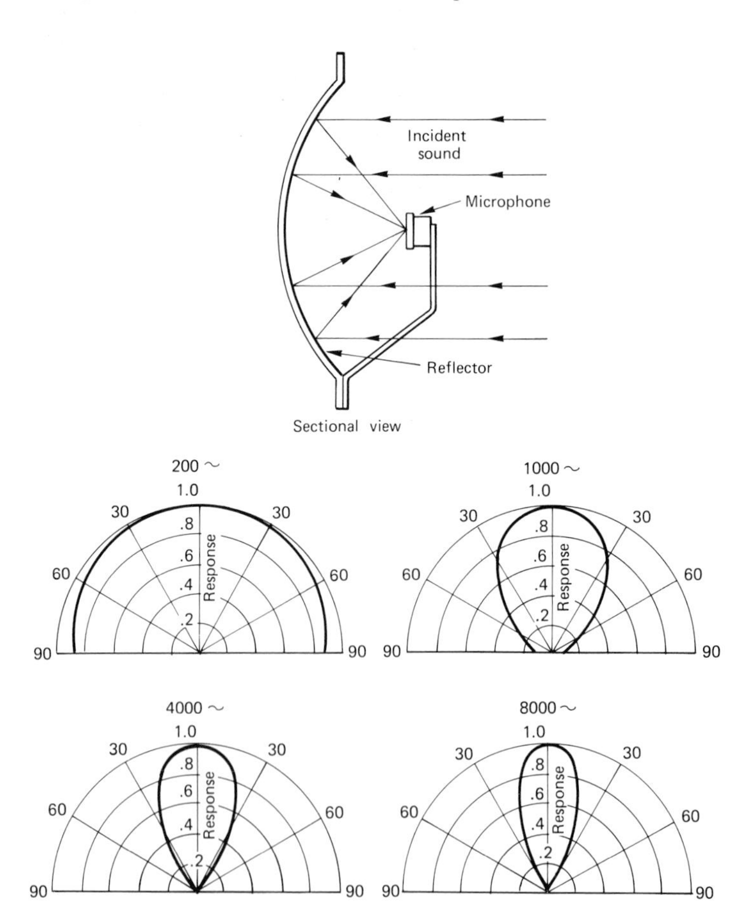

Figure 4.41 Parabolic reflector microphone and polar response at various frequencies (after Olson)

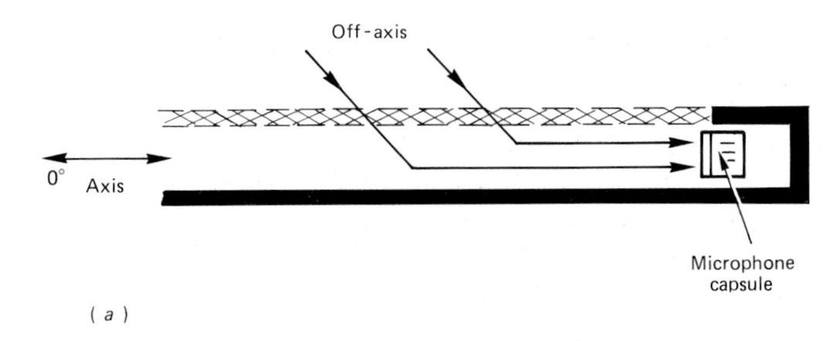

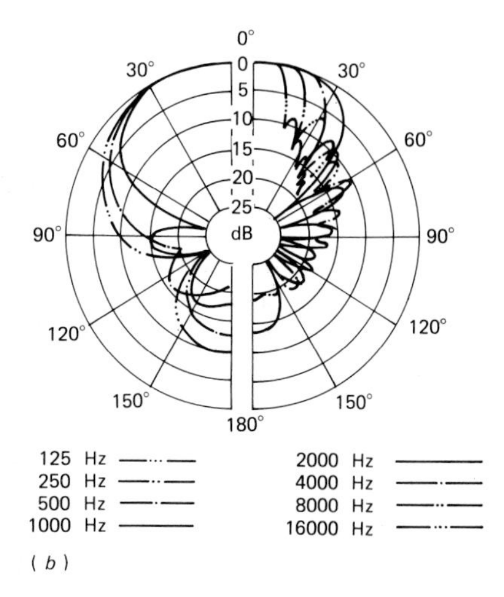

Figure 4.42 Shotgun (line) microphone. (a) Basic principle; (b) polar response of typical 555 mm long model (Sennheiser MKH 816)

acoustic line or pipe with equally spaced holes or a slit along its length and a pressure or (better) cardioid hypercardioid microphone element mounted at the rear end (Figure 4.42(a)). The desired high directivity results from the fact that only sound waves arriving on or near the axis produce additive pressures at the microphone diaphragm, whereas off-axis waves suffer varying phase-shift delays and tend to cancel. As with the parabolic reflector, the high-directivity action is restricted to middle and high frequencies, the bass extension being a function of the tube length. Figure 4.42(b) shows the polar patterns for a 555 mm long model.

In an alternative approach the single pipe can be replaced by a bundle of pipes of graded lengths. However, this design is now regarded as being too cumbersome, and a useful 15 dB rejection of sounds down to about 100 Hz for off-axis sounds can be achieved more elegantly by modern small-diameter single-line microphones. Designs combining second-order gradient elements with an acoustic line have also appeared.

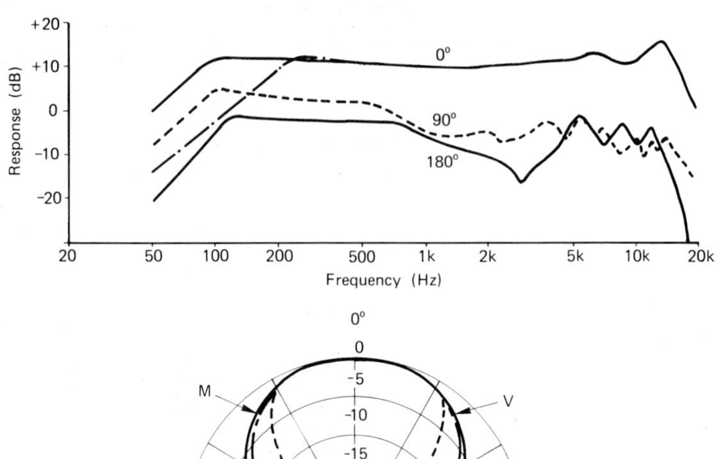

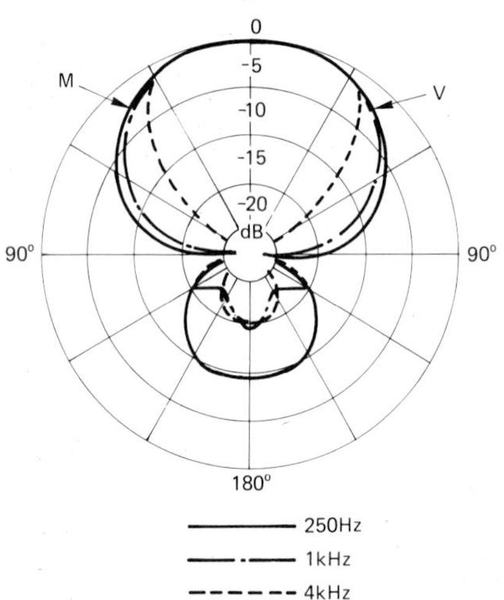

Figure 4.43 Sony ECM-672 304 mm shotgun microphone, shown mounted on the DXC-3000 video camera

Example of shotgun microphone design

Sony ECM-672 (Figure 4.43)
Transducer type: condenser (electret)
Polar pattern: highly directional to hypercardioid depending on frequency
Frequency range: 50–16000 Hz
Max. SPL: 114 dB
Dynamic range: 92 dB
Features: short shotgun construction (304 mm), windscreen, switched high-pass filter, internal AA-size battery gives 3000 h operation
Main applications: use on hand-grip or compact TV or video camera

In an interesting variant of this idea, a *zoom microphone* can be produced which is of special value for mounting on a TV camera. It is capable of providing a polar pattern variable from omni to highly directional (DF up to 2.7) as the TV scene demands. This uses two cardioid elements pointing forwards and spaced so as to give second-order gradient performance, plus a third cardioid pointing backwards which can be added in or out of phase at various relative levels and equalization settings. The zoom effect can even be controlled automatically by linking variable attenuators to the camera lens zoom mechanism.

Noise cancelling microphones have been developed for close-talking applications in noisy surroundings such as sports commentary, paging and special types of telephone. One classical lip-ribbon design dating back to the 1930s and still popular with radio reporters and sports commentators (Figure 4.44(a)) uses a ribbon transducer with a frontal guard-bar to ensure that the speaker's mouth is held at a fixed working distance of 63 mm. The proximity effect at this distance is naturally severe and must be equalized by means of a steep bass-cut filter which can be varied by means of a three-position switch to suit the individual voice. Thus filtered, the response to the voice approximates to the level curve shown in Figure 4.44(b). The degree of noise cancelling is also shown and can in some commentary situations be so effective in silencing random sounds that a separate

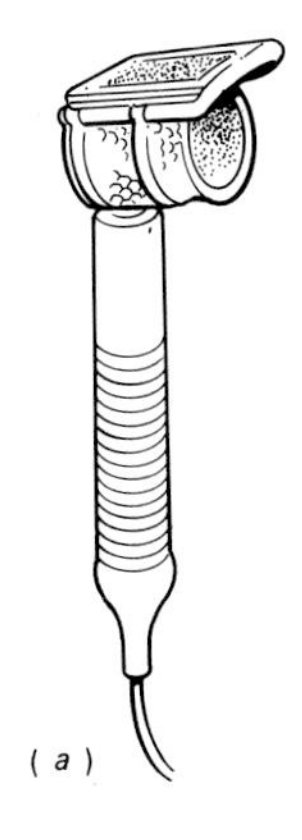

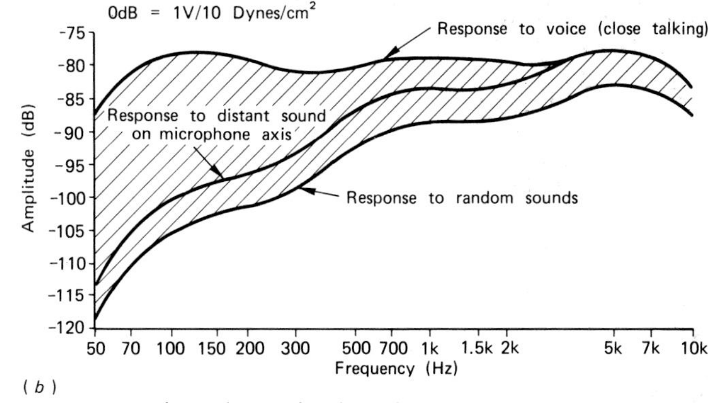

Figure 4.44 Noise-cancelling microphone. (a) S.T. & C. 4104 lip-ribbon microphone; (b) response to close speech and random sounds

microphone is rigged up solely to mix in the required amount of ambient atmosphere. Other noise-cancelling designs use a pair of cardioid or hypercardioid capsules in antiphase. The speech reaches the capsules at different time intervals and so cancellation is incomplete. However, the unwanted random sounds are almost completely cancelled.

Radio (wireless) microphones, which use FM radio transmission instead of the traditional microphone cable, are being increasingly employed for stage work and other applications where complete freedom of movement is needed. Typically, the radio microphone itself is small and attached to or concealed in the clothing or on top of the head, though handheld radio microphones are also common with the transmitter built into the handgrip to which a short aerial is attached. In the smallest versions, the transmitter is housed in a separate unit about the size of a pocket calculator and connected to the microphone by about a metre of cable which may also act as the aerial. As the special convenience of radio miking has been appreciated, its uses have multiplied and most manufacturers now offer complete systems with adaptors allowing any microphone type to be linked to the transmitter.

The microphone audio signal is frequency modulated onto the radio carrier. Frequency band allocations and licensing regulations vary from country to country, with both VHF and UHF bands used and elaborate design measures applied to overcome the practical problems of maintaining consistent signal levels and sound quality as artistes move anywhere on stage or throughout a televison studio. The transmitter aerial is basically omnidirectional, i.e. the signal falls off as the square of the distance, but the omni pattern is severely altered by reflections and shadowing by the performer's body, walls and other obstacles. This can lead to signal drop-outs, dead spots or buzzing due to multipath reception. The 8- and 2-metre VHF bands, with relatively long wavelengths, suffer less from the obstacle effect but require a proportionately longer aerial (up to 75 cm). The UHF band needs only a small aerial (10–20 cm) but is easily blocked by obstacles in its path. A final choice of waveband may also be influenced by local regulations and the risk of radio interference from known sources such as taxis, etc. in the given locality.

The receiver antenna should be placed where it will give line-of-sight reception over the performing area, well away from walls and preferably not more than about 60 m from the performers. Procedures used to enhance signal strength and consistency include treble pre-emphasis/de-emphasis to improve the signal-to-noise ratio as in normal FM broadcasting; limiting; companding (compression/expansion) for wide-band noise reduction; diversity reception using more than one receiver/aerial (see Figure 4.45) to overcome multipath fading, etc. Of course, each radio microphone and its receiver must operate on a different radio carrier frequency, which can call for racks of multiple receivers and careful initial alignment in case some channels cause interference. Large musical shows may call for a system made up of 30 or more microphones, with constant

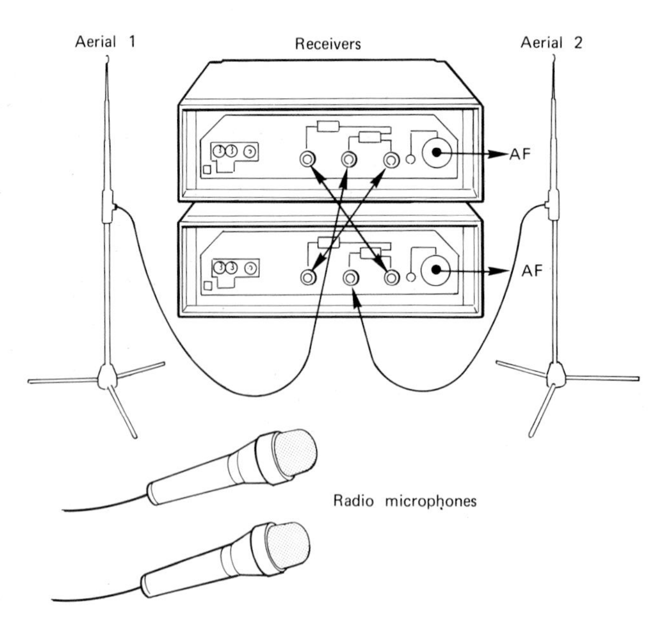

Figure 4.45 Radio microphone: showing equipment rig for diversity reception. (Courtesy Sennheiser)

switching between the available transmission channels as the show proceeds.

Example of radio microphone design

Sennheiser SKM4031 (Figure 4.46)
Transducer type: condenser
Polar pattern: supercardioid
Frequency range: 70–20000 Hz
Radiated RF power: 10 mW
Max. SPL for 1% THD: switchable 126, 136, 146 dB
Dynamic range: 96 dB
Features: cordless microphone with built-in battery-powered transmitter, attached 75 cm antenna, on-off switch, compander system, three-position sensitivity switch, 3.5 μm thick diaphragm, insensitive to handling noise, pop shield
Main applications: solo vocal, speech (close miking)

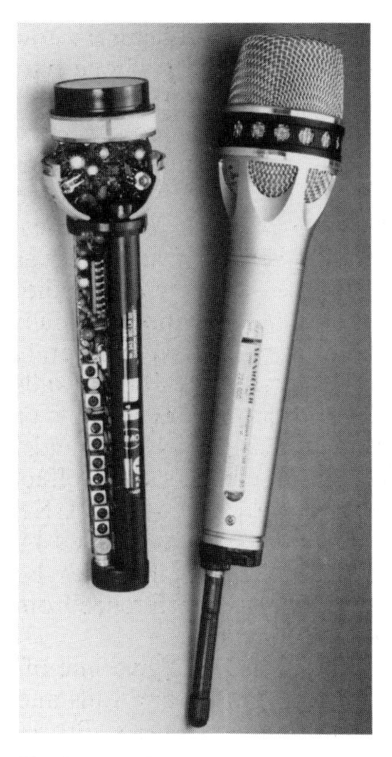

Figure 4.46 Sennheiser SKM 4031 radio microphone

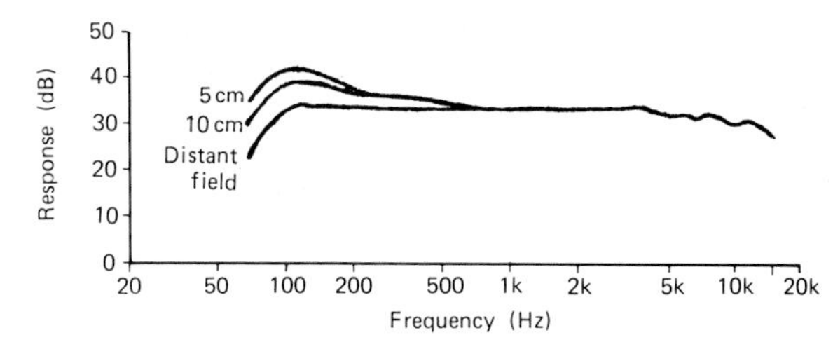

Not to be confused with the radio microphones just discussed is the *RF condenser microphone.* This was designed for optimum condenser operation in the days before high-quality head amplifiers were available. It produces a standard audio output but incorporates a built-in RF oscillator (8 M Hz), modulator and demodulator circuit. High stability and robustness are claimed, the usual high polarizing voltage being replaced by only about 10 V. Changing capacitance due to diaphragm vibration produces frequency modulation of the 8 MHz signal as in FM broadcasting and the resultant signal is demodulated to provide normal audio signals at the terminals.

Two-way microphones have been developed having separate diaphragm/transducer capsules to cover high and low frequencies, combined via a frequency selective crossover network as is commonly used in multiple-driver loudspeakers. The main objective is to produce a more consistent directivity pattern and high-frequency response than can normally be obtained when a single diaphragm has to cover the full audio bandwidth. In one popular cardioid design (Figure 4.47) the two diaphragms are, respectively, 15 mm and 30 mm in diameter. The HF system is located immediately inside the front protecting grille, with the LF unit behind it and the crossover frequency is set at 400 Hz.

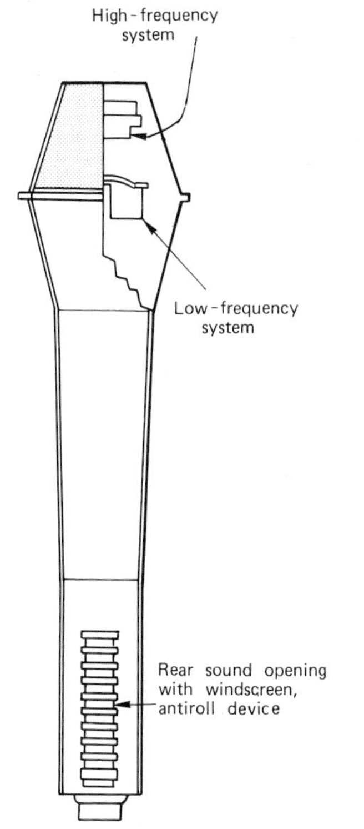

Figure 4.47 Two-way microphone, showing version with LF capsule mounted behind the HF capsule (AKG D202)

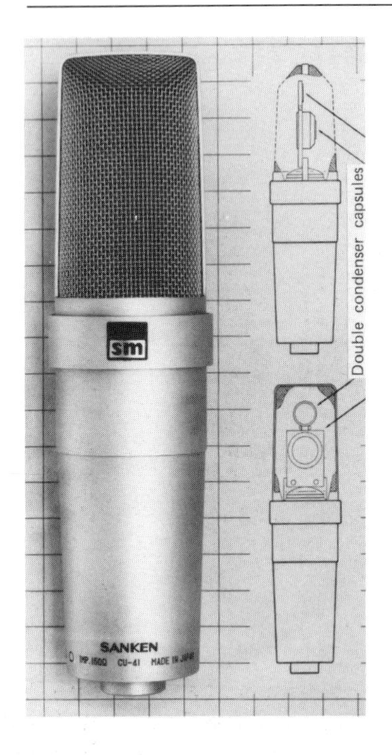

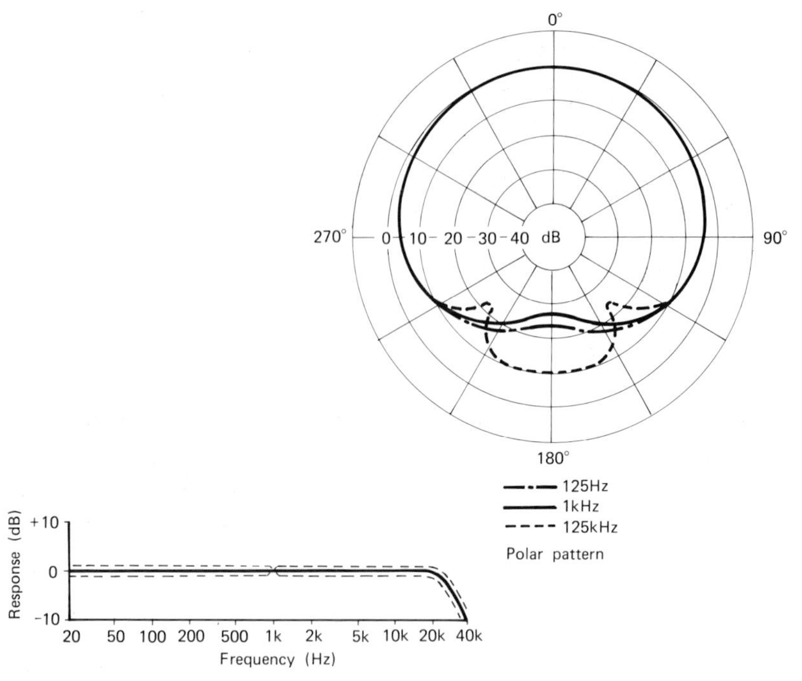

Figure 4.48 Sanken CU-41 two-way microphone

Example of two-way microphone design

Sanken CU-41 (Figure 4.48)
Transducer type: two-way condenser
Crossover frequency: 1000 Hz
Polar pattern: cardioid
Frequency range: 20–20000 Hz ± dB
Sensitivity: 0.7 mV for 74 dB SPL
Max. SPL for 0.5% THD: 134 dB
Impedance: 150 Ω
Equivalent noise level: 15 dB
Features: two separate capsules with 10 mm and 24 mm diaphragms of 1 μm thick titanium, vertically aligned
Main applications: music

Voice-activated (automatic) microphones are used in such varied situations as conference rooms, churches and security intruder monitoring. In the basic design, the microphone output is muted by a level-sensitive gating circuit until a predetermined threshold sound-pressure level is exceeded. Thus in a round-table discussion each talker can address a separate microphone, yet the confusion normally introduced by background noise and multiple-phase errors with all microphones constantly open is avoided. Each microphone will become live only when spoken to directly from within its frontal acceptance window. In a refinement of this idea, which avoids the need for careful threshold gating-level adjustment, two cardioid microphone capsules are placed back to back. The rear capsule continuously monitors the ambient sound level and the main front capsule is switched on whenever some differential level (say, 10 dB) is exceeded by direct speech.

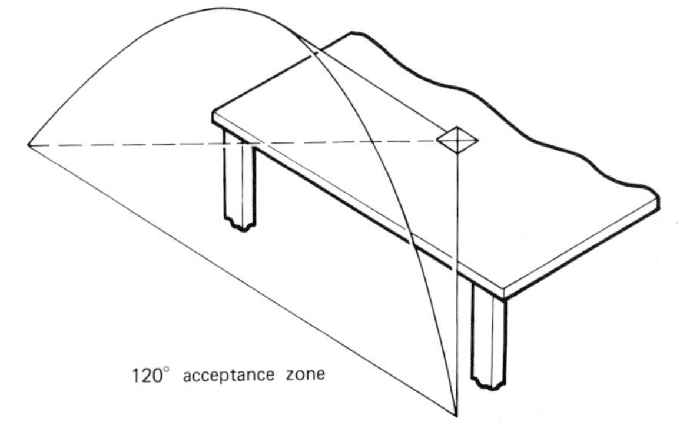

Figure 4.49 Shure AMS22 voice-activated microphone

Example of voice-activated microphone design

Shure AMS22 (Figure 4.49)
Transducer type: condenser (electret)
Polar pattern: hemi-cardioid
Frequency range: 50–10000 Hz
Sensitivity: −47 dBV
Equivalent noise level: 20 dB
Features: low-profile surface-mounting type to be used with the Shure Automatic Microphone System, automatic switch-on at 9.5 dB above ambient noise for sounds within 60°±10° of axis, Chairman muting, privacy or cough button
Main applications: discussions, courtrooms, conferences, churches, security monitoring

Contact microphones, as their name implies, are physically attached to the sound source and respond to mechanical vibrations rather than airborne pressure waves. Therefore they are not, strictly speaking, microphones but belong to the family of mechanical/electrical transducers which includes strain gauges, guitar pick-ups, throat microphones and accelerators used for vibration measurements. Any transducer principle may be used, but electromagnetic and, more recently, condenser/electret types are the most common. The unit can be fastened or taped to the soundboard or body of any acoustic musical instrument – string, wind or percussion – and of course must be small and lightweight. Exact positioning is critical to produce a pleasing timbre free of standing-wave resonance effects in the instrument structure. However, there are real advantages to be gained such as freedom of movement for the artiste, excellent separation from the sounds of other instruments or voices and rejection of ambient noise. One electret version is supplied in strip form only 1 mm thick and about 25 mm wide by 75 or 200 mm long. It is sufficiently flexible to conform (adhere) to a curved surface and may be battery or phantom powered.

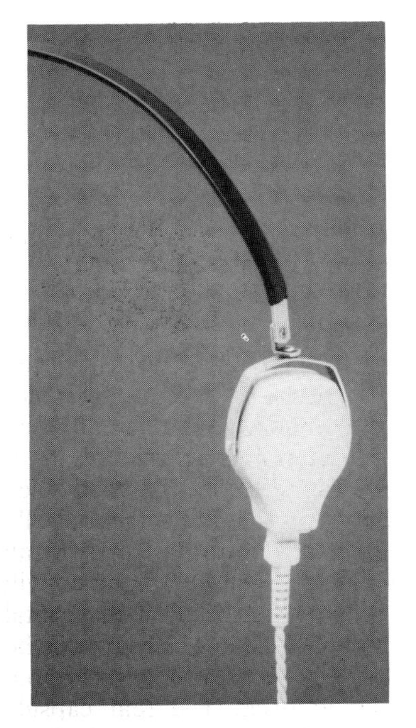

Figure 4.50 Contact microphone (Stanton MIT miniature inertial transducer)

Another recent magnetic type of contact microphone which is primarily designed to receive and transmit bone-conducted sounds (Figure 4.50) achieves intelligible communication in an ambient noise level of 120 dB. It may be worn on top of the head, pressed against the mastoid bone behind the ear or used as a throat microphone.

Underwater microphones (hydrophones) can be improvised by suspending a standard microphone, typically moving-coil, inside a waterproof cage. Professional versions are generally piezoelectric and supplied as complete corrosion-free kits with built-in amplifier and special waterproof cables capable of use to water depths of 1000 m. They also make robust all-weather outdoor microphones with a response up to about 15 kHz. The inertial (contact) microphone in Figure 4.50 is also available in an underwater version.

5 Microphones for stereo

From rather hesitant commercial beginnings in the 1950s, two-channel stereophony has largely taken over as the standard method of music recording and sound broadcasting (primarily in the VHF/FM waveband, though AM stereo also exists in the USA). Single-channel mono has certainly not disappeared completely, since AM radio in the medium and long wavebands is mainly mono and so is most television sound – though genuine stereo soundtracks on video and TV are beginning to grow in importance.

Before the 1950s everything was monophonic, conveyed along a single communication channel, and so we did not need a special name for it. Note that mono remains mono if the signals are restricted to a single channel at any point along the chain from microphone to loudspeaker. Any number of microphones may be mixed together in an attempt to produce a desired musical balance, or faded up in turn to follow the action in a play or documentary, but there will be no feeling of lateral spread when these are reproduced through the single (mono) loudspeaker. Adding extra loudspeakers, each fed with the same mono signal, does not help. Listeners still hear all the sounds coming to them from a single point in space located wherever the intensities from the various loudspeakers are effectively equal.

5.1 The stereo illusion

Stereophony, from the Greek word meaning 'solid', is an attempt to reproduce sounds in such a way that listeners have the same feeling of left–right spread in the frontal quadrant, and the same ability to locate the direction of individual voices or instruments, as they have at a live performance.

There are important differences between natural directional hearing, as described in Section 2.16, and the illusion of directionality created in stereophonic recording and reproduction. The listener to a stereo system of two spaced loudspeakers in effect hears everything twice, since the sounds from both loudspeakers reach both ears (see Figure 5.1). This simple fact sets a very real limit to the realism that two-channel stereo can or will ever achieve. The trick is to feed the left

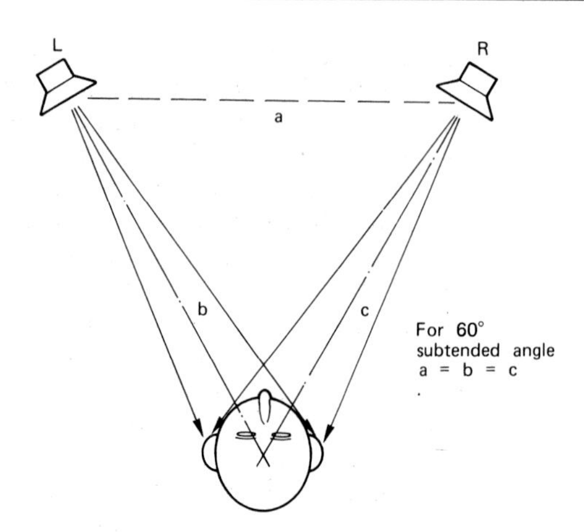

Figure 5.1 Loudspeaker stereo: in the conventional layout $a = b = c$ and is generally between 2 and 4 m. Note that the sound from each loudspeaker reaches both ears, with a slight time delay to the more remote ear

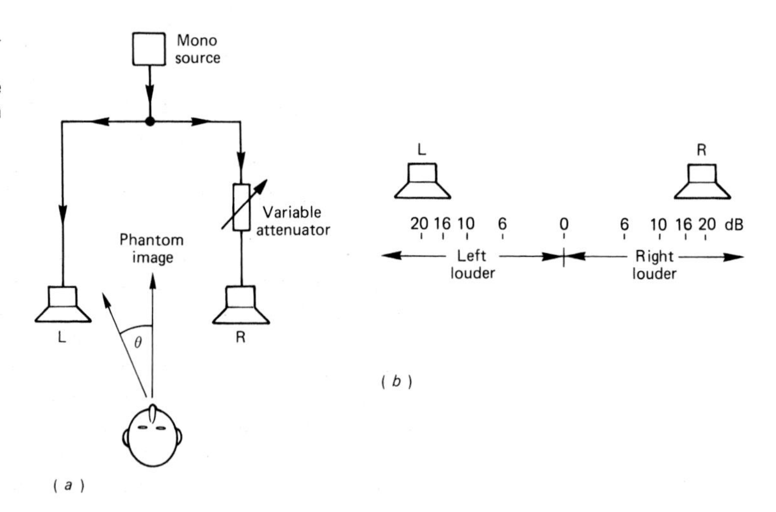

Figure 5.2 Intensity stereo. (a) Attenuating one loudspeaker causes the phantom image to move towards the other; (b) showing typical attenuation (in decibels) for image positions across the stero stage

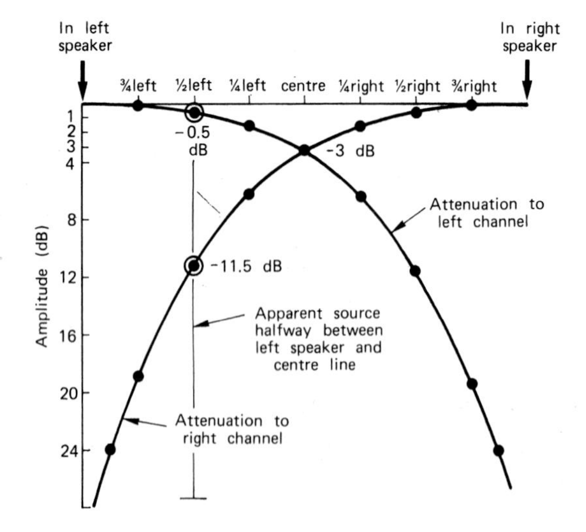

Figure 5.3 Panpot law

and right loudspeakers with signals which differ from each other in some way which will create the illusion of an extended stereo 'stage' across the imaginary arc joining the two loudspeakers. As Figure 5.1 shows, the standard layout for stereo listening puts the listener and the two loudspeakers at the corners of an equilateral triangle. In other words, the reproduction angle is set at 60°.

Researches by De Boer, Myer, Shodder and others have established the basic principles of two-loudspeaker stereo, showing that the required arc of phantom images can be established by making the signals from the loudspeakers differ by small amounts in intensity, time or both. Using a set-up as shown in Figure 5.2(a), where the signal level to one loudspeaker can be attenuated, the apparent location of the source from a central listening position will be found to shift progressively from the centre (when the levels are equal) towards the unattenuated loudspeaker (Figure 5.2(b)).

This is the principle of the familiar panpot (panoramic potentiometer), where a ganged pair of variable attenuators working respectively in the clockwise and anticlockwise sense can be used to shift a mono source to any required point between the loudspeakers. As Figure 5.3 shows, the attenuator law is chosen so that both channels are attenuated by 3 dB at the centre setting (to avoid the + 6 dB doubling which would otherwise occur) and graduated in such a way that the total intensity radiated by the two loudspeakers remains the same when the signal is panned all the way from left to right.

Similar tests with time delay to one loudspeaker produce the results indicated in Figure 5.4, with the apparent source location swinging from the centre towards the undelayed loudspeaker. Note, however, that the fact that both ears hear both loudspeakers already introduces an inherent pattern of level and time differences which interferes with any such differences in the left/right microphone signals. For so-called 'intensity stereo' these fixed dimensional differences simply reinforce the desired effect. However, for 'time difference stereo' the inter-loudspeaker differences introduce phase additions and cancellations (comb filter distortion).

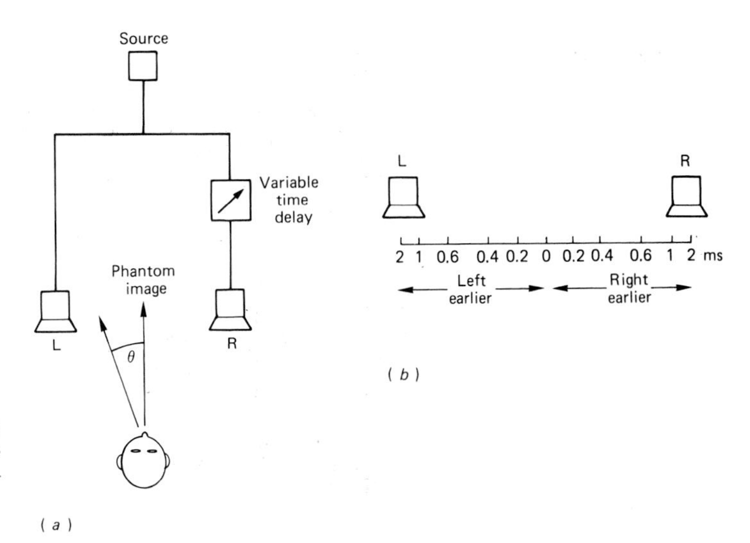

Figure 5.4 Time delay stereo. (a) Delaying one loudspeaker causes the phantom image to move towards the other; (b) typical time delays (in milliseconds) for image positions across the stereo stage

As might be expected, the deliberate introduction of both intensity and time differences together produces an additive (or subtractive) effect. For instance, if the sounds from the right-hand loudspeaker are both louder and earlier, the swing to the right will be greater than for each effect alone. On the other hand, a given image shift due to level difference can be diminished by introducing a time difference in the opposite sense.

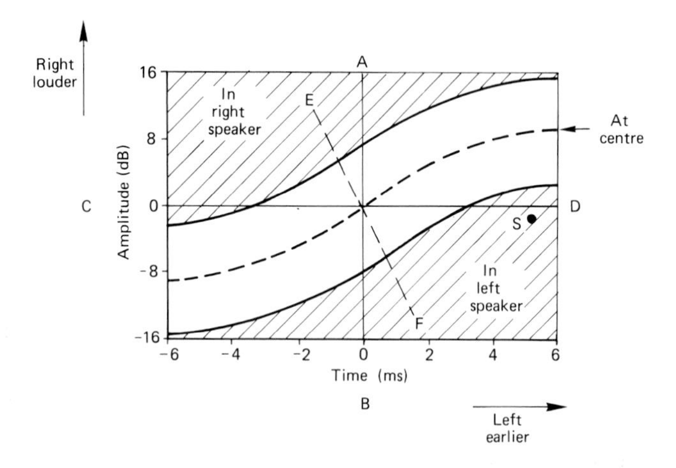

Figure 5.5 Combining intensity (vertical axis AB) and time delay (horizontal axis CD) effects. The shaded areas represent sound images located in either the left or right loudspeaker (after Franssen)

This combination of the two effects is summarized in Figure 5.5. Intensity differences produce shifts on the vertical axis AB, time differences on the horizontal axis CD. The two solid-line curves represent the boundary conditions when the image appears in either the left or right loudspeaker, and any further increase in level or time difference will produce no change. This figure will be referred to again in the following sections describing the various techniques currently used for two-microphone stereo recording.

5.2 Basic Blumlein

The most logical stereo pair configuration with which to begin our discussion is that named after A.D. Blumlein, whose comprehensive Patent No.394325 (14.6.1933) laid the foundation for the universal two-channel stereo system as we know it. Blumlein realized that producing simple level differences at the loudspeakers would in the event give rise to both intensity (IAD) and phase (ITD) directional cues to the listener's ears (as discussed in Section 2.17). His patent in fact covered a variety of microphone pairings to produce the desired 'intensity stereo', including the use of spaced omnidirectional microphones with a so-called 'shuffling' network and the M–S (Middle and Side) scheme described in Section 5.3 below.

However, the pairing generally regarded as basic Blumlein 'intensity stereo' consists of two bidirectional microphones arranged at ±45° to the front axis (see Figure 5.6(a)). The two diaphragms are placed as nearly as possible in the same point in space (coincident) – at least as far as the horizontal plane is concerned (i.e. one may be placed above the other, as in Figure 5.6(b)). Special stereo microphones give more precise coincidence. They have twin left/right

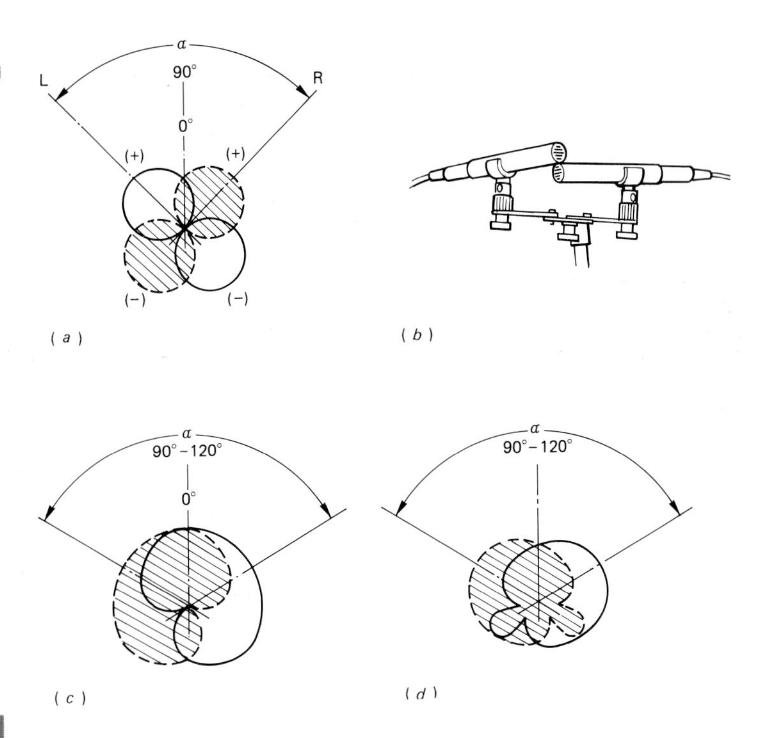

Figure 5.6 Coincident pair (intensity) stereo. (a) Basic Blumlein; (b) locating two diaphragms for coincidence in the horizontal plane; (c) crossed cardioids; (d) crossed hypercardioids

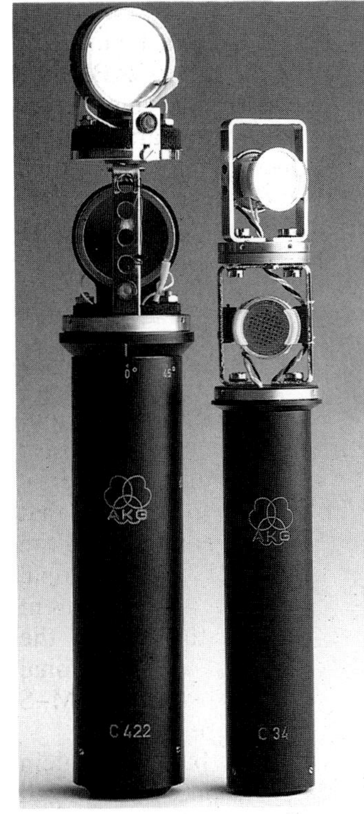

Figure 5.7 Stereo microphones. examples of designs allowing the mutual angle between the two capsules to be adjusted by rotating the upper capsule (AKG C422 and C34)

capsules mounted one above the other, with the included angle easily altered by rotating the upper capsule (Figure 5.7) Then, with time-of-arrival differences eliminated, the signals generated by the microphones will ideally differ in relative sensitivity only. The way this works to produce a phantom image at any desired point between a pair of spaced loudspeakers was described above and shown as the line AB in Figure 5.5.

The included angle between the forward axes of the two microphones is traditionally 90° as shown, and the maximum angle of pick-up (recording angle) for a left/right intensity difference of, say, 15 dB is slightly less than 90°. Remembering the cosine law as applicable to pressure gradient (figure-of-eight) microphones (Section 4.4) we see that the output from the left-facing microphone in Figure 5.6(a) is a maximum for a source at L ($\cos \theta = 0$) and 3 dB down at the centre ($\cos 45° = 0.7$). The same applies for the right-facing microphone. This gives the same smooth transition across the stereo stage as the panpot law shown in Figure 5.3. With modern high-quality microphones this arrangement of a pair of figure-of-eights at 90° produces excellent imaging in the frontal quadrant which is independent of the distance from the source. It also reproduces a uniform spread of reverberant sound between the loudspeakers. In some situations, however, it can appear to be lacking the sense of space present in some spaced-microphone systems. Among its main drawbacks is the fact that the outputs of the two microphones are respectively in antiphase in the two side quadrants, so that strong side-wall reflections may suffer phase distortion. Of more significance, sounds originating in the rear-right quadrant will be picked up by the rear lobe of the left microphone, and vice versa, and so will be radiated by the wrong loudspeaker (cross-channelled).

The need for the microphone diaphragms to be as 'coincident' as possible suggests that small-diaphragm types are best, and this is reinforced by the fact that smaller microphones also possess another esential ingredient for good-intensity stereo – a polar pattern which is maintained over the full frequency range. Where the off-axis response is variable with frequency, crossed microphones are liable to produce a coloured sound quality and the image may shift with changes in musical pitch. This is a serious problem, since so much of the sound source is edgewise onto both microphones. Indeed neither microphone is directed at the important centre soloist location. So again only high-quality microphones should be used.

As we have said, the Blumlein pair of bidirectional microphones set at an included angle of 90° gives a so-called recording angle of a little less than 90° and provides a smooth spread between the stereo loudspeakers arranged to subtend the traditional 60° angle at the listener's position. In practice, musical ensembles should be contained within this 90° frontal arc. The microphone distance is therefore critical. When this distance needs to be increased beyond the value which would be ideal for such other considerations as direct-to-reverberant sound balance, etc. some widening or narrowing of the recording angle can be achieved by altering the included angle between the microphones, but this means of control is limited (see Section 5.5).

Where a wider recording angle is esential, the usual practice is to replace the bidirectional microphones with cardioids or hypercardioids as suggested in Figure 5.6(c) and (d). A pair of cardioids at 90° will give a very wide theoretical recording angle up to 270°, though this is better interpreted as a useful angle of up to 180° (see Figure 5.8(c)) because of a subjective falling off in direct-to-reverberant sound ratio at wide angles. This allows a very close balance indeed, as well as efficient rejection of sounds from the rear, but a wider included angle is often chosen (say, up to 120°) to direct the microphones at the extreme edges of the musical ensemble.

Hypercardioids give an intermediate condition (Figure 5.6(d)). The 90° included angle produces a recording angle of 130° (Figure 5.8(b)) and helps to provide a more solid centre image. It also allows a more distant balance for a given ratio of direct-to-reverberant sound.

A special example of a 'coincident' microphone design is the Calrec Soundfield microphone, comprising four transducer elements closely arranged in a tetrahedron as described in Section 5.5.

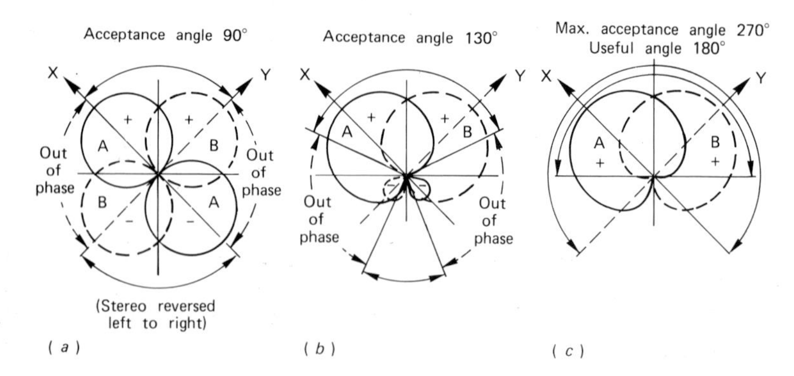

Figure 5.8 Stereo microphone recording (acceptance) angles for an included angle of 90°. (a) Bidirectional; (b) hypercardioid; (c) cardioid

5.3 M–S stereo

Another arrangement of two coincident microphones covered in Blumlein's 1933 patent consisted of one microphone (which can be of any directivity pattern) pointing to the front (Middle) to provide a (L + R) signal, and the other a bidirectional type at right angles (Side) providing a (L − R) signal. This Middle-and-Side or M–S pairing has a number of advantages over the Blumlein X–Y combinations of identical angled microphones so far discussed. However, it does require a matrixing sum-and-difference network to recreate the conventional Left and Right signals. In essence, this network sends the sum signal (M + S) = (L + R) + (L − R) = 2L to the left channel and the difference signal (M − S) = (L + R) − (L − R) = 2R to the right.

The choice of M component directivity pattern gives a range of stereo perspective possibilities, and this can be further extended by adjusting the relative sensitivities of the M and S channels. Figure 5.9 illustrates just nine examples based on an M component with figure-of-eight, cardioid and omni characteristics respectively and three M:S ratios. The resulting X–Y patterns are also shown. It will be seen that:

1. *An omni M component* produces back-to-back X–Y cardioids for equal M and S sensitivity (centre diagram). This changes towards back-to-back hypercardioids or subcardioids if the M:S ratio is increased or decreased (perhaps by remote control) as shown.

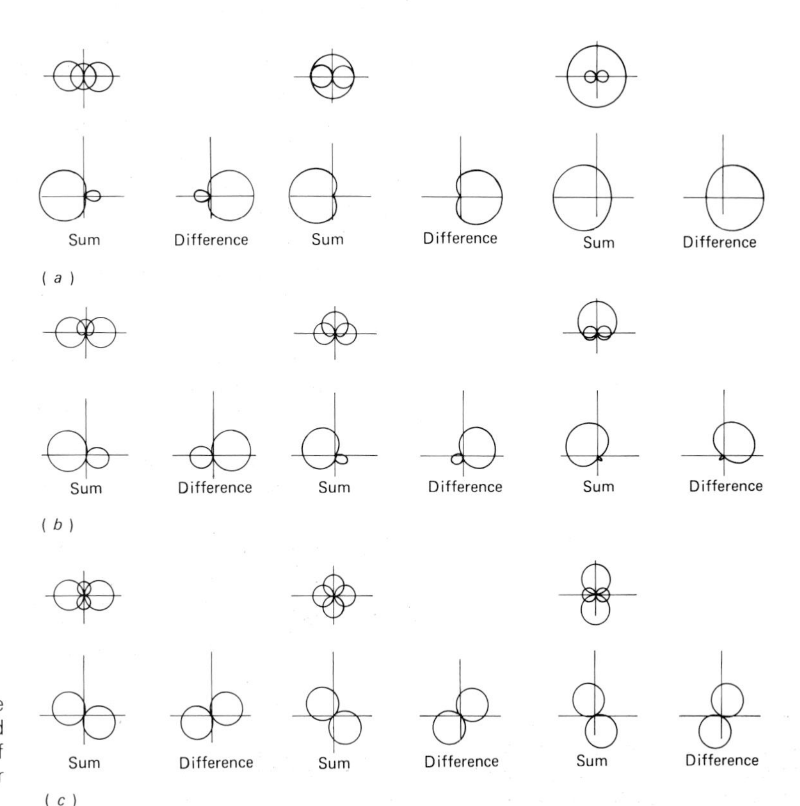

Figure 5.9 M–S stereo, showing combinations of the bidirectional Side component with (a) omni, (b) cardioid and (c) bidirectional for three ratios of M to S, 30:70, 50:50 and 70:30 (after Dooley and Streicher)

2. *A cardioid M component* produces crossed X–Y hypercardioids at an angle of about 127° for equally sensitive M and S units. Increasing or decreasing the M:S ratio not only narrows or widens the included angle but also changes the shape of the X–Y patterns making them respectively more nearly figure-of-eight or cardioid.
3. *A figure-of-eight M component* produces crossed X–Y figure-of-eights. When the M and S sensitivities are equal, the X-Y included angle is the basic Blumlein 90°. However, this can be increased or decreased by making the S component either greater or smaller than the M component, as shown.

M–S stereo is clearly a versatile option for all kinds of stereo balance and it has the further advantage that simple addition of the X and Y derived signals provides a fully compatible mono signal: i.e. (M + S) + (M − S) = 2M. This is attractive to broadcasters, who generally have both mono and stereo listeners to consider. The mono signal can be of the highest quality, not subject to the off-axis colorations and phase ambiguities present with other X–Y methods. It is also capable of fine tuning in terms of reverberant sound balance and degree of mono compatibility.

When a pair of microphones (or one of the special double-capsule stereo models) is used with remote control of the directivity patterns the engineer can make adjustments to image width and ambience to take account of changes in performer numbers or layout during a concert or between the no-audience rehearsal situation and the acoustically less lively performance with audience present. Altering the apparent image width is accompanied by a subjective change in the reverberation and therefore in the feeling of distance. This is equivalent to a zoom lens and is incorporated into some M–S stereo microphones used in TV (see Figure 5.11).

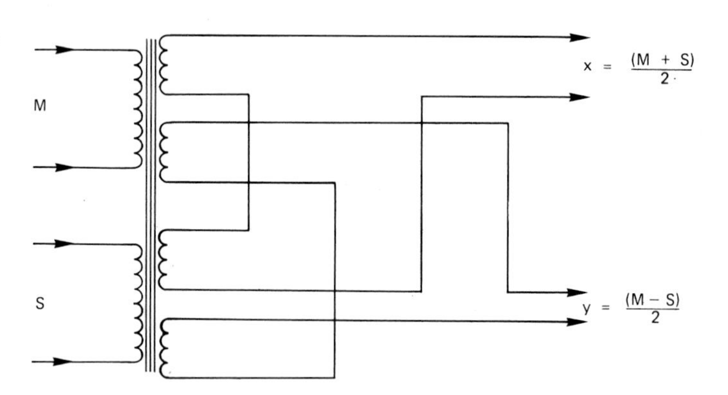

Figure 5.10 Basic sum and difference circuit used to derive the equivalent X–Y signals from M and S microphone inputs

Separate EQ or processing of the M and S components can be carried out prior to matrixing, or the M and S signals can be sent to separate tape tracks, allowing final decisions on balance to await the later mixdown stage. It is of incidental benefit that a centre soloist is on-axis for the forward-facing M microphone but on the null-axis of the bi-directional S microphone, so that the latter mainly picks up reverberant sound and gives direct control of reverberant balance.

Various matrixing networks to derive the equivalent X–Y signals and route them to the left and right channels are now available separately or as built-in console features. In essence they involve splitting both the M and S signals, for example by using Y-adaptors or hybrid transformers with one primary and two independent secondary windings (Figure 5.10), and then combining one pair of outputs in phase and the other in antiphase. Thus the left channel becomes $(M + S)/2$ and the right one becomes $(M - S)/2$. Each matrixing method can lead to increased noise and distortion, though later designs which take M and S to independent preamplifiers and matrix at line level have reduced this to a minimum.

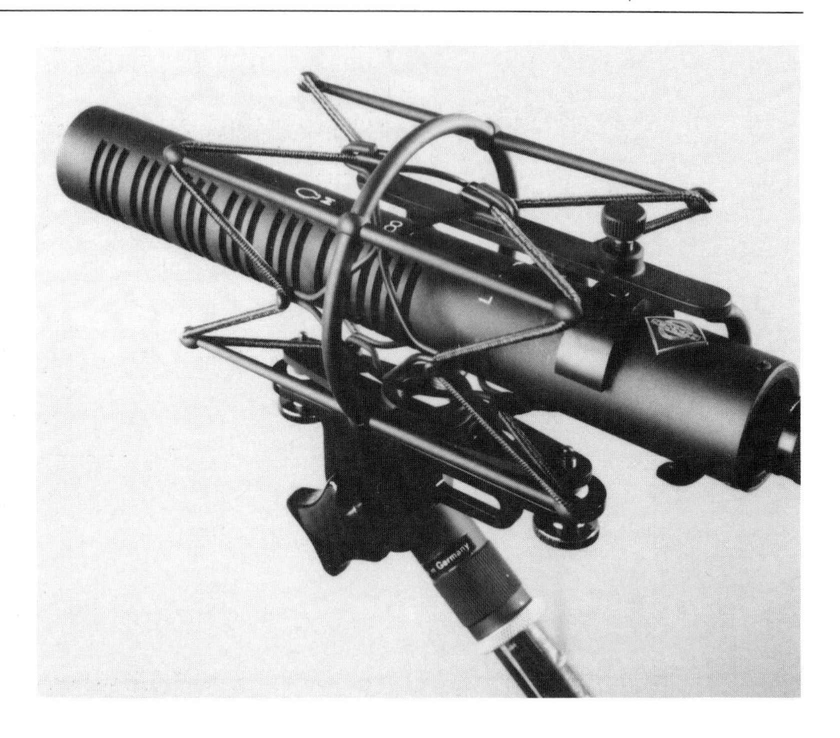

Figure 5.11 Neumann RSM190i stereo microphone

Example of stereo microphone design

Neumann RSM 190i (Figure 5.11)
Transducer type: MS stereo condenser
Polar pattern: hypercardioid (Middle) plus figure-of-eight (Side) giving variable M–S and X–Y patterns
Frequency range: 40–18 000 Hz
Sensitivity: 23 mV/Pa
Max. SPL for 0.5% THD: 134 dB (= 100 Pa) or 144 dB with pre-attenuation
Impedance: 50 Ω
Equivalent noise level: 25 dB (M), 31 dB (S)
Features: comprises two separate capsules, one short shotgun hypercardioid (M) and one pressure gradient (S), remote-control six-position level switch on S capsule changes the width of the stereo image for a 'zoom' effect, built-in M–S to X–Y matrix for left/right stereo, transformerless circuit, 40 Hz high-pass filter, polyurethane windscreen
Main applications: stereo newsgathering, TV and film where stereo width control is needed during recording or later in post-production.

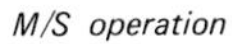

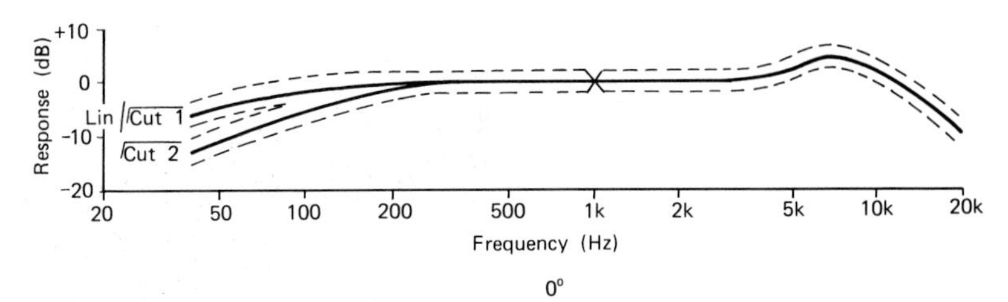

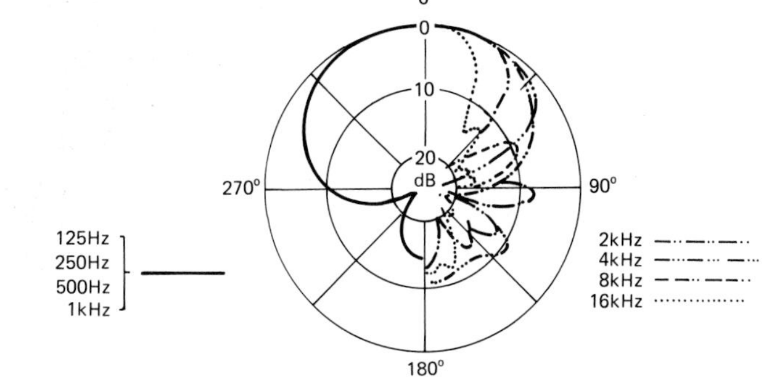

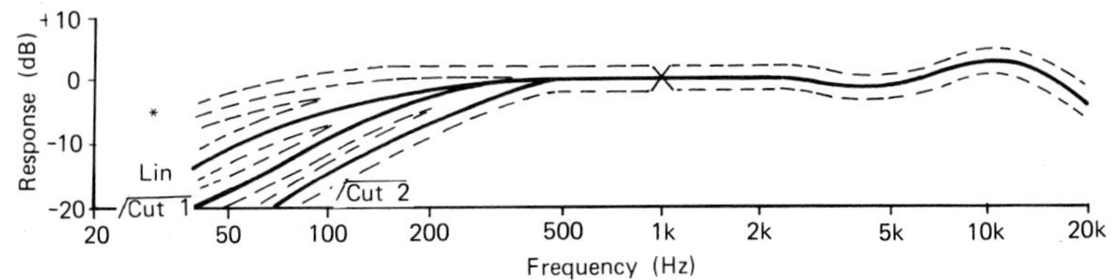

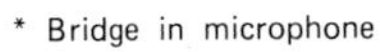

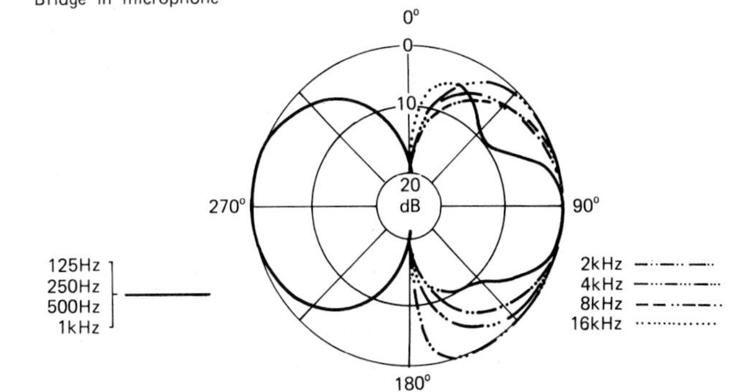

X/Y operation

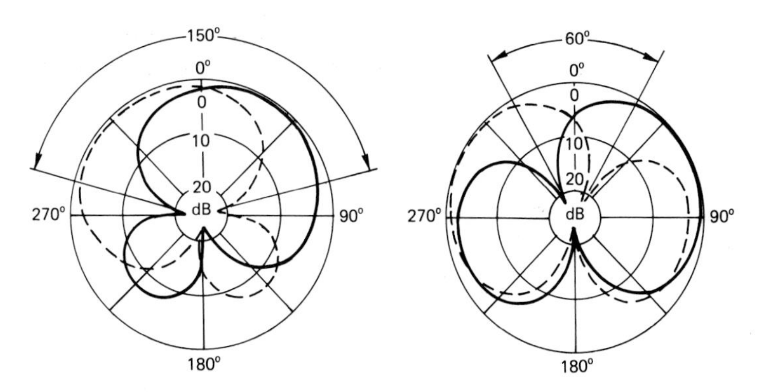

5.4 Non-coincident (spaced) microphones

The use of spaced microphones for stereo has at least as long a pedigree as Blumlein's 'intensity stereo' coincident-pair method, as can be seen from the experiments at Bell Laboratories (with the active cooperation of the famous conductor Leopold Stokowski) in the early 1930s. The Bell researchers set out to recreate in the listening area the whole soundfield of the original, and began with the notion of a line of microphones across the stage-front linked by separate channels to an equal number of loudspeakers arranged behind a curtain in the remote auditorium (Figure 5.12). It did not

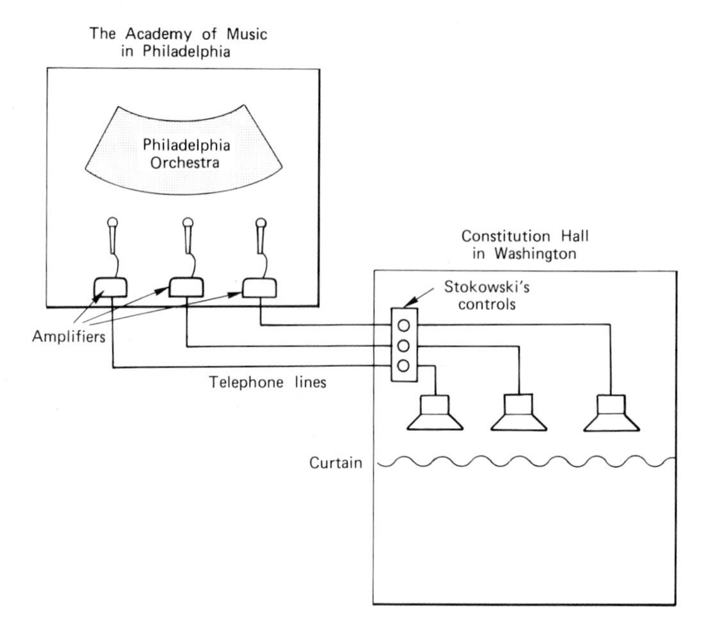

Figure 5.12 Set-up for one of the Bell Laboratories three-channel 'broadcast' experiments

take long to see this as economically impractical and a reduced scheme with two or, at most, three channels was studied. It is worth noting that they worked with a narrower subtended angle at the listening position of only 35° compared with the 60° now universally adopted, and this may have caused them to underrate the importance of lateral imaging compared with the followers of Blumlein.

As a historical aside, it is salutary to remind ourselves that the idea of using a long line of microphones along the front of a real stage actually goes back more than a hundred years to 1881, when Clément Ader arranged two groups of five telephone transmitters in the footlights of the Paris Opéra (Figure 5.13). The transmitters may be regarded as primitive telephone-type microphones, consisting of carbon rods and induction coils, and were arranged as left and right pairs relayed to the left and right earphones of banks of 80 telephone receivers located at the International Exhibition of Electricity, 3 km away. Visitors queued nightly for the chance to put the appropriate receivers to their left and right ears. The *Scientific American* issue for 3 December 1881 quoted its contemporary *L'Electricien* as follows: 'The singers place themselves, in the mind of the listener, at

Figure 5.13 Array of carbon telephone transmitters placed in the footlights of the Paris Opéra by Clément Ader in 1881 (*Die Elektricität in Dienste der Mensheit*)

a fixed distance, some to the right and others to the left. It is easy to follow their movements and to indicate exactly, each time that they change their position, the imaginary distance at which they appear to be.'

The first thing to be understood when non-coincident microphones are placed in front of a sound source is that the signals reaching the left and right microphones will not only differ in intensity but also in time of arrival or phase. This gives less consistent positional information with greater source distances than pure 'intensity stereo'. It can also introduce such problems as low-frequency comb filter effects on wider spaced sources, vague centre imaging and unpredictable mono compatibility.

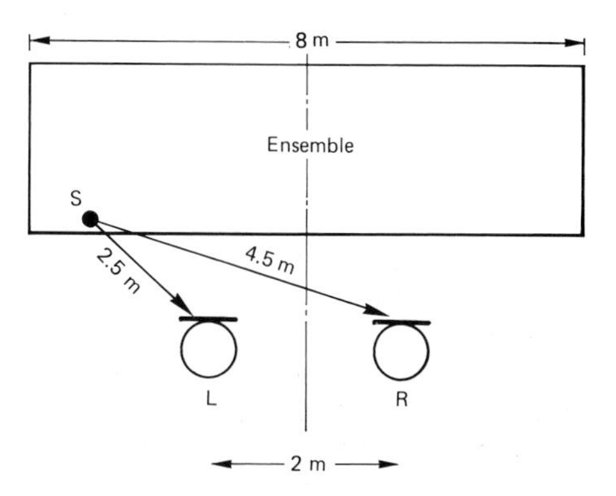

Figure 5.14 Spaced microphones produce both time and intensity differences. (See also Figure 5.5)

By way of example, Figure 5.14 shows a pair of omnidirectional microphones placed 2 m apart in front of an 8 m wide ensemble. An instrument at S is, respectively, 2.5 m and 4.5 m from the left and right microphones which will produce a time difference of $(4.5 - 2.5)/344 = 5.8$ ms. Also, treating S as a point source so that the inverse square law of attenuation with distance applies, there will be an intensity square ratio of (4.5/2.5), which corresponds to 2.1 dB level difference. The combination of these two effects places source S firmly in the left-hand loudspeaker at the point shown as S in Figure 5.5. In fact the microphone spacing in this example is wider than the ideal, and produces a 'hole-in-the-middle' effect with the musicians

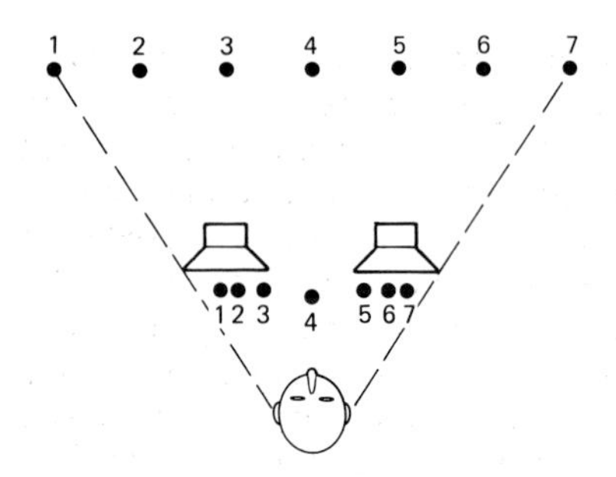

Figure 5.15 The hole-in-the-middle effect; showing crowding of sources in the two loudspeakers if the microphones are too far apart

Figure 5.16 Tree of spaced microphone pair with a third microphone at the centre (Decca photo, with Sir Georg Solti conducting)

on either side of centre seemingly pushed outwards to crowd together in the loudspeakers, as shown in Figure 5.15.

Microphone spacing should, in general, be no more than about one-third of the source width, and values between about 1 m and 3 m are usual, depending on the source width and the working distance. Spaced omnis are most common, available omnis having superior frequency response smoothness and, in the best small-diaphragm models, consistency of off-axis response with frequency. The larger diaphragm omnis should be angled outwards slightly to point their axes to left and right of centre. As always, when making subtle stereo adjustments it can be very helpful to use a stereo indicating meter or oscilloscope as a check on microphone separation and phasing.

Note that omnis pick up the full reverberant sound and so a more forward position may become necessary, causing the centre musicians to be louder than the rest. Cardioids, hypercardioids and even bidirectional microphones can be used instead, allowing a more distant balance and overcoming some of the problems of exaggerated perspective. A spaced pair of PZM-type boundary microphones is also a possibility, giving a good sense of the ambient space. If the PZMs are laid on the floor, they do, of course, favour the front desks of the orchestra, which to some extent mask the players beyond them. As an alternative, PZMs can be mounted on a pair of vertical Perspex panels to good effect.

When the hole in the middle cannot be avoided, a third central microphone is often introduced to enhance the central image, as in the Bell Laboratories experiments, and partly operates as a 'width control'. This type of two- or three- microphone spaced technique is very common, and at least one recording company regularly sets up a 'tree' of three microphones for many of its acclaimed classical recordings (see Figure 5.16). However, spaced microphones, unless used carefully, suffer from a number of shortcomings compared with the coincident systems, which rely solely on intensity differences. For one thing, the predicted stereo spread produced by time difference applies only at low frequencies and for impulse sounds with time differences of up to about 3 ms. At middle and high frequencies there is a general loss of imaging focus, with the inter-loudspeaker time differences introducing frequency-dependent comb filter peaks and troughs, leading to a 'phasey' confusion. Against this it should be said that the spaced technique does produce a more 'open' sound, albeit lacking in pinpoint image location, and is preferred by many engineers and record buyers.

5.5 Near-coincident microphones

Broadcasting and recording engineers in various countries have developed a number of stereo microphone configurations which come somewhere between the coincident and the widely spaced methods so far discussed. As a rule, they behave very like a coincident pair at low frequencies but also give clear, unclouded directional cues at higher frequencies. The left/right spread effect falls broadly on the line EF in Figure 5.5.

Any of the standard polar patterns may be chosen and, by varying the spacing between the microphones as well as the angle between them, an infinite number of possibilities is presented. In practice, the final decision will depend on individual taste regarding the inevitable compromise between achieving a coherent stereo image, free of angular distortion, and the preferred ratio of direct-to-reverberant sound.

A valuable analysis by Williams summarizes the possible choices of microphone spacing (up to 50 cm) and angle (up to 180°) for pairs of each of the basic polar patterns. These are calculated to provide the standard ±30° reproduction angle from ensembles subtending angles from ±30° to ±90° (the recording angle). His results for a pair of cardioids are shown in Figure 5.17. They take into account purely physical theory and the physiological (subjective) estimates of the intensity and time differences needed to produce a 60° reproduction angle, as described by Simonsen (broadly confirming the results summarized in Figure 5.5).

The curved lines in Figure 5.17 represent the maximum recommended recording angle, and it will be seen that any required recording angle can be obtained using various distances and angles between the microphones. For example, an ensemble subtending a half-angle of 40° (80° total) would be covered using a pair of cardioids at 12 cm 160°, 17 cm 145°, 22 cm 125°, 30 cm 90°, 40 cm 50° and 50 cm 20°.

Examples of near-coincident microphone arrays currently in use will now be described, and it will be seen that two of them (ORTF and NOS) are marked in Figure 5.17.

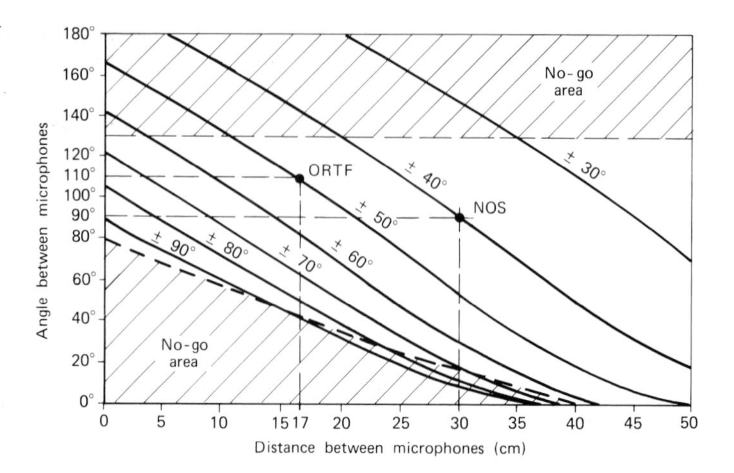

Figure 5.17 Recording angle for a pair of cardioid microphones at various spacings and angles (after Williams)

(1) *Binaural*. This is a special case, a pair of microphones being arranged physically to imitate the locations of a pair of human ears to produce recordings or broadcasts primarily for headphone listening. The result can be uncannily realistic, and has been employed, for example, in radio drama and documentaries both by the BBC and West German radio. Its adoption more widely has been handicapped by the fact that the impressive localization effects heard on headphones, when each ear receives exclusively the signal from one channel, do not translate ideally to loudspeakers, where both ears receive the signals from both loudspeakers. For best results, it is now realized that the feed to headphones should include special equalization to correct for the acoustic effects of the ear canal. Adaptor networks to optimize binaural signals for loudspeaker listening have also been proposed.

A binaural microphone pair usually consists of two pressure omnis placed ear-distance apart, i.e. about 14 cm, on either side of a dummy head or a baffle designed to introduce the same time differences as a real head (Figure 5.18). Indeed, some of the most impressive results have been obtained with small probe microphones worn in the ears of a real person. (When the wearer then listens on good-quality open-backed headphones the unique agreement of the head diffraction dimensions possibly reproduces the most true-to-life stereo images available.) The Sennheiser MKE 2002 is a proprietary version of this idea. Next in verisimilitude comes the building of suitable microphones into a dummy head, as in the Neumann KU80, but essentially

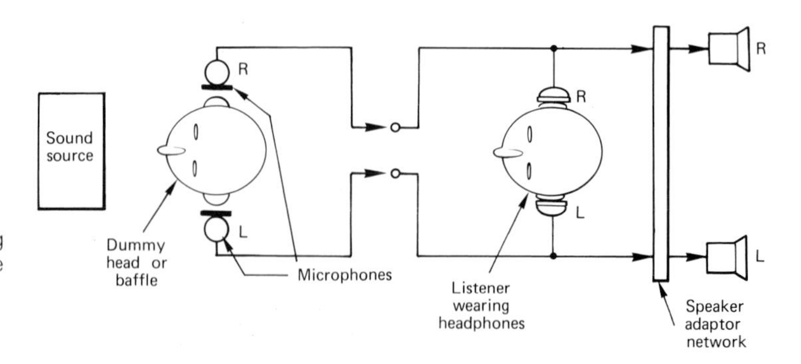

Figure 5.18 Binaural recording. Using a dummy head or baffle can produce realistic stereo on headphones, but needs an adaptor network for loudspeaker listening

similar results can be obtained using a simple baffle as a stylized head. For example, the BBC have been successful with a 25 cm (10 in) diameter Perspex disc and a pair of small electret microphones mounted on a rod pushed through the disc a little off-centre. Other recordists have used an absorptive baffle with omnis or figure-of-eights fixed at 7.5–10 cm on either side.

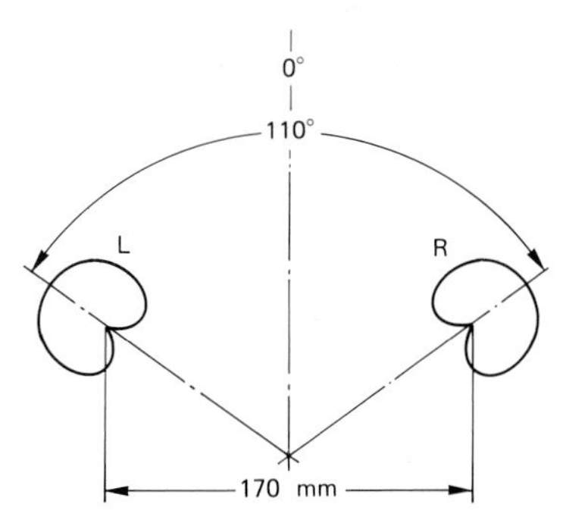

Figure 5.19 ORTF near-coincident pair

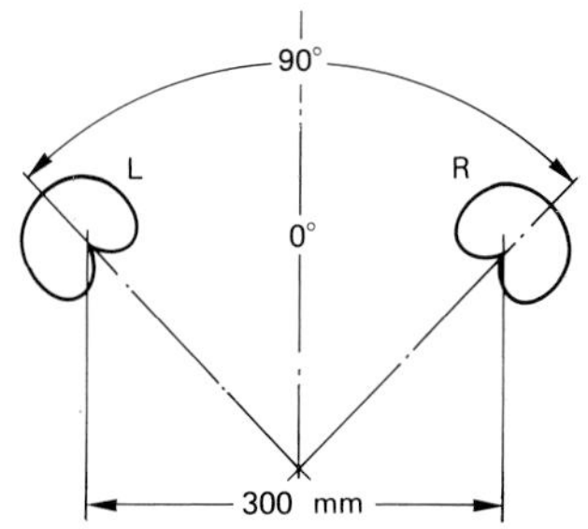

Figure 5.20 NOS near-coincident pair

(2) *ORTF*. This near-coincident technique accredited to French National Broadcasting (Office de Radiodiffusion Française) uses two cardioids spaced 17 cm apart and pointing outwards with an included angle of 110° (Figure 5.19). Its claimed virtues include a sense of 'airiness' combined with good imaging. As shown in Figure 5.17, it most closely suits a recording angle of ±50°.

(3) *NOS*. This variation on the ORTF approach has been used by Dutch Broadcasting (Nederlandsche Omroep Stichting) and features cardioids at 90° spaced 30 cm apart (Figure 5.20). The ideal recording angle suggested by Figure 5.17 is ±40°.

(4) *Jecklin/OSS*. Here two pressure omnis are placed on either side of an absorptive circular baffle in what Swiss Broadcasting call the 'Optimum Stereo Signal' arrangement. Microphone separation is 16.5 cm and the baffle diameter is 28 cm (Figure 5.21). The system is said to work best for internally well-balanced ensembles, i.e.

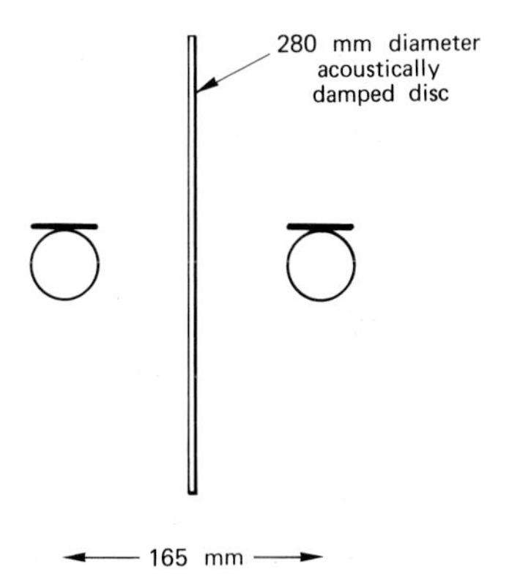

Figure 5.21 Jecklin/OSS near-coincident pair using a circular baffle

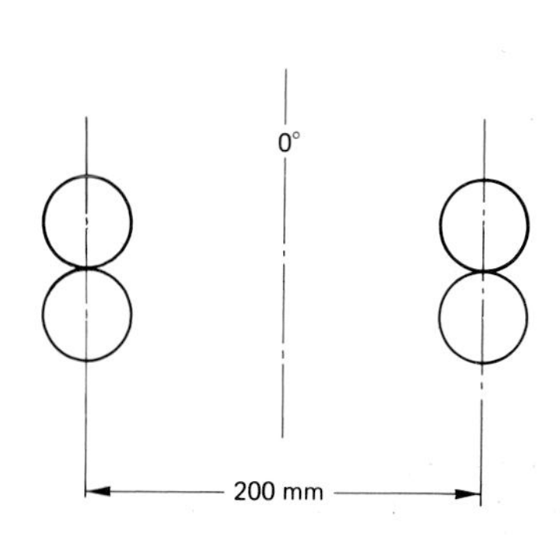

Figure 5.22 Faulkner near-coincident pair

classical music, in a room having a normal or short reverberation time. In too-reverberant conditions a semi-cylindrical screen can be fixed behind the assembly to give simulated cardioid response. The assembly is sold commercially.

(5) *Faulkner*. The British recording engineer Tony Faulkner has been successful with a simple arrangement of two bidirectional microphones pointing straight ahead and spaced 20 cm apart (Figure 5.22). Though this might at first appear to provide only minimal stereo information, in fact its mixture of pure Blumlein with a short spacing can produce coherent imaging plus an open feeling.

5.6 Surround sound

We began this chapter by describing how man's ingenuity has succeeded in creating 'the stereo illusion'. This advanced beyond the old single-point single-channel recording and reproducing system we now call 'mono' (Figure 5.23(a)) to the stage where a believable spread of sound images can be established across a 60° arc in the horizontal plane between a pair of spaced loudspeakers (Figure 5.23(b)).

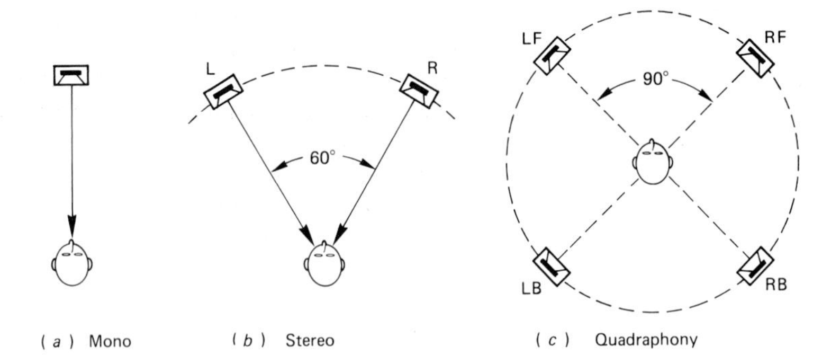

Figure 5.23 Comparing (a) point image mono; (b) line image stereo; (c) plane image quadraphony

Of course man's ambitions have not stopped there. Since sound recording began there has been more or less continuous research and development into the much more complicated problem of creating a full-scale three-dimensional 'surround sound illusion'. This begins with the realization that the sounds reaching the ears of a listener at a live concert, or indeed in any real-life situation, arrive from all directions. At all times we are immersed in a three-dimensional soundfield and, until this can be satisfactorily recorded and reproduced in our homes, true high fidelity cannot be said to have arrived.

In the 1970s, uncountable investigative man-hours and vast sums of money were expended on solving this problem. The idea was to double up again on the two-channel stereo scheme. To replace 'stereophony' with 'quadraphony', elaborate systems were put together based on four microphones, four recording channels and four loudspeakers, with the listener sitting in the middle (Figure 5.23(c)). In the event, quadraphony proved to be a commercial failure and after only a few years the record companies collectively decided to cut their losses and go back to stereo.

However, no one should doubt that surround sound will come back again. The enhanced realism is still a worthwhile target to aim at, and the surround sound effect, whether real or synthesized, is already seen as a desirable feature for the cinema, home videos, pop concerts, public relations promotions, etc.

The main reason usually given for quadraphony's failure in the 1970s was the existence of several incompatible systems, which confused the buying public and inhibited the marketing of recordings and playback equipment. These systems were trying in various ways to solve the problem of mixing (matrixing) the original four signals down onto two channels so that they could be issued on normal stereo discs and cassettes, and broadcast on stereo FM radio (to give what is called 'backwards compatibility' with existing home systems). For each competing system the listener needed a different decoder unit, and for most consumers any interest in quadraphony died there.

A more potent cause of quadraphony's downfall, however, was the fact that each of the simplistic encoding systems proposed relied on separate signals fed to the four loudspeakers arranged into a square around the listener. As Figure 5.23(c) shows, each pair of loudspeakers subtends an angle of 90° to the listener instead of the standard 60° used for stereo. In practice, this means that the sound stage in each front, back and side quadrant suffers from a 'hole-in-the-middle' effect with sounds tending to crowd together near the loudspeakers.

The essential ingredient missing from these early experiments in simple quadraphony – which, it will be noted, were in any case flawed in that they merely sought to reproduce a two-dimensional sound stage in the horizontal plane, with no 'height' information – was a true sense of the timing, phase dispersal and direction of the reverberant sound waves.

The British system known as Ambisonics addresses this problem and offers a range or 'hierarchy' of encoding schemes from a basic stereo-compatible UHJ format, which some record companies are already using, all the way up to 'periphony', giving full-sphere playback from a suitable array of loudspeakers.

Central to the Ambisonic system is the use of a special cluster of four microphone capsules, and the Calrec Soundfield microphone offers this feature in a single unit (Figure 5.25). The four diaphragms are arranged as a regular tetrahedron array, sampling the soundfield as if on the surface of a sphere and capable of reproducing the sound pressure at the effective centre. The four outputs represent one omnidirectional (pressure) element plus three bidirectional (pressure gradient) signals in the left/right, front/back and up/down planes.

The Soundfield microphone's control circuitry not only provides an ideal source for Ambisonics surround sound but also a versatile coincident stereo microphone. Axial directivity is continuously steerable over 360° and the close diaphragm spacings produce an effectively coincident pair whose included angle can be set at any value between 0° (mono) and 180° with any polar pattern between a circle and figure-of-eight. Effective tilt and source distance (zoom) are also adjustable electronically, either by remote control at the time of the recording or during post-production processing. So far, the non-existence of suitable domestic four-channel reproducers has prevented Ambisonics from establishing its surround sound potential. As digital processors proliferate, however, extra channels will

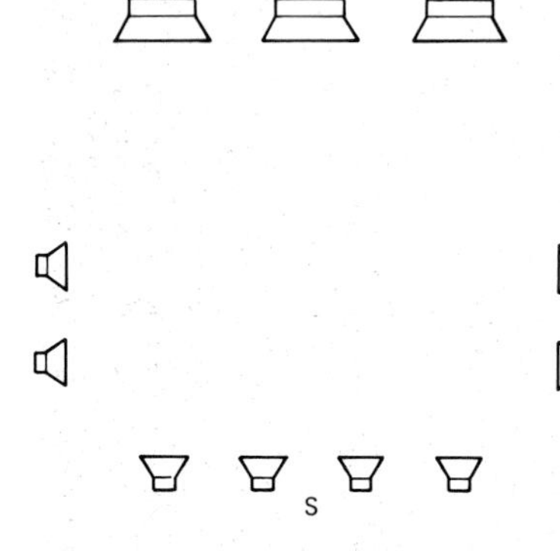

Figure 5.24 Typical cinema loudsspeaker arrangement for Dolby Surround soundtrack reproduction

be easy to accommodate and surround sound in the home could become commonplace.

Surround sound in the cinema is already technically successful. The Dolby Surround cinema system exists in two standard recording formats both intended for reproduction via three full-range behind-the-screen loudspeakers and a U-shaped array of 'surround' loud speakers positioned around the rear half of the theatre (Figure 5.24) Additional bass loudspeakers may be used. The 70 mm wide-screen format employs four magnetic soundtracks to give discrete recording of the three front signals and a fourth surround signal. There are two additional magnetic tracks mainly used for bass enhancement. The simpler 35 mm format has two optical tracks carrying a matrix-encoding of the original four tracks which is decoded at suitably equipped cinemas and is the basis of the home consumer Dolby video playback surround sound format.

Microphone techniques for surround sound films are basically similar to those for mono or stereo. Post-production processing makes use of quadraphonic panpots having a joystick control for steering voices and sound effects. Most dialogue is concentrated on the front centre channel, while music is rendered in a three front tracks plus single surround-array track mode. The rear track is delayed slightly to maintain the impression that most sounds are coming from the area of the screen.

Example of surround sound microphone design

Calrec Soundfield (Figure 5.25)
Transducer type: condenser (four identical capsules mounted very close together)
Polar pattern: continuous adjustment to any pattern
Frequency range: 20–20 000 Hz
Sensitivity: 1 mV/μbar
Max. SPL for 0.5% THD: 140 dB
Impedance: 100 Ω approx.
Equivalent noise level: 16–17 dB
Features: control unit provides separate muting, gain setting or fine level control of all four capsules, four built-in preamplifiers, LED to indicate main reference axis, switchable for Ambisonics B-Format or stereo operation, totally steerable with variable zoom
Main applications: Ambisonics surround sound, stereo recording and broadcasting.

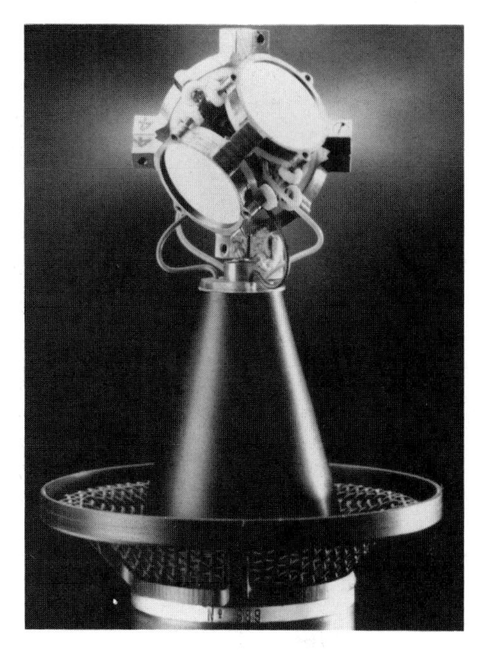

Figure 5.25 Calrec Soundfield microphone

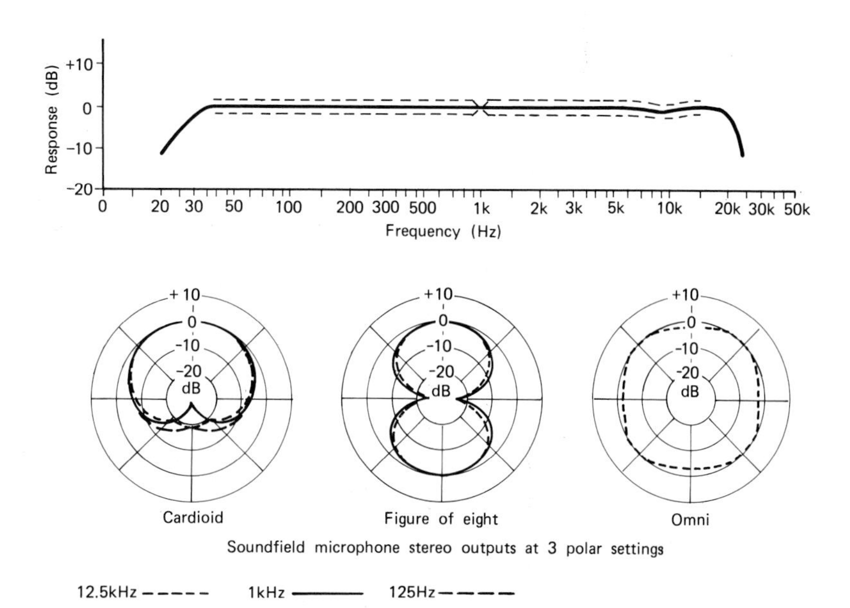

6
Microphone connectors and accessories

Nowhere is the lack of industry standards more in evidence than in the realm of connectors and accessories. However, recent years have seen a growing unanimity on the part of manufacturers, and new microphones for professional use are tending increasingly to conform to the practices outlined here.

6.1 Plugs and sockets

Microphones for communications or amateur use are often fitted with relatively short, fixed cables terminating at the equipment end in a plug to the three- or five-pin DIN standard or the familiar 6.3 mm (0.25 in) jackplug. These may be wired for unbalanced or balanced line working. In the main, however, professional microphones will be designed primarily for balanced line and will have a built-in socket to be used with detachable plug-in cables for added flexibility. The built-in socket will usually have some screw-collar (e.g. DIN or Tuchel type) or spring-loaded latchlock device (e.g. XLR type) to prevent the cable/plug from working loose.

Figure 6.1 illustrates typical internal wiring and built-in socket arrangements designed either for connection to an unbalanced cable or a standard balanced line, i.e. comprising two identical signal conductors within a grounded braiding or screen as described in Chapter 4 (see Figure 4.26). The systems shown are:

1. A standard three-pin DIN type socket with the transducer signal leads wired symmetrically to pins 1 and 3 and pin 2 earthed. This would be suitable for a low-impedance microphone (30–50 Ω) using up to 200 m of two-conductor screened cable. To match the usual high-impedance amplifier or mixer input, a step-up transformer may be needed at the equipment end of the cable.
2. Unbalanced DIN connection with pins 1 and 3 both taken to the single live conductor and pin 2 to the screen of a coaxial cable.
3. Unbalanced jackplug connector.

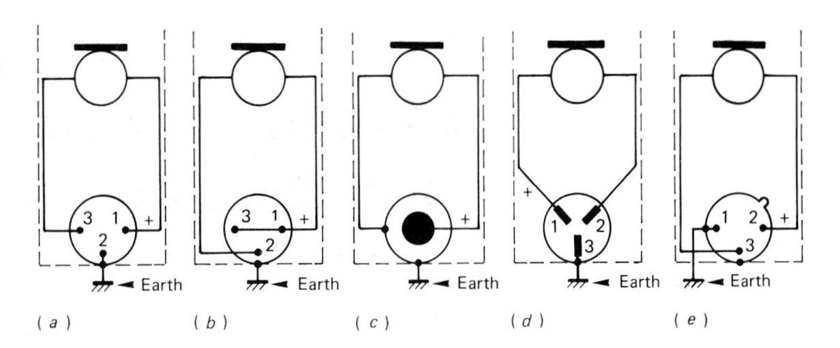

Figure 6.1 Microphone internal wiring. Examples of socket connections

4. A basically similar arrangement to (1) but representing the Tuchel type of socket which, by convention, has pin 3 earthed.
5. The XLR type of socket, now virtually a recording and broadcasting industry standard and discussed in detail below.

6.1.1 XLR connectors

The most widely used universal type of connector is the XLR, and this is now the preferred choice when wiring up a new studio. The studio specification should allow for a generous number of microphone circuits wired to wall sockets distributed round the studio at convenient positions which will help to keep cable lengths short. The sockets can be arranged in groups of four, eight or 12, and parallel-wired groups on opposite walls, labelled 1A, 2A, etc., will reduce the clutter of cables criss-crossing the studio floor. In small talks studios a variation on this idea of parallel sockets sometimes has alternative sockets located side by side, the second one wired with a bass-cut network. This socket will be chosen by the balance engineer if room resonances are troublesome or close-miking is producing excessively boomy speech due to the proximity effect (normally a problem only with pressure-gradient microphones). The traditional floor-level skirting-board position for mounting microphone (and mains power supply) sockets can with advantage be changed to a height of about 1 m to facilitate the insertion and removal of latchlock cable plugs and the reading of socket number labels.

Figure 6.2 illustrates three versions of the three-pin XLR connector. Note that XLR code numbers identify the number of pins followed by the type, the odd type numbers indicating a socket (female) and the even numbers a plug (male). The most common socket and plug types are the 11C/12C in-line cable connector; 13/14 round panel mount; 15/16 right-angle cable connector; 31/32 rectangular panel mount. Thus the three three-pin versions shown in Figure 6.2 can be recognized as (1) XLR-3–13 round panel mount socket, (2) XLR-3–12C plug and (3) XLR-3–11C cable socket. Note that the panel and cable socket connectors are fitted with a push latchlock to retain the plug, though this can be removed if not required.

The majority of studio microphones are now constructed with a built-in three-pin XLR plug (male) to conform to one of the unwritten laws of XLR connectors that the pins should point in the direction of signal travel. Similarly, a microphone cable will normally have XLR-3–11C (female) at the microphone end and XLR-3–12C (male) at the other, while wall sockets and mixer inputs will be of

Figure 6.2 XLR three-pin connectors. (a) Round panel mount socket; (b) plug; (c) cable socket

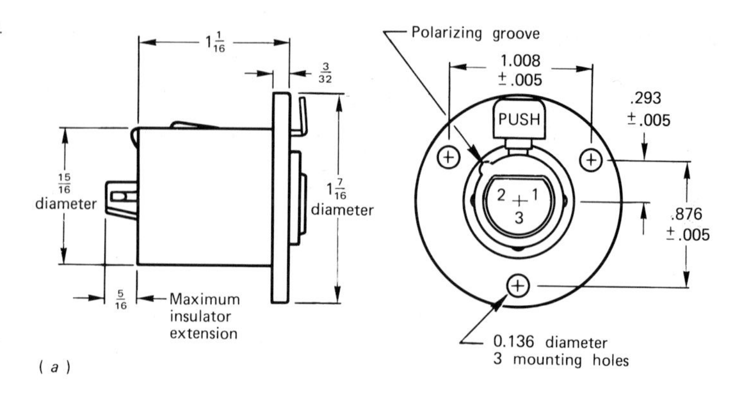

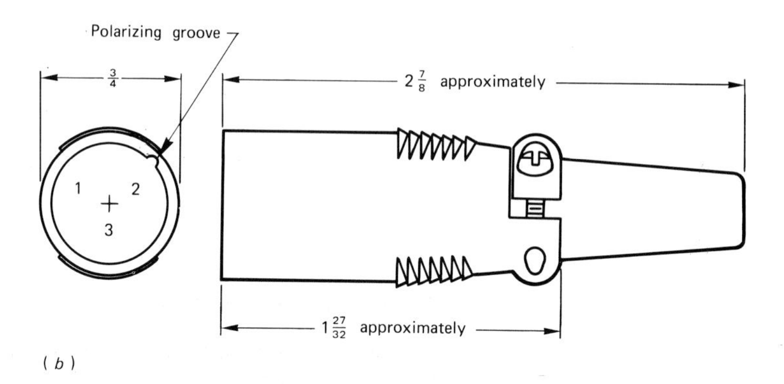

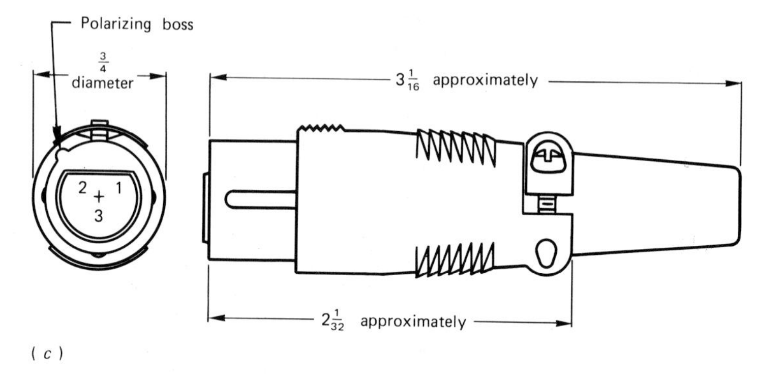

the circular or rectangular XLR female type. The same law should apply to inputs and outputs on tape machines, processors and monitoring loudspeaker systems where XLR connectors are in use.

In case non-standard XLR connections have to be used (for example, power amplifiers and loudspeakers often use reverse-sex XLRs) it is useful to keep a supply of 'sex-reversal' adaptors having the same type of connector at each end (both male or both female). These adaptors are available commercially (see Figure 6.3(a) and (b)) or can be made up locally on short lengths of cable and colour-coded for easy identification.

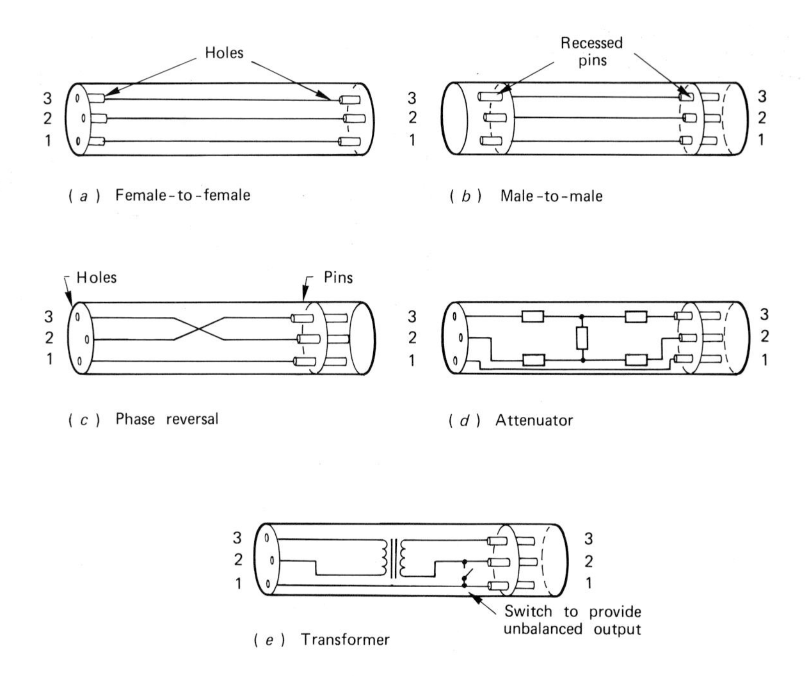

Figure 6.3 XLR in-line adaptors

6.1.2 Phasing or polarity

There is a second unwritten law related to XLR connectors which sets down the convention to be followed in pin numbering. This is already implied in Figure 6.1(e), and shown in more detail in Figure 6.4. It will be seen that the recommendation is for pin 1 to be grounded with the positive or 'hot' signal connection made to pin 2 and the return signal to pin 3. Connecting pin 1 to the screen, isolated at the microphone end but taken to the centre tap of the equipment input transformer or a suitable transistor circuit, has always been the universal practice. This makes sense when it is realized that the metal sides of socket hole 1 on XLR connectors are longer than the others to ensure that the screened connection is made before the signal when the plug is inserted.

Until recently, however, some studios and manufacturers have adopted a non-standard attitude to which way round they wired pins 2 and 3, i.e. some connectors were wired in antiphase or opposite polarity to others. At first this may seem to be of only academic interest, remembering the balanced symmetry of the signal circuits, and could go undetected except when two or more microphones are

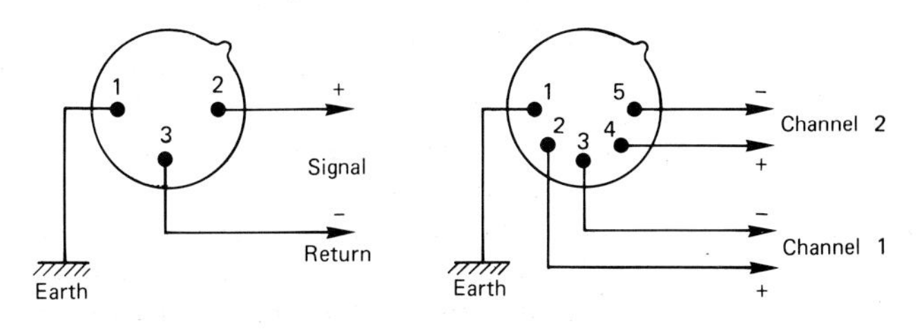

Figure 6.4 XLR three- and five-pin wiring conventions

faded up together. Then the results of mixed polarity connections can be disastrous, showing up as severe low-frequency attenuation and particularly damaging when the cables to a stereo microphone pair are accidentally wired in antiphase.

Today the above convention of wiring pin 2 to 'hot' and pin 3 to 'return' is becoming not so much a vague recommendation as a hard-and-fast rule. This is because of new awareness of the importance of audio signal polarity – the idea of 'absolute phase'. This relates to the desirability of preserving the phase of the original sound wave throughout the entire chain from microphone to loudspeaker – and the realization that, unlike pure sinewaves, many music and speech signals are asymmetrical, with transients, for example, exhibiting a distinctly positive or negative swing in sound pressure on the initial attack. Ideally, for the same sensation of attack to be reproduced, the loudspeaker diaphragm should follow the same forward or backward motion that the microphone diaphragm performed at the outset.

Preservation of absolute phase throughout the entire chain may be a pipe dream, and is surely unattainable in a multi-microphone sound balance where so many direct and reflected waves contribute random phase elements to the final mix. However, the attempt is worth making and is being set down in the relevant international standards (e.g. EBU Recommendation R50–1988). Some links in the chain, notably microphones and loudspeakers, are already covered by IEC Publication 268. An audio signal is taken to be positive when it results in an increase in acoustic pressure on the microphone diaphragm and therefore a displacement of the diaphragm towards the rear. A microphone is correctly wired internally when such a displacement produces a positive voltage at pin 2 with respect to pin 3. Again, a loudspeaker has been correctly phased when an instantaneous positive voltage at its red or 'positive' terminal produces an outward

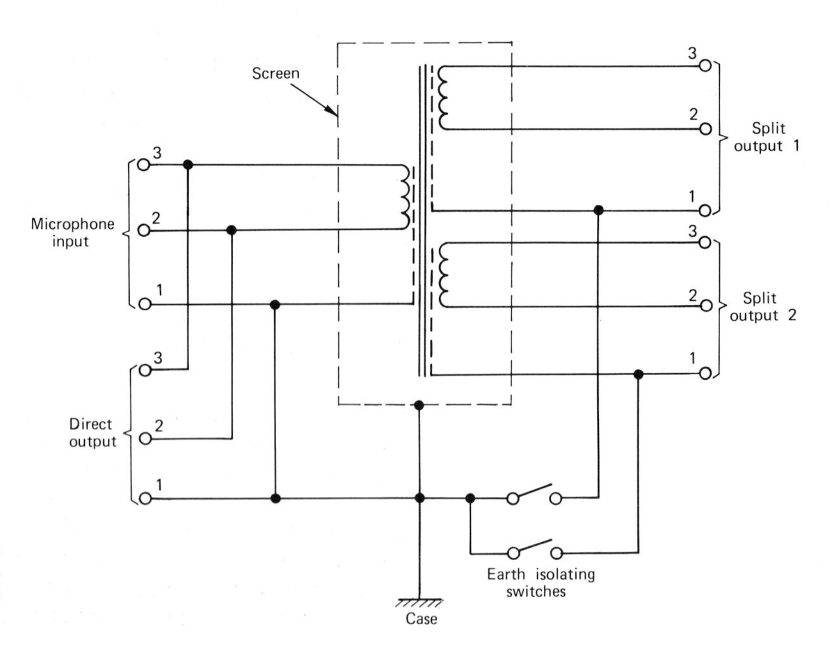

Figure 6.5 Microphone splitter: basic circuit deriving two independent outputs plus a direct output from a single balanced input

motion of the cone or diaphragm. It remains for manufacturers and studio workshops to apply similar logic to plugs, sockets and cables (as in Figure 6.4), analogue tapes (a positive audio signal should correspond to magnetization in the same direction as that of the motion of the tape), amplifiers, processors, digital systems, etc.

A phase-reversal adaptor can be useful in case non-standard polarity situations arise. As mentioned in respect of sex-reversal adaptors, commercial phase-reversal adaptors are available (see Figure 6.3(c)) or can be made up locally. Other convenient XLR adaptors exist, including on/off switches, bass-cut filters to reduce the proximity effect and line attenuators or microphone loss-pads (Figure 6.3(d)). These may be rated from −10 dB to −60 dB and can be used, for example, to avoid overloading of microphone circuits when close-miking percussion instruments, etc. The switchable attenuator built into some microphones may not reduce the output level sufficiently, and is in any case designed primarily to protect and avoid distortion in the microphone internal circuitry. The −60 dB microphone pad is convenient for obtaining a microphone-level signal from a line-level source prior to mixing. Figure 6.3(e) shows an in-line transformer adaptor, which is a convenient method of impedance matching. A step-up version, for example, might be used to match a low-impedance microphone to a high-impedance input, when pins 1 and 2 could be wired together to provide balanced-to-unbalanced conversion.

A microphone splitter box is another popular XLR accessory (Figure 6.5). This allows the signal from a microphone to be routed to two independent destinations. A typical application occurs when a live performance using sound reinforcement is being simultaneously recorded or broadcast. To make life easier for the performers and keep the stage area reasonably clear, the cables from selected microphones may be taken to the splitter inputs and from there sent on to both the public-address circuits and the mixer console. Simple Y-connectors would not provide a sufficient degree of electrical isolation between the two destinations in this application, or in various broadcast programmes where independent feeds are essential. The figure shows how the isolating microphone splitter derives its two completely independent outputs using a transformer with a single primary winding and two separate secondary windings. Note that condenser microphones requiring phantom power must be supplied via a direct circuit as the splitter transformer acts as a block to DC.

6.2 Microphone cables

Clearly, at the local studio level, cables and connectors should be wired to an agreed polarity convention. For example, if the twin conductors in the given cable have, say, red and white insulation it could be made a rule to wire red to pin 2 and white to pin 3. There then comes the question of cable quality. It is clearly false economy to choose a type of cable which in any way degrades the signal quality or fails to stand up to the rigours of professional use – a factor which is even more important when choosing cables for location recording.

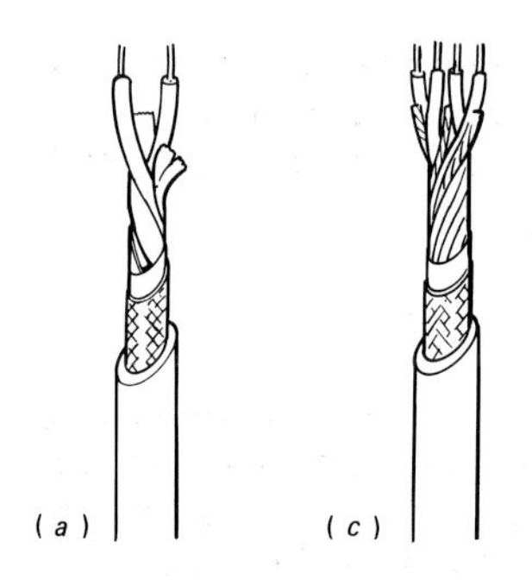

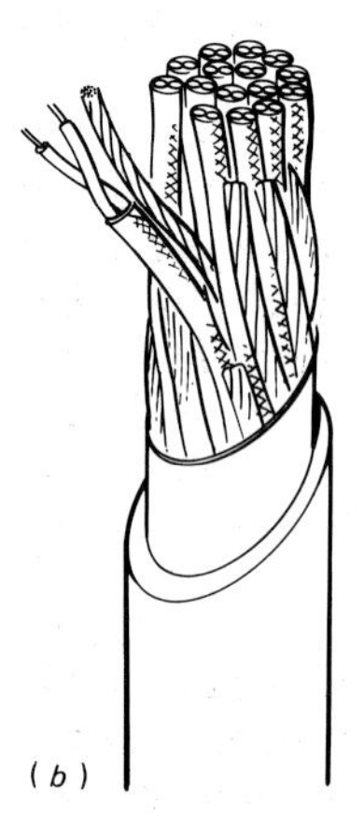

Figure 6.6 Microphone cables. (a) Two-conductor type; (b) multichannel type; (c) Star-Quad type

6.2.1 Two-conductor shielded cables

Figure 6.6(a) illustrates the basic features of a good-quality two-conductor microphone cable. Typically, each conductor will be about 0.7 mm in diameter (22–gauge AWG) and might comprise 60 strands of very thin (0.08 mm) copper or tinned cadmium bronze wire to provide greater flexibility than the equivalent single thick wire would give. The insulation and rubber or PVC outer sheath should also be designed for good flexibility even at low temperatures. The braided shield should be high density for efficient screening against stray hum fields, electrostatic noise and radio frequency interference.

6.2.2 Multichannel cables

Multichannel 'snake' cables containing, say, eight, 12 or 16 pairs of colour-coded conductors simplify the task of setting up location sessions. They can be terminated at the console end with a multi-pin plug/socket while the microphone end can consist of a junction box with the appropriate number of female panel-mount or in-line three-pin XLR sockets ready to accept short cables from the individual microphones. As Figure 6.6(b) shows, each signal channel has its own shielded two-conductor cable. Their shields have no separate insulation so that a common ground exists.

6.2.3 Star-Quad cables

Figure 6.6(c) shows a four-conductor Star-Quad type of cable which is capable of ten times the rejection of stray hum and noise achieved by conventional two-conductor types. Diagonally opposite conductors are connected together at both ends of the cable and all four conductors are tightly twisted within the screen to ensure that any induced interference is exactly the same in each leg of the balanced circuit for effective cancellation at the console input transformer. Star-Quad cable is particularly suitable in televison studios, theatres or anywhere that lighting dimmers or other notorious causes of electromagnetic noise are found. Multichannel 'snake' Star-Quad' cables are also available.

Cable reels can simplify the transportation of long cables and speed up the setting up and removal of large microphone installations (Figure 6.7). A winding device is often fitted and the cable may be permanently wired to XLR sockets on the reel side, requiring only a short extension cable to the mixer/console.

Figure 6.7 Typical microphone cable reel. (Courtesy Keith Monks)

6.2.4 Speciality (audiophile) cables

In addition to various grades of microphone cable in the configurations described above, new kinds of exotic cable are now on offer at relatively high prices. These may employ complex interleaved constructions or special low-oxide copper wire to produce a claimed improvement in subjective sound quality. In a properly assembled system, where cable lengths and matching at the terminations are already up to high professional standards, these claimed benefits may be too small to detect easily by ear, and will almost certainly not show up in conventional measurements. The best advice in such cases is probably to stay with high-grade conventional cables until a scientifically planned A/B comparison test demonstrates that a change to one of these speciality cables clearly justifies their high cost.

6.2.5 Cable losses

Losses in signal level and sound quality are inevitable when long cables are used but these are normally insignificant. The primary electrical considerations are the resistance, capacitance and inductance (too small to worry about) presented by the cable. This total impedance appears as an addition to the load as seen by the microphone.

Studio microphones are generally operated in approximately 'no-load' conditions, i.e. with the load impedance Z_L at least five times greater than the microphone source impedance Z_S. As was seen in Chapter 4, most microphones have a Z_S in the range 150–300 Ω (nominally 200 Ω) so that Z_L is typically 1000 Ω or higher. The cable resistance R_C is a series effect which increases with cable length taking the whole loop formed by the send and return conductors into account; C_C is the parallel or shunt capacitance which also increases with cable length.

Typical values for good-quality cables are about R_C = 0.06 Ω/m and C_C = 70 pF/m. The effect of the purely resistive element in the cable impedance is simply to introduce a small loss in signal transfer level. In practice, this can be ignored except for extremely long cables. For example, R_C = 0.6 Ω for a 10 m cable and is still only 6.0 Ω for a 100 m one. This makes only a small difference to the total load, normally at least 1000 Ω as we have said, and reduces signal level by just a fraction of 1 dB.

As for the frequency-dependent reactive elements, the tiny inductance can be ignored, but the shunting effect of C_C at high frequencies where reactance falls to relatively low values is certainly measurable and, in a worst-case situation, may result in audible high-frequency loss. Taking a 70 pF/m cable as an example, C_C will increase to 7000 pF for a cable length of 100 m. This suggests that a 100 m cable has an insignificant (because large) shunting reactance of about 23 000 Ω at 1 kHz (calculated from $X_C = 1/2\pi fC$) but that this falls to only 1138 Ω at 20 kHz, introducing a loss of almost 3 dB.

6.3 Microphone supports and stands

All steps must be taken to ensure that the performance of a microphone is in no way degraded by the way it is supported or suspended. In general, some form of shock-insulation will be needed, since all microphones react to vibrations transmitted to them along the direct physical path through cables, slings, stands or booms. Yet there is often a need to fix a microphone's position with almost millimetre accuracy, and so too compliant a mounting, which may sag or twist with time or changes in temperature, must be avoided. Some 'universal' spring-loaded clips and supports exist capable of holding a range of microphone types and dimensions but, where dedicated accessories are available for particular microphones, these will be the first choice and will hopefully provide a precise match to the microphone's dimensions, weight, etc.

Apart from considerations of secure vibration-proof mounting, there is always the question of safety. All connectors must be firm and the whole assembly must be put together so as to minimize any risks of being knocked over or falling on artistes, instruments or audience. The first security measure is to route cables for minimum risk of accidental contact. They should be taken along close to walls and taped up and over any doors through which artistes or audience may walk. When the cable must cross the main floor or stage area, a neat route should be chosen with generous fixings of gaffer tape to prevent loops forming. At the microphone stand, the cable should again be firmly taped or clipped but with enough slack to allow later changes in microphone angle or position as may be necessary.

When microphones are to be slung overhead (often a preferred arrangement in crowded halls and, of course, theoretically capable of providing a high degree of stability compared with tall stands or booms) safety becomes paramount. A minimalist approach is to suspend the microphone by its cable alone as shown in Figure 6.8, either pointing straight down or held at an angle by means of a clip. This works well enough with lightweight microphones, though even then it is advisable to tape the microphone across to the cable so that it cannot become dislodged, and not rely solely on the XLR latchlock. One nuisance with the angled arrangement is that it is very difficult to secure the top of the cable so that the microphone points along the desired compass bearing and, even when this has been achieved, the cable may twist after a time.

The answer is to add a tie-cord attached to some other convenient part of the building (Figure 6.9(a)). Using the microphone cable as the main vertical support, the cord is anchored to the cable just above the microphone itself and then tied back to a support which will keep the microphone at the desired location and angle. Alternatively, the microphone can be suspended on a separate guy wire between two main side supports and pulled back to the desired position by a tie-cord as before (Figure 6.9(b)). The supports must be strong and well anchored, thin nylon cord or wire being better than rope, with the

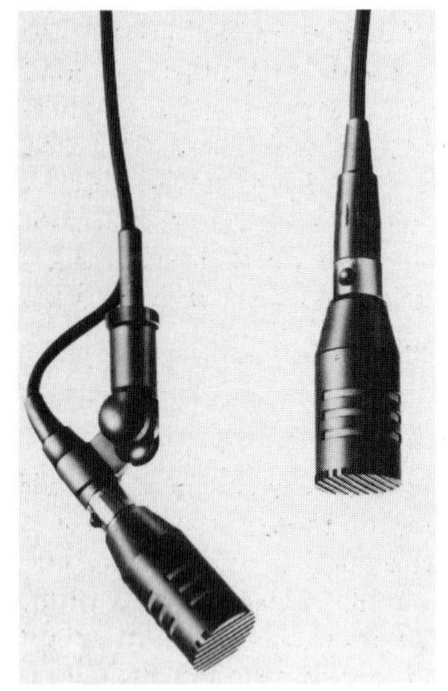

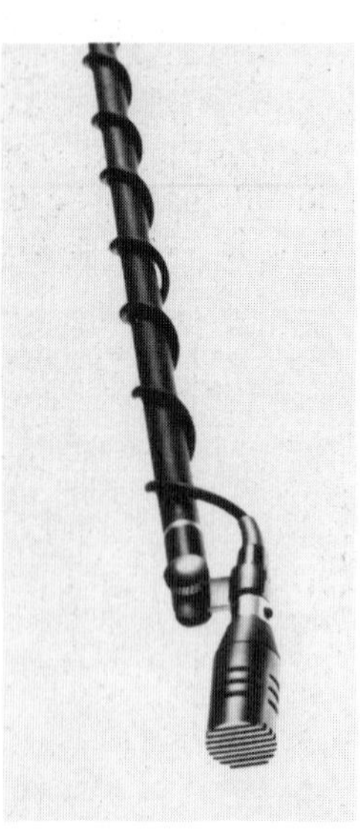

Figure 6.8 Simple microphone suspensions. (a) Angled clip; (b) pointing straight down; (c) using a fishpole. (Courtesy AKG)

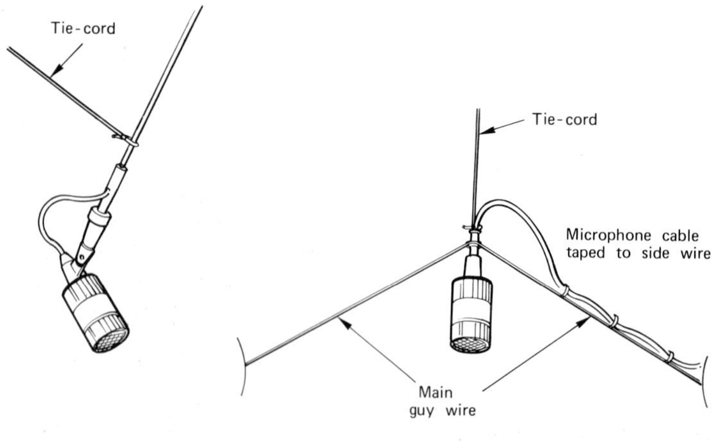

Figure 6.9 Microphone suspensions. (a) A single tie-cord; (b) a more controllable three-wire system

microphone fixed at the centre or off-centre as required. The microphone cable can then be taken upwards as before or taped along one side of the supporting cord as shown. Of course, such a crossed-sling arrangement can be used to support more than one microphone, perhaps forming a spaced or coincident stereo pair. A fourth guy wire can be attached to the microphone clamp or bar to give control over angle of tilt. For a more permanent installation, a system of winches and alternative anchor points is often assembled to combine versatility with maximum safety.

6.3.1 Microphone clamps and suspensions

Most manufacturers supply clamps or adaptors which make a safe fitment between the microphone and stand or gooseneck suspension, etc. These will usually have an inner threaded collar for screwing onto the stand top and an angle joint to allow microphone tilting (Figure 6.10(a)). As well as these dedicated clamps which fit only a specific microphone body, for which an elastic suspension cradle may be included, there is a 'universal' type which can grip a range of microphone diameters. This has spring-loaded V-shaped jaws, preferably rubber lined, and a shaft containing the threaded collar. There is no single thread standard, though the two most popular are of 9.5 mm (3/8 in) or 15.9 mm (5/8 in) diameter with either Whitworth or 27 turns per inch thread. Suitable male and female adaptors are available, as shown in Figure 6.11.

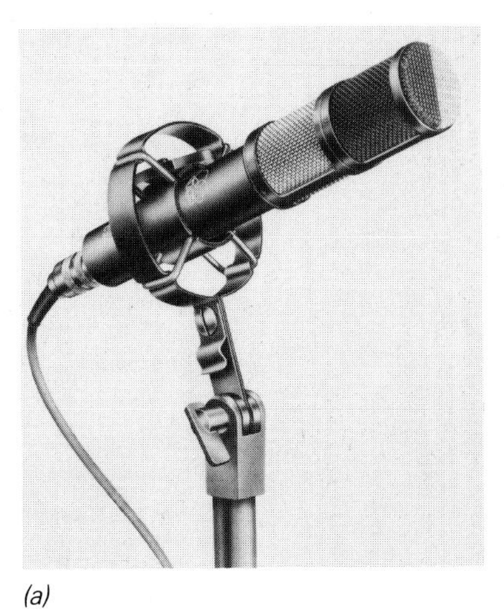

(a)

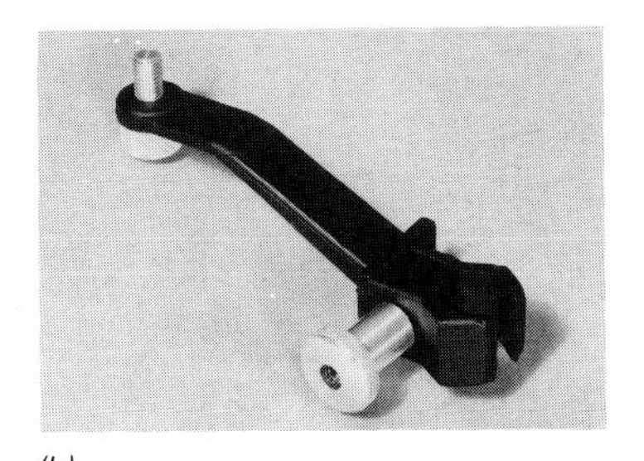

(b)

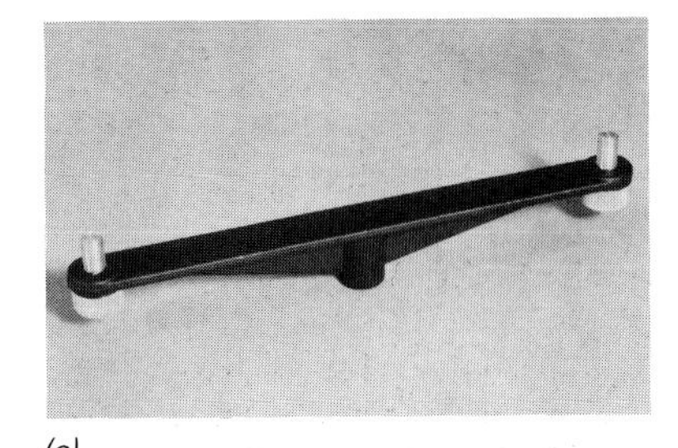

(c)

Figure 6.10 Microphone clamps. (a) Cradle type; (b) vertical shaft clamp; (c) stereo bar

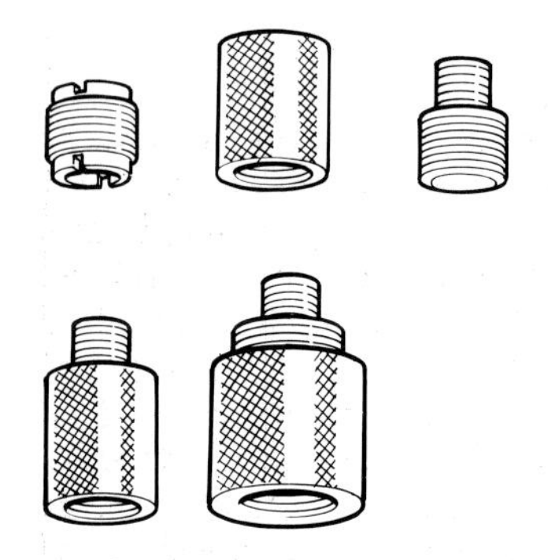

Figure 6.11 Examples of internal and external thread adaptors

More elaborate shock-mounts also exist and, of course, should be chosen where floor vibrations are a likely hazard. There are also simple clamps for attaching to the vertical shaft of a microphone stand or indeed to any suitable structure (Figure 6.10(b)). Gooseneck adaptors can be obtained either with threaded inserts at both ends or with a pre-wired XLR socket at one or both ends. These are spring wound and can be set to any required angle. Stereo bars are quite common, having captive sliding screws for attaching the clamps of two microphones, with a stand adaptor at the centre (Figure 6.10(c)). These simplify the setting up of coincident or closely spaced microphones for stereo.

6.3.2 Table stands and floor stands

Numerous designs exist for desk or lectern mounting of microphones, with or without built-in on/off switches or flexible gooseneck attachments. These work best when they are a good match for the particular microphone in terms of dimensions and weight. Again the microphone cable should be dressed neatly to provide maximum security and safety. The need for the user to adopt a proper voice tone and to address the microphone directly, avoiding breath and popping effects, is discussed in Chapter 9, along with tips on avoiding desk-top reflections.

Floor stands also come in various forms. One common type covering heights from about 1 to 2 m has a vertical telescopic column and three or four screw-in or fold-away rubber-ended legs. The top of the column carries a male threaded connector onto which any of the microphone clamps discussed above can be fixed directly. Very often this type of stand is made more versatile by adding a boom arm onto which the microphone clamp is screwed (Figure 6.12). This allows the microphone to be extended towards an instrument or singer. In its simplest form, the boom arm is about 1 m long and set in an adjustable T-clamp which permits it to be extended and adjusted at any angle.

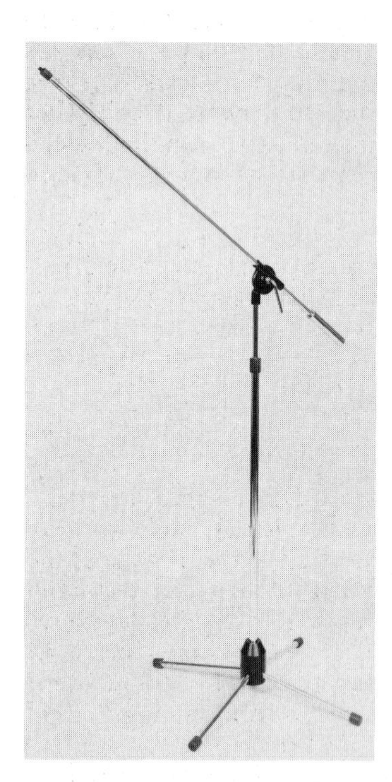

Figure 6.12 Floor stand with boom attachment. (Courtesy Keith Monks)

Larger tripod stands giving height extensions up to 3 or 4 m need extra care. When they are fitted with telescopic boom arms to reach over obstructions or towards back-row musicians there is a real risk of overbalancing with a heavy microphone or stereo pair fitted. In most cases a counterbalance weight is attached to the rear end of the boom arm, and, of course, this helps to keep the forces on either side of the T-clamp about equal. However, in all such cases where there is a danger of the system becoming top heavy it is worth working out the approximate height of the overall centre of gravity and ensuring that this cannot fall outside the area of floor-support. Test tilting of the stand at various angles should confirm the system's stability. With a tripod stand it is always good practice to line up one of the legs in the same direction as the microphone as an insurance against falling forwards.

When extra height is used, the legs must be splayed to their widest extent and weighed down with special weights or sandbags if necessary. If absorbent pads are placed under stands to reduce floor-transmitted vibrations, the system must be checked for any residual rocking motion.

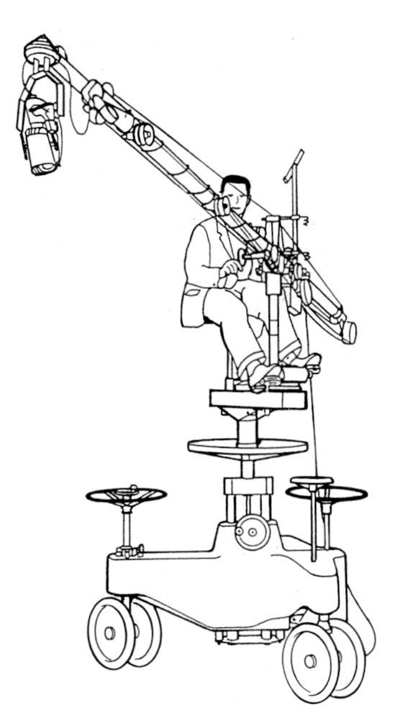

Figure 6.13 Typical dolly-mounted microphone boom

6.3.3 TV booms and fishpoles

When the microphone is required to follow TV or film action, or move over an audience to pick up individual voices, an advanced form of boom is needed. A popular type has a long telescopic arm with the microphone suspended in a complex cradle at one end and a counterweight at the other (Figure 6.13). The arm is supported on a pivoting pillar which is built onto a three-wheeled trolley or pram.

The operator stands on a platform on board the trolley and an assistant propels the whole boom assembly around the floor as necessary. The operator turns a handle held in the right hand to extend or retract the boom arm, and this action simultaneously shifts the counterweight through the smaller distance needed to maintain balance. Control of the microphone position in all three planes is carried out by the left hand. The operator can swing the arm round for horizontal position, tilt the arm on its pivot for height, turn a lever to rotate the microphone cradle for direction and squeeze the two grips of the lever to change the microphone's angle of tilt.

Simpler mobile booms are available and also more elaborate designs with a seat for the operator, boom extension up to 5 m and a wheel-lock arrangement allowing sideways or 'crab' tracking as well as the normal back-wheel steering.

A skilled operator can control all this manoeuvrability to keep the microphone continually at just the right position and angle to pick up the artistes with the required sound balance in relation to each other and to any accompanying instruments or effects. In TV drama this is no mean feat, and is complicated by the need to keep the microphone and its shadow out of the picture. Operators need to know all about lenses and camera angles so that they can raise or retract the boom for long shots and move in for close-ups – a necessary procedure if visual and aural perspectives are to be kept in synchronism. Headphone monitoring may not give them an ideally clear indication of the sound balance, and so talkback to and from the sound mixer is a further requirement.

Where less manoeuvrability is needed, a lightweight microphone can be mounted on a fishpole. This is a single telescopic arm with rubber handgrips and three or four extension sections reaching up to, say, 4.5 m.

6.4 Windshields

Air movement can lead to turbulence around the diaphragm and the phase-shift vent-holes (in pressure gradient microphones). This can have a devastating effect on a microphone's performance or, at the very least, introduce unpredictable bursts of overload distortion (blasting). Much of the energy in wind noise is at low frequencies or even subsonic, so that a high-pass filter designed to remove all frequencies below, say, 60 Hz can be quite effective. In addition, for close speaking or singing and outdoor work, some form of pop shield or windshield is often needed.

As with clamps, many microphones can be fitted with specially designed 'dedicated' windshields, but there are some 'universal' models which will fit a range of microphone types. These are usually simple foam push-on pop shields and may be brightly coloured, and easily seen by a close-singing vocalist, or non-light reflecting and almost invisible. The best examples have been carefully designed using wind-tunnels and the like, and can reduce wind noise by 12–20 dB helped by a small amount of bass-cut. Others are little more than cosmetic and capable of protecting the microphone from moderate breath effects. Equivalent results can be achieved using an improvised metal frame with nylon or silk stretched over it.

For serious outdoor work, particularly using highly directional microphones to pick up sounds at a distance, complex basket-weave windshields forming a double cage with a thin acoustic screen are needed to keep wind interference away from the phase-shift vent holes. The dedicated versions provide about 20–30 dB improvement.

Part 2
The technique

7
Principles of microphone balance

7.1 Basic guidelines – classical music

It is often said that microphone balance is a subtle art for which there can never be hard-and-fast rules. This assertion certainly applies in the field of pop music recording, where the ultimate sounds and the music itself are totally wrapped up together in a combined creative process and should not be inhibited by any arbitrary set of rules. We might almost call pop music 'microphone music'.

In classical music recording, where the engineer's role may be more fairly described as interpretative rather than creative, there are again no strict rules, but certain clear conventions have been evolved which all recordists would do well to follow. The balance of solo instruments, voices and ensembles in classical music must establish for the listener as many as possible of the sensations and impressions experienced at a live concert. This is often no easy matter and requires engineers to begin by deciding which of the two quite distinct basic impressions they want to create:

1. Listeners feel transported to idealized seats in the concert hall – and this can be an imaginary 'perfect' hall or some real-life venue, as when the announcer for a broadcast concert says, 'And now we take you over to the Royal Albert Hall';
2. Listeners feel that the musicians have been transported to them and are performing there and then in their own rooms or in an extended space just beyond the room walls. This latter impression is virtually impossible to create when recording large forces, but may be very effective for solo guitar, a song recital, etc. In practice, listeners are also aware of the acoustics of their rooms, but this is usually no more than a mild distraction, though there is a case for recording engineers to choose a slightly dry balance to allow for this small additional reverberation, since many living rooms are slightly more reverberant than professional studio control rooms.

That first locational decision having been made, some or all of the following features need to be borne in mind if the 'perceived' musical performance is to match up to the real thing:

1. Sound quality or timbre of voices and instruments
2. Musical balance between voices and instruments
3. Apparent distance and depth of perspective
4. Apparent left/right (stereo) spread
5. Apparent acoustic environment
6. Dynamic range or apparent loudness variation.

Each of these perceived characteristics of a sound balance will now be discussed in some detail, but it must be realized that they are not mutually exclusive. Decisions about one will often affect decisions about another. Note that this section treats classical music as a separate grouping, with pop music following in Section 7.2. However, this division into categories – and, of course, there are others which might have been singled out, such as jazz or rock – is no more than a broad attempt to illustrate the principles. Many music-recording situations will demand their own unique form of treatment within these guidelines.

7.1.1 Sound quality or timbre

Each voice or instrument should be reproduced as faithfully as possible. This means, for example, that too close a balance is ruled out if odd mechanical noises not normally heard by an audience are to be avoided. It also suggests that the microphone itself, and indeed the whole of the recording/playback chain of equipment, should be up to high-quality standards. It is also necessary to position the microphone carefully in relation to the known directional properties of each instrument (or any directivity peculiarities noted at rehearsal) to ensure fidelity to its expected timbre. These properties are discussed in Chapters 8 and 9, but, by way of example, it may be noted that the violins in an orchestra are played at quite an oblique angle to the ears of the audience. Placing a microphone vertically above the violins is likely to produce a more shrill sound due to the greater concentration of high-frequency energy on the instrumental axis.

7.1.2 Musical balance

The relative loudness balance between instruments and voices should appear natural and correspond faithfully to the composer's markings in the score – as interpreted by the artistes and conductor. When the particular work or artistes are not well known, it becomes very necessary to spend time listening to them 'live' to assess the true balance before making a final decision as to the microphone positions and gain settings. All temptations to impose a standard balance between parts on what is a unique interpretation or to highlight soloists, for example, should generally be avoided. When the performers appear to be opting for what seems like a wrongly judged balance, this should be discussed tactfully.

As far as possible, the balance between parts should be achieved with the performers taking up their normal preferred layout. This will give them maximum comfort and confidence, since they will hear each other and be able to adjust their own relative loudness balance just as at a normal concert or rehearsal. In a studio recording where the acoustic environment is proving difficult, or the work to be performed is unusually complex or the musical forces available are non-ideal (e.g. too few string instruments or male voices) an unconventional layout or microphone technique may be the best answer, if all the performers agree.

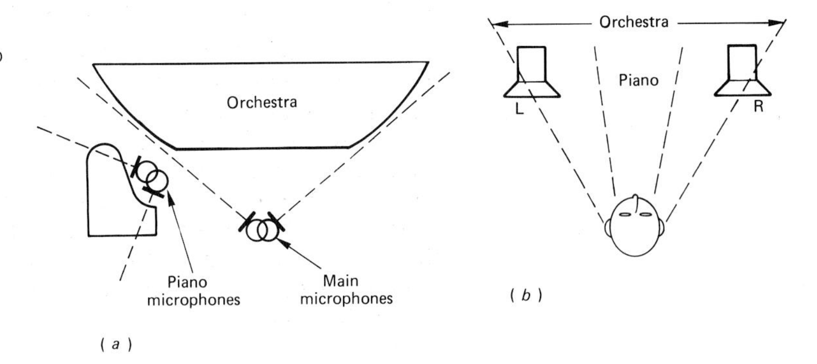

Figure 7.1 Microphone panning to simulate ideal stereo layout. (a) Piano awkwardly placed at the side: (b) piano microphones panned to produce a centre image

Note, however, that subtle balance shifting will be needed to recreate the normally expected layout as heard by the listener. For example, if a cramped studio makes it necessary to locate the piano in a concerto along a side wall, a special microphone (or preferably a stereo pair) should be used and panned as necessary to recreate the normal central image (see Figure 7.1).

7.1.3 Distance and depth of perspective

No performers should appear to be nearer than they would from a normal audience listening position. Indeed, since proper use of microphones can give a considerable degree of control over this 'apparent distance', the engineer might as well have in mind the ideal seat in the auditorium, having a more or less idealized range of effective distances from the front- and back-row performers, and seek to simulate that situation.

For many reasons, including microphone directivity patterns and room acoustics, the perceived distance from performer to listener is not the same as the actual performer/microphone distance. This is easily confirmed by a simple experiment. Choose a good seat in a concert hall and place a microphone (or stereo pair) there. Almost always the sounds picked up will be of disappointing quality and make the musicians appear much too far away, with the front-to-back perspective (between violins and percussion, for example) unnatural.

It follows that to capture the correct apparent distance may require considerable experiment combined with experience. For example. as was discussed in Section 4.5, a microphone with a figure-of-eight polar pattern picks up only about one-third of the total reverberant sound whereas an omnidirectional microphone responds equally in all directions (except at high frequencies, where the polar pattern generally narrows). Therefore to reproduce the same apparent ambience (ratio of direct to reverberant sound) the figure-of-eight

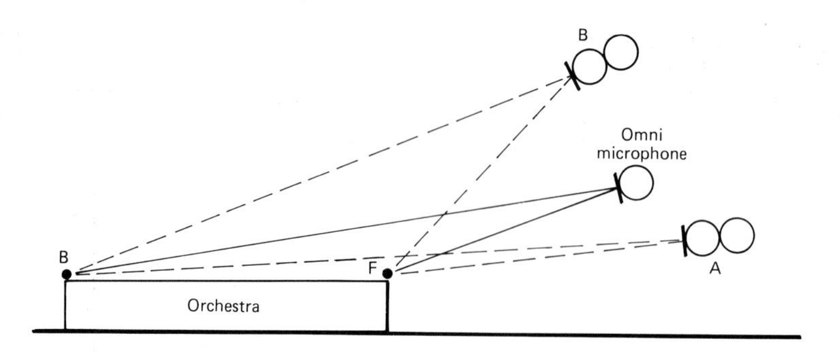

Figure 7.2 Maintaining depth of perspective with different microphone types. The omni and fig-8A microphones reproduce the same apparent direct/reverberant sound ratio but the latter has less depth of perspective. This can be increased by raising the microphone to position fig-8B

microphone should be placed about 1.7 times further away than an omni (see Figure 7.2). However, this may seriously alter the apparent front-to-back perspective; i.e. the effective distance between the microphone and the front- and back-row musicians. Therefore a change of microphone type may call for an adjustment not only in the working distance but also in the height, if the chosen depth of perspective is to be preserved. In the case of stereo, there is the further problem of maintaining the optimum recording angle when the physical distance is substantially altered.

Table 7.1 Factors restricting the choice of microphone distance

1. Existence of background noise
2. Proximity of other instruments or voices
3. Length of true reverberation
4. Tonal quality of the reverberation
5. Smoothness of the reverberation
6. Directivity of instruments
7. Directivity of microphone
8. Mechanical noise of instrument
9. Proximity effect (with pressure-gradient microphones)

Finding the best microphone position to meet these interdependent distance/perspective/stereo width requirements is certainly as much an art as a science. In practical situations there are many factors which restrict a free choice of actual microphone distance. Some of these are listed in Table 7.1.

7.1.4 Stereo spread

Except for listeners using some form of pseudo surround sound playback system, and, of course, mono installations, the assumed loudspeaker layout is the standard stereophonic one, with the two loudspeakers and the listener located at the corners of an equal-sided (equilateral) triangle. This makes the loudspeakers subtend a 60° angle to the listener, as described in Section 5.1 (Figure 5.1) and fixes the maximum width of the listener's left/right stereo stage accordingly.

It is therefore standard practice for engineers and producers to monitor their recordings and broadcasts in this way (see Figure 7.3). The perceived stereo spread within this imposed 60° restriction

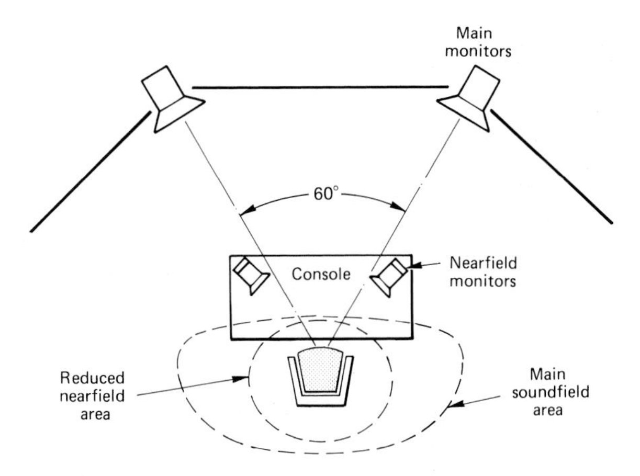

Figure 7.3 Conventional control room layout for stereo monitoring. Using the alternative near-field monitors gives a reduced area in which the stereo effect can be properly assessed

should be appropriate to an idealized concert location. Thus a full orchestra will usually be made to appear spread across the whole 60° arc. This would simulate, for example, a seat about 20 m back from the platform for an orchestra with a physical spread of 25 m. In fact the orchestra subtends a much narrower angle when viewed from most seats in the auditorium, but the enhancement of the width to 60° can be regarded as a legitimate cheat in that it helps to compensate stereo listeners for their lack of visual clues by giving them a 'privileged seat', dead centre and near the front.

In the case of smaller ensembles, such as a string quartet, the real-life angular spread presented to the audience is usually much less than 60°, and the microphone balance should reflect this situation. The reverberant sound may indeed spread over the whole speaker-to-speaker arc, but the musicians should appear at a natural lateral distance from each other.

The same convention of naturalness applies to all classical music balances, with the additional requirement in many cases that 'best use' of the stereo stage should be made. For example, it generally sounds wrong or even perverse to locate the voice in a solo recital anywhere other than dead centre. Again, in oratorio, it is usually best to follow convention and locate the soprano, contralto, tenor and bass voices in that order from left to right, even when another physical layout has been used for some reason, and to make them occupy about the same restricted angular arc as they would present live.

7.1.5 Acoustic setting

The performance should appear to be taking place in appropriate, if possible ideal, surroundings. For example, a large concert hall acoustic must be used, or simulated by appropriate microphone technique, for a romantic or twentieth-century symphonic work; most church music sounds best in reverberant surroundings; an idealized 'music room' acoustic suits most solo recitals and chamber works, and so on.

It is clearly advantageous to record in a hall or studio where the given acoustics are already suitable for the works to be performed.

Fine tuning of the acoustic effect will still be needed for every assignment, but at least the problems and compromises in microphone technique will be minimal. Such ideal recording venues are surprisingly few in number but, once identified, they tend to be used regularly. It is not enough for the hall dimensions and basic reverberation time to be right. The reverberant sound must be evenly diffused and consistent over the whole frequency spectrum. Traffic, aircraft and air-conditioning noises must be low and such convenience factors as an adequate monitoring room, catering and car parking are all very important.

Whatever the prevailing acoustic environment, the apparent ambience as heard by the listener can be varied to a considerable extent by the recording technique used. Obviously, for example, too dry an acoustic can be opened out by judicious use of artificial reverberation. This used to be a crude hit-or-miss procedure, feeding the microphone signals to special hard-walled echo chambers or reverberant springs or plates. Today's digital processors give much more subtle control of acoustic sweetening both tonally and in terms of added reverberant warmth.

Even so, the engineer is well advised to get as close to the desired acoustic effect as possible by microphone technique alone. The perceived effect, as mentioned earlier, is basically a function of the ratio of direct-to-indirect sound reaching the microphone. Simply increasing or decreasing the microphone/musician distance therefore produces a considerable change in apparent acoustic liveliness as the direct sound falls or rises in value while the indirect sound remains virtually unchanged. This is illustrated in Figure 7.4, where a given microphone distance in (a) produces a direct sound plus reverberant sound curve as shown at (b). If the microphone is then moved closer, as in (c), its output will appear as in (d) with a higher level of direct sound but essentially an unchanged amount of ambient sound. Then reducing the channel gain to produce the same peak level reading as

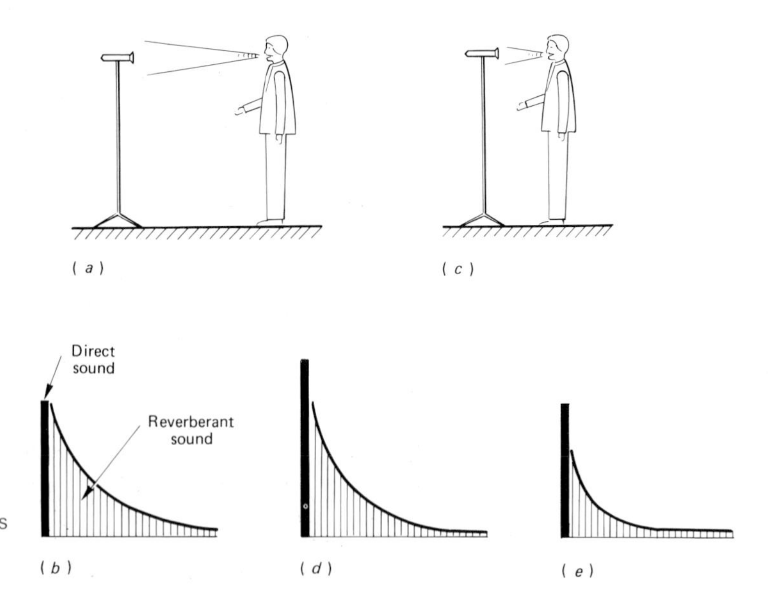

Figure 7.4 'Apparent reverberation' is a function of microphone distance (see text)

before will proportionately reduce the contribution of reverberant sound to the whole, as shown at (e).

A convenient method of adding reverberation when the real or apparent acoustics resulting from a close microphone balance (perhaps imposed by considerations of extraneous noise) are too dead is to introduce extra 'atmosphere' microphones (see Figure 7.5). These can be some distance from the musicians and angled away from them to pick up mainly reflected sounds, or they can comprise a coincident or near-coincident stereo pair adjusted for a narrow recording angle. Mixing in just a small contribution of this ambient sound avoids the complications of external processors and has the advantage that the quality of the added sound matches the indirect sound picked up by the main microphones.

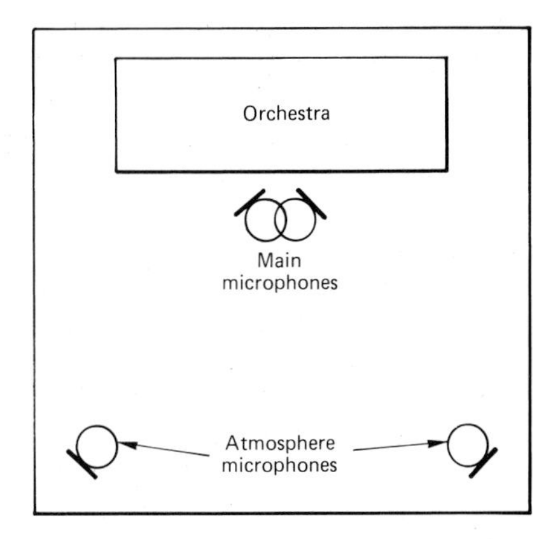

Figure 7.5 Atmosphere microphones: used to add a controlled amount of reverberant sound

As in all cases where extra microphones are introduced, it becomes important to present all the musicians as if they are performing not only in a proper musical, perspective and left/right balance (as already discussed) but also in the same acoustic setting. For example, where a solo voice is given a separate microphone, the relative level of this must be carefully adjusted to conceal the fact – otherwise the singer may appear to be enclosed in a smaller (and less reverberant) setting than the accompanying instruments.

7.1.6 Dynamic range

The measured dynamic range for typical musical forces and media are shown in Figure 2.16. Classical music in general covers a wide range, up to about 80 dB, from the quietest pianissimo which the performers feel will be just audible to the audience under the given ambient noise conditions to the full power which the particular piece of music seems to demand. In the case of the technically superior media such as digital recording and FM radio, all but the most demanding works for large forces can be accommodated. It would therefore seem a simple matter for engineers to determine accurately the optimum gain setting before the recordings begin (perhaps by asking the orchestra to play a few bars from one of the loudest passages) and set the faders for maximum reading on the level meters. They can then record with 'hands off', making no further adjustments.

Within reason, this technique is now quite commonly used for works of moderate dynamic range – up to, say, 60 dB. Yet the domestic listening situation is so different from that in the concert hall that listeners may actually prefer something less than the full dynamic range enjoyed at a live concert. For one thing, members of a concert audience are located at a considerable distance from the loudest instruments and can actually see the brass or percussion players preparing to play in loud passages. Second, the sound peaks build up more rapidly in the confines of a small room and give a subjectively greater impression of loudness than in the greater spaces of a large hall. Third, there are restrictions in the highest sound levels tolerable in most home settings due to considerations of neighbours and the ability of the given reproducing system to handle peak levels without clipping, i.e. running into severe distortion. Also the lowest audible sound levels may be restricted by the presence of a higher ambient noise 'floor' in a typical urban home.

This imposition of the need for some degree of level compression has been traditionally recognized by recording engineers. In the case of the vinyl long-playing record, for example, the limitations of the medium itself have obliged the engineers to limit the peak level and/or raise the quietest passages to conform with inherent groove and surface noise restrictions. The more limited media such as AM radio, telecommunications and background in-store music have always demanded quite severe compression to a mere 20 dB or even less.

The greater freedom theoretically offered by compact discs and FM radio cannot usually be enjoyed in many listening situations (for example, in-car, where ambient noise levels are unpleasantly high). Even listeners with high-quality home systems have been known to complain when the full dynamic capabilities have been exploited on CD. Unless they are already familiar with a particular composition, they may find it difficult to set their volume control for the opening bars and later feel obliged to change the setting as the music proceeds. Some tempering of the maximum spread in levels may therefore be desirable. This can sometimes be kept to a minimum by choosing a microphone layout which avoids undue extremes in sound level. Otherwise, sensitively applied compression can be introduced automatically using a modern limiter/compressor unit.

Although it goes against the purist 'hands-off' philosophy, experienced engineers even today will often resort to small amounts of manual compression by carefully shifting the gain settings in anticipation of uncomfortably quiet or loud passages in the music. Figure 7.6 illustrates how this might be introduced during a piece of music where a strong peak in level is followed by a long diminuendo sinking below the expected noise floor. During the run-up to the peak, the setting is gradually reduced as shown by the broken line to a point where the peak itself can be handled without clipping. These small adjustments will normally go unnoticed by the listener, partly due to the limited ability of human hearing to detect very small shifts in average level or to make accurate assessments of level at any given instant. They are certainly preferable to a situation where the peak is clipped and thereby severely distorted. After the peak has been successfully negotiated, the engineer can slowly restore the gain

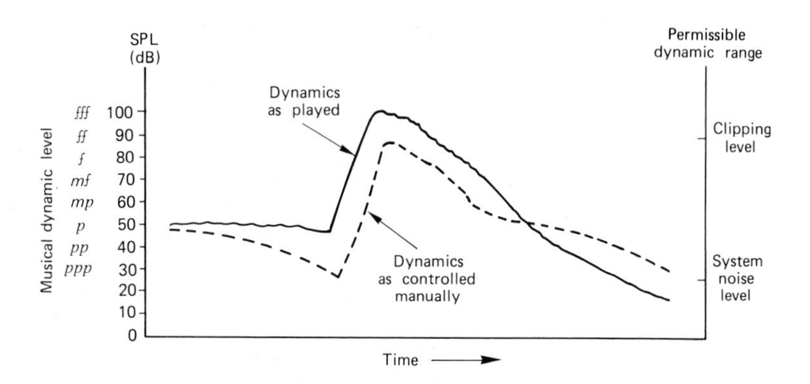

Figure 7.6 Manual compression of dynamic range (see text)

setting to prevent the next quiet passage from getting lost in noise, as shown in the figure.

7.2 Basic guidelines – pop music

Microphone technique for pop music is less restricted in its objectives than that for classical music. The need to conform at least to some extent to the expectations of a natural 'live' concert hall experience does not apply: freedom rules. However, even to work in an unrestricted field demands a positive attitude based on knowledge of what can and cannot be achieved. It is therefore possible to outline a number of basic pop music concepts using the same six perceived characteristics which formed the headings in our guidleines for classical music.

7.2.1 Sound quality or timbre

It is very important to work out and then work to capture the precise sound quality of each voice and instrument that the particular song or album seems to demand. Indeed, many hours may be devoted to getting the 'sound' right both during and after the recording of the individual tracks. This is not to say that the desired tonal quality is necessarily the singer's or player's true natural sound: much more often it will be a carefully engineered mixture of equalization, processing (sweetening), artificial reverberation, etc.

It is a major plus feature that the widely used technique of recording and overdubbing individual tape tracks for a solo voice or instrument, often with no other performer playing, can produce a standard of high-fidelity purity of sound that many classical recording engineers, wrestling with ensembles of instruments, might envy. Yet microphone positioning is often arrived at from other considerations of attack or dynamics rather than timbre fidelity. Even when the microphone signal is a faithful replica of that voice or instrument, the production will often call for further treatment, as mentioned above.

In the early days of one-performer one-track-at-a-time overdub recording, close-miking was invariably used as a kind of hangover from the time when a very tight balance on each instrument or voice, working in acoustically dead studios, was the only way to get good separation from other instruments playing at the same time. However, fashions change, and the cleaner, more open sound that can be picked up with a slightly more distant microphone technique has now become popular. This has led to studios being designed more and more with a 'live room' or at least a live-end/dead-end construction.

For established vocalists and instrumentalists, of course, there will be a known sound which should normally be the aim on later recordings – or deliberately departed from only when the new production suggests this. Artistes often use a particular type of microphone and technique as one sure way of keeping their sound the same.

Electronic keyboards, guitars, synthesizers, samplers and instruments using contact microphones, whose signals are amplified and reproduced through loudspeakers, are a special case as far as timbre is concerned. The recording engineer has a number of options:

1. *Ignore the speaker sound*, using no microphone at all but simply taking a direct feed of the electrical output signal. This is called direct injection (DI), and at least has the advantage that the studio acoustics and outside noises are completely excluded. Indeed, DI instruments can be set up in the control room for maximum liaison with the engineer and producer as the final balance is worked out. Of course, DI gives a very dead sound, except where the instrument has built-in reverb, but this can easily be livened up to taste using artificial reverberation and other processing such as double-tracking or digital delay.
2. *Concentrate on the speaker sound*, though this is often of indifferent quality and suffers from inherent amplifier noise and variable directional effects, calling for special care and experimentation when close-miking.
3. *Use both DI and live microphone sound*, either mixing the two immediately or recording onto separate tracks for later rebalancing at mixdown.

7.2.2 Musical balance

Freedom reigns here too. A pop music arrangement is invariably put together with microphones and processors in mind. Therefore quiet instruments can be made to dominate; a solo flute can compete on equal terms with a full brass section or a whispered vocal can be made louder than everyone else. Backing vocals can be made to sound loud and yet avoid masking of the soloist, for example, by introducing a measure of middle-frequency attenuation. (In any case, as we shall discuss in Section 7.2.6, dynamic range is usually deliberately restricted.) Slightly boosting the 'presence' frequency band, from about 2 kHz to 8 kHz, has the opposite effect of emphasizing or bringing forward voices. In short, every musical line can be treated as a separate entity and either made to take the lead or be set to some chosen accompaniment level to taste.

This freedom of choice in perceived loudness balance is more difficult to achieve when several performers are playing or singing together on stage or in the studio unless each is individually and tightly miked. The major requirement becomes 'separation', with as little spill as possible of unwanted sources onto each microphone. Keeping the drum sound out of a vocal or acoustic guitar microphone, for example, may be virtually impossible even when they are spaced far apart and screened or put in a special booth.

It was this need for total separation which led to the popularity of multitrack recording (as opposed to multimicrophone ensemble recording) mentioned in the previous section. Each vocal or instrumental contribution (usually treating vocal groups, strings or brass sections as one) can then be performed 'solo' and recorded onto a separate track or stereo pair of tracks. Previous tracks, including a click-track or guide vocal, can be listened to on 'foldback' headphones or carefully placed loudspeakers, and the new track recorded at peak level for later balancing and individual treatment at the mixdown stage (see Chapter 11).

7.2.3 Distance and depth of perspective

Deep perspective effects do not feature in the majority of pop recordings – nor are they used much at live pop concerts. Everyone tends to be close-miked and balanced purely on subjective loudness terms, with a more or less flat perspective. The spatial effect, when it exists, is achieved by subtle use of EQ, stereo panning and reverberation added to what is initially a very close and up-front sound.

7.2.4 Stereo spread

Though many people listen to pop music in mono, from AM radio and demonstration videos, for example, the added stimulus and involvement which stereo reproduction can provide is a significant factor in all pop recording. Again a purely naturalistic sound is only occasionally adopted. For example, a two-microphone balance onto two tracks is very common on piano but the microphones may be only a few centimetres from the strings or soundboard, making a truly natural stereo effect impossible, and the final stereo balance may rely solely on panning to provide a bass/treble divide. Similarly, the component parts of a drumkit are often recorded onto separate tracks and subsequently panned individually to produce the required left/right spread. Excessive width can be a mistake, however, if it becomes too obvious that the drummer could not possibly reach such a wide spread.

That said, planning the perceived stereo 'stage' for any song can be a long and involved creative process. With each music line on a single mono track (or occasionally laid down in genuine stereo) the engineer can experiment with any left/right placement and even, for example, pan a voice to left of centre and its reverb or echo to the right. The best starting point is a rough plan of the kind illustrated in Figure 7.7. A trial mix could position the musicians roughly to correspond to the plan and then lead to successive refinements as other decisions relating to musical balance, etc. are reached (see Chapter 11).

Mobile panning, which makes vocals or instruments swing about between the loudspeakers, has occasionally been used to good effect

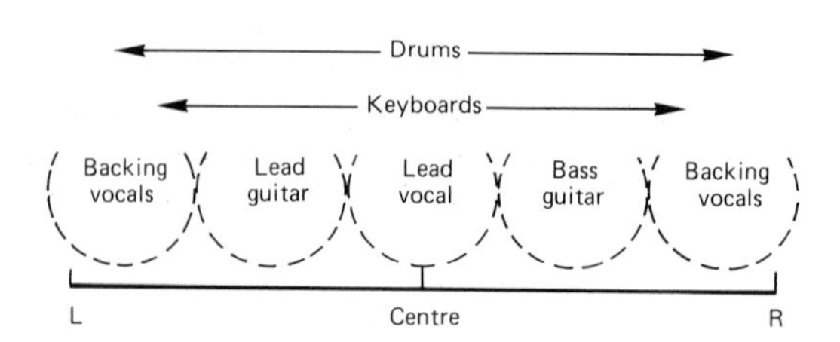

Figure 7.7 Typical stereo location plan for a pop recording

but can easily be overdone and may cause phase problems in mono listening. Now that so many listeners are using headphones, it may be advisable to monitor on headphones at some stage during a sensitive stereo mixdown to see whether any exaggerated spread produces undesirable effects.

7.2.5 Acoustic setting

The prevailing fashion for close-miking in pop music has meant that, even now, studio design normally concentrates on a dead, resonance-free and low ambient-noise environment. The creation of an apparently reverberant ambience on the recording is therefore a matter of introducing echo chamber sounds or relaying some of the direct sound through reverberation plates or digital processors. Vocals will almost always need added reverb and this is also true for strings, woodwind and brass, which can sound thin and edgy when close-miked without some added sweetening or acoustic opening out. Of course, this is less true of the modern studios with a 'live room' capability in which a well-integrated acoustic balance can be obtained naturally.

In the same way as for studio recordings, live albums generally call for some degree of treatment after the event to recreate a sense of space and the full experience of being in an audience. In addition, special microphones can be rigged at the concert venue to record atmosphere or audience reaction tracks. None of these stratagems should be regarded as unethical, provided the principal recording content is the performance itself. They are simply steps taken to give depth and width to a recorded event at which, on the night, there were problems enough in ensuring that all the performers were captured on tape at something close to their correct level and sound quality. (Live recording is discussed in more detail in Section 7.3.6.)

7.2.6 Dynamic range

A major difference between the philosophies applied to classical and pop music relates to the question of dynamic range: what should be the maximum extent of the level differences between the quietest and loudest passages? For classical music, as we have seen, the ultimate aim is to let the perceived dynamic range approach as close to concert hall realism as the medium and the expected listening environment will allow.

In the main, however, pop music is seen as an accompaniment to dancing, or for enjoying on the move on headphones, in-car, on AM radio – all situations where wide swings between quiet and loud passages are unwelcome. Live pop shows make much use of amplification and may often produce higher sound-pressure levels throughout the audience area than the unassisted instruments of even the largest symphony orchestra. Yet they seldom move down the level scale to match the true pianissimos met in classical music. Indeed, the rhythm section habitually registers the same peak levels from start to finish, which therefore ties the vocals and other instruments to a similarly restricted range of dynamics.

All this can be confirmed by watching the level meters during a pop concert or recording session when no more than a 20 dB or even 10 dB range is usually indicated. Limiters, compressors and noise gates are frequently employed to help contain the music within this prescribed target range. Early versions of these dynamic range processors produced a saturation or crushed effect on peaks, with pumping or 'breathing' of the residual noise as levels were automatically restored after any brief cut-back in gain as the circuit ducked down for each loud peak. Fortunately, modern dynamic processor units are more sophisticated, and the required degree of compression can be achieved with very little in the way of telltale side effects.

7.3 Practical matters

7.3.1. Before the artistes arrive

The parable of the wise virgins has a direct relevance to the art of microphone balance. Even when the exact numbers of the performers and their individual layout and microphone preferences are known, it pays to anticipate all possible hardware requirements in advance of their arrival. The studio or stage should be prepared and keyboards and other heavy instruments put in approximate positions along with chairs and music stands. Microphone cables should be run out, allowing for a few spares and looped to permit further extension if necesary. The same applies to foldback headphones or loudspeaker cables, cue lights, intercom telephone, etc. (see Figure 7.8).

The microphones should be plugged in and set to their approximate positions. If some are to be slung high or are otherwise difficult to reach, they should be included in a thorough identification and quality check while still accessible. This is sometimes called the 'scratch' test, because it traditionally consists of getting an assistant to scratch the grille of each microphone in turn and speak into it some identification such as 'piano mike 2'. This procedure is physically rough and ready and hardly appropriate for some of today's more sensitive microphones. Instead, the gentlest tapping should be enough to tell the engineer at the console that he has faded up the correct microphone (whereas simply speaking an identification could be misleading if picked up at about equal level on more than one microphone). Then a few words spoken at a consistent distance will provide a brief check that microphone quality and sensitivity are satisfactory. Gentle cable movement will reveal any plug/socket problems and a quick listen will show if hum or interference are present. When large distances are involved, as when one engineer is adjusting microphone slings from a gallery or through a false ceiling and another is setting the precise microphone position and angle, two-way radio intercoms may be necessary.

Figure 7.8 Before the artistes arrive. Typical studio scene prior to start of a film score recording. Note all microphones and foldback headphones already put in position. (Courtesy CTS Studios, London)

When all microphones have been checked and suitably labelled on the mixer board, a general phase check may be carried out (see Section 6.1.2), fading up microphones in pairs and listening for maximum bass as the phase-reversal switch on the microphone channel is clicked in and out. This phase check may be unnecessary when all the microphones are of often-used types, but it becomes of increasing importance when new microphones are in use and for stereo pairs – as is the proper identification of the left and right channels.

7.3.2 Preliminary balancing

As soon as even a few of the musicians begin to take their seats and tune up, it becomes possible to save valuable session/rehearsal time by carrying out preliminary balance and quality checks – listening to the sound of various instruments on their 'wanted' microphones as well as the degree of unwanted spill onto all other nearby microphones. This can be followed by some judicious relocation and angling of microphones, plus the adjustment of channel level presets, panpots, etc.

When all the players are in position, it will pay dividends later if they can be persuaded to play short key passages of music for preliminary balance purposes. For example, the optimum gain setting (at least as important in digital recording as in analogue) can be determined by asking the musicians to play one of the loudest passages. Internal balance of a brass or saxophone section can be set by asking them to play a built-up chord. Vocal groups and soloists can similarly be balanced before the performance starts in earnest, so that the first run-through can already be approximately balanced with all major problems anticipated and, to some extent, solved.

7.3.3 Monitoring

The monitoring function is central to the art of microphone balance, yet there are many factors which combine to make it almost impossible to monitor reproduced sound quality with a high degree of accuracy or consistency. Even such psychological factors as tiredness can affect one's judgement. Engineers and producers must therefore make the best use they can of the given monitoring situation, and this means being continually aware of the main causes of inaccuracy. These include:

1. Loudspeaker frequency response on axis
2. Loudspeaker directionality
3. Listening level
4. Room reverberation time/frequency response
5. Room resonances (standing waves).

This is not the place to discuss the choosing of loudspeakers and other monitoring hardware in any detail. However, a proper attitude to the monitoring function does suggest a few basic rules. For example, it has been proved that engineers and producers carrying out balance or mixdown operations with non-linear loudspeakers or acoustically coloured control rooms will produce tapes with a mirror-image response – overbright loudspeakers (or rooms) will produce dull tapes and vice versa.

Therefore Rule 1 would seem to be: use loudspeakers with as flat and smoothly extended a frequency response as possible. Locate them correctly in a well-designed room free of response anomalies, and sit in a central position where best judgements can be reached as to balance and stereo effect. This presupposes that the engineer and the producer will normally sit close together or change places when some critical point is being checked – otherwise they must trust each other's ears.

Rule 2 must surely relate to the monitoring level – and a realization that marked changes in tonal balance and the extent to which one instrument masks another can result from any major change in the monitoring loudness setting. It is well known that the great majority of sound engineers and producers monitor at a much higher sound level than home listeners. The reasons for this are clear enough. The professionals have a complicated job to do. They must feel confident that they will detect any brief musical or technical blemishes during each take, and receive early warning of any intrusion of unwanted noise or interference – whether from aircraft, vehicles, footsteps, air conditioning, a system fault, etc. They may also need to make split-second adjustments to controls or processors on a musical cue while being distracted by their colleagues moving around, setting up tapes or other pieces of apparatus. Therefore high monitoring levels are a necessary evil, always remembering that excessive levels over long periods can adversely affect judgements and even lead to irreversible ear damage.

By contrast, home listeners can generally relax and just enjoy the music. Therefore they do not need high sound levels, and in any case their playback equipment may not be able to reproduce at high levels without distortion. There is also the subjective increase in apparent loudness in a small room, and nervousness about causing annoyance to neighbours, as mentioned earlier.

Given that professional monitoring levels are higher than domestic ones, it is enough to glance at the equal loudness curves shown in Figure 2.15 to appreciate that choosing a 100 dB peak SPL compared with 80 dB will substantially increase the relative loudness of both extreme bass and treble frequencies. Therefore a carefully adjusted tonal balance may appear lacking in bass and treble when heard by the consumer. This loss of bass and treble impact will be most serious in passages where these aspects are critical to the desired effect. It is good practice, therefore, to monitor sample passages from time to time at a lower level, say −20 dB, listening specifically for signs of reduced impact or upsets in musical balance. If the amount of loss is unacceptable it may be decided to lift the bass and treble slightly in the interests of communicating the desired effect to the domestic listener.

This is perhaps a suitable point at which to warn against the use of excessive monitoring levels for long periods of time. Legislation aimed at protecting employees from working conditions which might lead to irreversible hearing damage already exists or should soon appear in many countries. As a guide to permissible upper limits, the figure of 90 dBA/L_{eq} is sometimes quoted. This means any exposure to loud sounds which is the equivalent of 90 dBA SPL over an 8-hour working day. For example, this would mean listening at 120 dBA for no more than 30 s.

Rule 3 relates to the considerable differences between the actual playback equipment, as well as the environment, used by professionals and consumers. Professionals should bear in mind the fact that they are in a favoured position. They are listening to the direct sound from the console or to a first-generation tape through state-of-the-art electronics and, usually, massive loudspeakers capable of very extended bass and treble with little or no overload distortion. They are also listening in specially dedicated rooms which, even if not acoustically perfect, will almost certainly have better insulation against extraneous noises than the average home.

To make at least partial allowance for any anomalies that these very different listening situations will introduce to the final product, the 'domestic level' check suggested above is often best performed on small loudspeakers located on top of the mixer or elsewhere close to the monitoring position. These near-field monitors are shown in Figure 7.3 and can be special mini speakers or typical domestic units. Placed only about 1 m from the listener, they are less subject to the diffuse soundfield reflection components of the room but provide a smaller soundfield area for correct monitoring.

The above discussion relates to the ordinary run-of-the-mill consumer set-up. Yet it should not be forgotten that, perhaps ironically, an important segment of the domestic market includes high-fidelity equipment which is equal or even superior to that

Table 7.2 Quality criteria

The main technical aspects of sound quality to be continually assessed during monitoring are as follows:

1. *Frequency range*: are the bass and treble well extended and in a natural balance?
2. *Dynamic range*: are the minimum and maximum signal levels as widely spaced as the music requires and as the medium will permit?
3. *Distortion*: are the tonal qualities of voices and instruments lifelike and free from harshness on peaks etc?
4. *Noise*: is background noise below audibility?
5. Wow and flutter (applicable only on listening to recorded material): is the pitch of notes true and free from fluctuations?
6. *Internal balance*: is the relative loudness of instruments and voices properly balanced?
7. *Perspective*: do all instruments and voices sound at the correct relative distance?
8. *Stereo spread*: are the instruments and voices properly spaced across the stereo stage?
9. *Ambience*: does the acoustic environment sound consistent and appropriate for the type of music?

employed in many studios and location-recording installations. Designers of esoteric high-end systems for the dedicated audiophile are continually seeking perfection, and their best designs, freed from the professional insistence on ultra-high levels and maximum robustness, do sometimes set new standards of fidelity. Added to this is the increasing availability to the consumer of digital formats like compact discs and digital audio tape, whose signals are almost indistinguishable from the master tape.

Every stage in the recording, processing and manufacturing chain now needs scrupulous monitoring to minimize the intrusion of chair creaks, pages turning, too dry recording (previously a common approach to allow for subsequent loss of definition in the analogue chain), noisy equalizers and processors of all kinds.

Often working at high speed, with no chance of a retake, sound engineers have to use their ears primarily to respond to time cues requiring some manual adjustment to the controls. At the same time they have to switch their attention continually between various quality criteria, subconsciously checking that each is satisfactory. Some of these criteria are listed in Table 7.2.

7.3.4 Checking takes

The various procedures relating to mixdown sessions, editing, post-production and manufacture of vinyl discs, CDs and cassettes are beyond the scope of this book. However, the balance engineer should be aware of them and plan microphone techniques as far as possible to avoid later problems with peak levels, dynamic range, equalization, etc.

An important opportunity for checking such details occurs when the artistes join the producer in the control room to listen to individual takes. Most of the attention will be given to artistic matters such as late entries, fluffed notes or tempo errors. Then an immediate decision can be reached on whether a whole passage needs to be re-recorded or just a few bars, while the musicians are still assembled. However, of equal urgency is the technical acceptability of the various quality criteria listed in Table 7.2 – recorded ambience, perspectives, stereo placement and the rest. These must be checked very carefully during this playback session and remedied before it is too late. Thus it may be necessary to ask for a retake for purely technical reasons after some small microphone or level adjustment has been made, even when everyone is happy with the particular take from the musical performance point of view.

7.3.5 Keeping a log

It is standard practice to keep a careful log of take numbers and timings at recording sessions. This may simply consist of basic details of track numbers and start/stop timings, with or without comments on the good and bad points within each take, as an aid to subsequent editing or playback operations. It may even be short enough to be written on the tape box label or an insert sheet.

Alternatively, the session may be of enough complexity to warrant the logging of full information on mixer control settings and studio layout. A computer-assisted mixing console avoids the need for handwritten logging by providing instant recall of panel set-ups. With less exotic mixers it is a great time saver to print up standard log-sheets which can be entered up in a few seconds. Figure 7.9 shows a typical example covering up to 16 channels and four group faders. The studio floor plan log-sheet shown in Figure 7.10 is a further memory aid which can be filled in very rapidly. The completed forms can subsequently be put on file and are then invaluable when a later session takes place, using the same musicians and microphone set-up, perhaps with a different team of engineers. Systematic log-keeping using standard forms is particularly important on location recordings, as discussed in the next section.

7.3.6 Location recording

Recording or broadcasting assignments away from base will call for all the same practical considerations just discussed. In addition, however, there is the important task of drawing up a checklist of all items of equipment that will be needed to cover the assignment. This will also need to anticipate problems with equipment failure, and all known characteristics of the venue which will affect the provision of power supplies, lengths of cable runs, monitoring room acoustics, etc.

Many professional recording organizations keep a file of log-sheets and floor plans of the type described above covering all past location sessions. These will be used as a reference by engineers and producers whenever a previously visited venue is to be used again. When a new venue is scheduled, a special reconnaissance visit should be arranged if time and distance permit. Blank log-sheets and layout floor-plans can be taken along, including the special location recording log type illustrated in Figure 7.11. Then, armed with this survey information or as much of it as can be obtained on the spot (or by telephone) from the venue caretaker, a full-scale plan of action can be drawn up covering the items listed in Table 7.3.

Table 7.3 Planning a location recording

1. Study the venue	– acoustically – electrically – geographically – other services, food, etc.
2. Study the music	– the scores (durations) – special instruments or effects – musical objectives
3. Study the artistes	– good and bad points – times available – transport and accommodation
4. Plan rehearsal/recording schedule	– for minimum waste of artistes' time and energy – for clearest cueing/editing of tapes
5. Order materials	– scores – special instruments – screens or rostra
6. Plan equipment	– mixer, mains supply – loudspeakers, headphones – tape machines – tape – microphones – stands/slings – cue-lights, talkback – closed-circuit TV camera/monitor – cables – transport

7.3.7 Microphone care and testing

Most microphones give every appearance of being extremely robust and, of course, some of them really are – particularly those designed for stage use, handheld vocals, etc. Yet it should never be forgotten that they are precision instruments and liable to produce poor quality or fail completely if they are badly treated. Moving-coil (dynamic) microphones are the most robust among professional types, and so they are a first choice where physical rough-handling is a possibility or unduly high sound levels inside a bass drum. The fact that they contain a powerful magnet means that they should be kept a few inches apart when possible and certainly not placed next to tape boxes. Ribbon microphones have the same magnetic features and are in fact relatively fragile. The ultra-light ribbons cannot withstand rapid movement or physical shocks, and so some care is needed in rigging and transporting them. Condenser microphones have particularly sensitive diaphragm/capsule assemblies and, of course, built-in electronics which demand a certain amount of care.

It is best to set up all stands, pianos and other heavy items before putting the microphones in position, and to return the microphones to their protective cases as soon as the session is finished. On location this quick gathering up of all microphones makes good sense from the anti-theft point of view as well as avoiding possible damage. Keeping microphones in their cases when not in use is a protection against dust and accidents. There can still be problems with

(*Overleaf*)
Figure 7.9 Typical recording log sheet
Figure 7.10 Typical studio floor plan sheet
Figure 7.11 Typical location recording log sheet

Recording Log

DATE :

STUDIO :

ENGINEERS :

TITLE :

ARTIST :

TAPE Nos.:

CHANNELS		1	2	3	4	5	6	7	8	9	10	11	12	13	14	15	16	Groups	1	2	3	4
MIC. TYPE/LINE																		Fader Setting				
LOCATION (see over)																		Limiter 1/2				
PRE-SET GAIN																		Threshold				
TREBLE SETTING																		Recovery				
PRESENCE	FREQUENCY																	Compressor 1/2				
	SETTING																	Ratio				
BASS	FREQUENCY																	Threshold				
	SETTING																	Recovery				
HIGH PASS FREQUENCY																		Echo Send 1/2				
ECHO	PRE / POST																	Echo Gain				
	GAIN																	Echo Return				
TO GROUPS																		Echo Gain				
SIGNAL PAN.																		Echo Pan				
SIGNAL FADER SETTING																		Plate Time				

TAPE MACHINES :

TRACKS :

SPEED :

EQ :

STUDIO FLOOR PLAN	MICROPHONES (Height, Angle Etc.)	
	1	
	2	
	3	
	4	
	5	
	6	
	7	
	8	
	9	
	10	
	11	
	12	
	13	
	14	
	15	
	16	

LOCATION RECORDING LOG

LOCATION ADDRESS	
RECORDING DETAILS	DATE: ARTISTS: ENGINEER:
HALL/CONTACT	NAME: TELEPHONE:
ACCESS/PARKING	
HALL	DIMENSIONS: OTHER DETAILS (Stage seating, etc.)
ROOM FOR MONITORING	DIMENSIONS: OTHER DETAILS:
MAINS FACILITIES	
MIC. POINTS	
OTHER FACILITIES	
ACOUSTIC PROPERTIES	
CABLE ROUTING	
MICROPHONE PLACEMENT (RESTRICTIONS RE STANDS OR SLINGING)	
THE RECORDING	BRIEF DESCRIPTION OF SPECIAL PROBLEMS OR DIFFICULTIES: ASSESSMENT OF SUCCESS OR FAILURE:

condensation or moisture on close vocals, and electronic failure is always a remote possibility. Therefore regular check-ups should be carried out.

The 'scratch test' described earlier as a means of identifying all the microphones set out for a recording or stage show is also designed to check for gross losses in sensitivity or sound quality, but more detailed test procedures are needed from time to time. These could clearly be left to maintenance personnel and involve test equipment and perhaps an anechoic test chamber. However, many situations can arise where balance engineers should run their own tests not just to see if each microphone is working properly but also to make instant comparisons between one type and another or evaluate new models. Basic guidelines for mono and stereo testing are as follows:

Checking microphone quality (mono)

1. *Visual inspection*: examine microphones for any signs of damage, extending such examination to all cables and connectors, where 90% of all problems arise.
2. *Specification check*: study the manufacturers' literature if necessary to see what performance to expect.
3. *Microphone set-up*: position at least two microphones side by side at head height, but not so close together as to introduce problems with reflections/diffraction. A two-microphone comparison is recommended even when only one microphone is being tested, and the other is used as a 'control' reference. The control microphone does not need to be a model of perfection as long as it is known to be performing to specification.
4. *Choice of test sound source*: the spoken word makes an excellent test signal, therefore it may be enough to ask an assistant to speak or read at, say, half a metre from each microphone in turn, or from an equidistant point while the channels are switched or faded up one at a time. Otherwise a solo instrument could be played, a bunch of keys shaken or a loudspeaker used to reproduce suitable programme material – speech, music or white noise.
5. *Sensitivity test*: watching the level meters will quickly establish whether the microphones have identical sensitivity or differ by an amount which agrees fairly closely with the values quoted in their printed specifications. If the microphone under test is significantly below specification it should be handed over to maintenance for internal inspection and cleaning or returned to the manufacturer.
6. *Built-in switches test*: a similar routine can be used to check the performance of any built-in switches such as attenuators, bass-cut filters or polar pattern changeover. As well as checking meter readings, a listening test should be made on all aspects of sound quality.
7. *Directivity test*: the source position or microphone angle should be changed in, say, 30° steps while the meter readings and sound quality are checked.
8. *Special applications test*: additional checks can be carried out on the microphone's suitability for particular applications. For example, a microphone intended for handheld close-miking on vocals should be checked for its ability to reject breath noise, popping and handling noise.

Checking microphone quality (stereo)

It is often more convenient to check and experiment with different stereo pair configurations away from the real-life situation where the artistes are usually anxious to get on with rehearsal or recording. Therefore a logical test procedure is described below, and can be applied either to two suitably disposed mono microphones or a twin-capsule stereo model.

1. *Individual microphone test*: where necessary, carry out the relevant checks outlined above on each microphone capsule.
2. *Monitor set-up*: make sure that the monitor loudspeakers subtend a 60° angle to the listening position and are correctly balanced in terms of output level. This is best checked by feeding a mono signal to both monitor lines and adjusting the gains for a true centre image.
3. *Studio set-up*: position screens or curtains to make the studio as acoustically dead as possible. Choose and mark a central spot for the microphone stand and draw a 2-m radius chalk circle on the floor around this spot.
4. *Microphone set-up*: position the microphones at about head height and select the polar patterns for the stereo configuration to be tested: e.g. coincident XY, M–S, spaced or near-coincident.
5. *Phase test*: check that microphones and connectors are in-phase; for example, by aiming both forwards and speaking centrally to the pair, preferably with (A + B) sent to just one loudspeaker. An out-of-phase connection (A − B) will be severely attenuated, especially at low frequencies.
6. *Channel balance*: identify the microphones respectively connected to the Left and Right channel faders, for example by setting them at 90° to each other and speaking directly to each in turn. Then balance the gain on the two faders by speaking centrally to the microphone pair and either producing identical peak level readings or a clear central image. Alternatively, switch one channel to reverse phase, send both channels to one loudspeaker and set the faders for a null (minimum signal).
7. *Recording angle test*: the maximum acceptance angle which produces the standard reproduction angle of 60°, using the particular stereo configuration being tested, can be checked by asking an assistant to speak or rattle keys, etc. at the central position and then as the assistant moves round the marked circle to left and right until the image just moves into the centre of the appropriate loudspeaker. The angle through which the assistant has moved is the efffective recording angle. It should line up with theory for the basic coincident pair or spaced pair systems outlined in Chapter 5, but other microphone types and angles can be investigated and logged for future reference.
8. *Sound-qualty test*: as well as checking the basic directivity of a stereo pair, this set-up can be used as a subtle test of the sound quality to be expected on random sound at oblique angles outside the intended recording angle. Any serious loss of high frequencies or unnatural phase effects will usually show up very clearly, whereas they may be difficult to identify in an actual session.

8
Musical instruments and the microphone

8.1 The instrument as a sound source

Musical sounds can in general be distinguished from random noise by the fact that they are organized in terms of one or more of the properties pitch, rhythm, intensity and colour or timbre. Opponents of contemporary (*avant-garde*) music may complain that some recent compositions possess none of these qualities in any recognizably organized way, but we can widen our definition to include all kinds of music by stating 'musical sounds give pleasure (at least to someone) and noise annoys (nearly everyone)'. It follows that musical instruments are designed to produce pleasing, or at least recognizable, sounds and that the player has some control over the various properties of the sounds that emerge.

The essential components of a musical instrument are one or more vibrators/resonators, some means of exciting the vibrators into motion and a radiator capable of efficiently transforming the vibratory energy into airborne sound waves. In a few cases the vibrator and radiator are one and the same: e.g. tubular bells and cymbals. More often, the vibrators are separate entities such as strings or air columns, and the radiator is a composite of the body of the instrument and sometimes an attached soundboard. The vibrator will usually possess a natural resonant frequency which will determine the fundamental pitch of the note being played, and will simultaneously generate a family of harmonics or overtones, as discussed in Section 2.6. However, the radiator component of the instrument is seldom a passive transmitter of this original mix of fundamental and harmonics; it tends to be frequency selective and to respond more energetically to certain bands of frequencies. It is therefore constructed in such a way as to reinforce the middle register of the instrument's compass rather like a broad bandpass electrical filter. Where a specific relatively narrow band of frequencies is boosted, this region of pitch is called a formant, and is specially relevant to the human voice (see Chapter 9) and many string instruments.

Of course, as described in Chapter 2, the numbers and relative strengths of the harmonics – as modified by the characteristics of the radiator – contribute much to the feeling of colour or timbre of the instrument. (If two instruments played the same note without harmonics, the result would be an identical sinewave in each case and the sounds would be indistinguishable – not to say musically uninteresting.) It should be noted, however, that the mechanism of sound generation is more complicated than this simple description would suggest. For one thing, the overtones are often non-harmonic. This is the case for all string instruments, for example, since the effective length is not constant but varies as the string itself stretches on being struck, plucked or bowed. Also the string supports, and especially the bridge, are not completely rigid but must flex to some extent to communicate the vibrations to the rest of the instrument. Again, because the stiffness of the string has a direct bearing on the frequency of vibration (and stiffness increases with frequency), the higher harmonics become increasingly sharp with respect to the fundamental. In wind instruments, too, the effective length for the various harmonics alters with frequency due to the so-called 'end correction', which is a fixed function of the radius at the end opening.

A further factor which increases the richness of real-life sounds is that there is an initial transient as each individual note is sounded – even for the sustained notes of a bowed string or steadily blown clarinet. During the initial period when this transient vibration remains loud enough to be heard, it adds appreciably to the character of the instrument. Listeners often rely on their ability to detect this initial sound component to follow the presence of relatively quiet instruments in a large ensemble. Recreating this highly individual start-up sound component by electronic means is quite difficult, and this explains why many synthesizers or electronic organs cannot ideally reproduce particular instrument timbres.

8.2 Characteristics affecting microphone technique

Listeners at a live performance have the subconscious ability to make continuous allowances for physical imbalances in the sound reaching their particular seats, and can mentally replace many of the sound elements that may have become lost on the way. When microphones are placed in position as surrogate or substitute ears, not only are the listeners to the reproduced sounds from loudspeakers robbed of any visual clues but they are forced to make do with a fixed aural viewpoint which may convey only a restricted and therefore unrealistic pattern of sound cues. There are several principal characteristics of a musical instrument's behaviour which need to be taken into account when planning a microphone balance:

1. Frequency range
2. Directivity
3. Dynamic range
4. Mechanical noise.

Each of these will be examined briefly before the individual families of instruments are discussed.

8.2.1 Frequency range

The nominal pitch range or compass of each instrument tells us the range of fundamental frequencies which it can radiate (see Figure 2.9). To this must be added the often considerable upward extension contributed by the harmonics or overtones. String instruments, for example, have important harmonic content up to the limits of human hearing and beyond. Percussion instruments certainly owe much of their impact to their extended high-frequency components.

To capture these sounds with absolute fidelity, the chosen microphone should possess as flat a response as possible, at least up to the generally accepted upper limit of 20 kHz. Also, the microphone should be positioned with special attention to the short wavelength directional properties of both the microphone and the instrument in question (see Section 8.2.2 below). At the bass end of the spectrum it will be seen from Figure 2.9 that most musical instruments do not extend much below 60 Hz. Exceptions include the tuba, contra-bassoon, piano and, of course, the organ, which can in some cases produce tones down to 16 Hz. Many microphones, whatever their nominal directivity pattern, become approximately omnidirectional at very low frequencies, and studio screens are similarly ineffective as barriers to long-wavelength sounds. Therefore in any multi-microphone situation it can be assumed that bass instruments will spill over onto the unwanted as well as the wanted microphones, and bass cut may be needed in some of the microphone channels.

To a certain extent, the need for the microphone and the subsequent reproducing chain of equipment to respond to these very low frequencies is relieved by one of the more subtle subjective properties of human hearing. This is the ear's ability to synthesize and thus 'hear' a low fundamental tone, which may in fact be completely missing from the reproduced sound, by a process of recognizing the characteristic pattern of the harmonics. The ear is helped in this by its non-linearity, which generates the fundamental as the difference-tone between any two adjacent harmonics. The effect is called the *subjective fundamental*, and explains why quite small loudspeakers can often sound better at the bass end than their physical size would suggest possible. Of course, true audiophiles will accept no substitute, and may build enormous labyrinthine loudspeakers to get down to that elusive 16 Hz, though in fact many of the discs and tapes they buy will have been rolled off at 30 Hz or higher to clean up unavoidable motor rumble or other subsonic interference.

The spectrum analysis of sound-pressure levels contained in octave or narrow one-third octave bands during several minutes of a piece of music are a good indication of the frequency range covered and the relative amplitudes of low, middle and high frequencies. Figure 8.1 shows a typical result for a rock band and it is found that classical music also contains reduced energy in the high-frequency region. It is also common knowledge that, as musical sounds decay, the upper harmonics tend to die away at a faster rate than the lower components. This happens in piano tone, for example, because the string itself smooths out as it comes to rest. Reverberant sound similarly loses treble with time, due to air losses and the predominantly high-frequency absorption in furnishings, etc. However, it

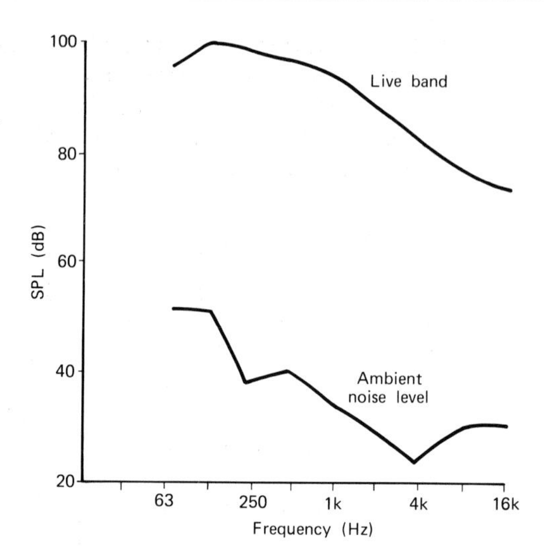

Figure 8.1 Typical octave-band spectrum for a rock band, measured on the dance floor, showing relative fall at high frequencies (after Cabot, Center and Lucke)

should not be deduced from this that the high harmonics are comparatively unimportant. Their contributions to transient impact, timbre, instrument recognition and 'musical fidelity' are vitally important. By a selective hearing mechanism the final tail of sound decay remains audible even when it falls below the level of the ambient noise 'floor'.

8.2.2 Directivity

Because of their often complex shapes, many musical instruments radiate sounds in a directional manner which varies so much with frequency as almost to defy analysis. Considerable experience is therefore needed – plus a live balance rehearsal – if the angular relationship between a solo instrument and the microphone is to be adjusted for optimum pick-up of the peak energy and the characteristic timbre.

An intelligent guess as to an instrument's basic directional radiation behaviour can, of course, be made from a knowledge of the range of fundamental frequencies (and therefore wavelength λ) which it produces in relationship to its physical diameter D. As discussed in Chapter 2, so long as D/λ is less than about 0.1, the instrument will radiate almost equally in all directions. As D/λ increases, for higher frequencies, the radiation pattern progressively narrows. For the highest frequencies the radiated energy will be confined to a narrow beam centred on the main axis. Positioning a microphone on this axis may seem the obvious approach for best balance. However, some instruments such as the violin comprise twin radiating surfaces (the front and back) separated by a time lag, and so the true axis is not the physical one but is skewed at some angle which varies with frequency.

As a rule, putting the microphone precisely on-axis does introduce problems since the highest overtones and such unwanted accompaniments as finger squeaks from a guitar or wind noise from a trumpet may appear on-axis at a much higher relative level than they are usually heard by the audience. It also presupposes that the performer will keep the instrument virtually motionless while playing: otherwise any slight movement may make a considerable difference to both level and timbre, particularly in the case of close-miking. For these various reasons a slightly off-axis microphone position is usually best, at least where classical music is concerned.

8.2.3 Dynamic range

Musical instruments are naturally limited in the loudest and quietest sound levels that they can produce under normal playing conditions. Figure 8.2 shows the dynamic ranges for typical instruments at a distance of about 3 m (10 ft). As a very broad guide, it can be said that woodwind instruments can sound about twice as loud as strings (+ 10 dB) and brass instruments about 5–10 dB louder still. This accords well with the normal orchestral layout placing the louder instruments progressively further from the audience. Most instruments can be made to sound quieter than shown, but, in practice,

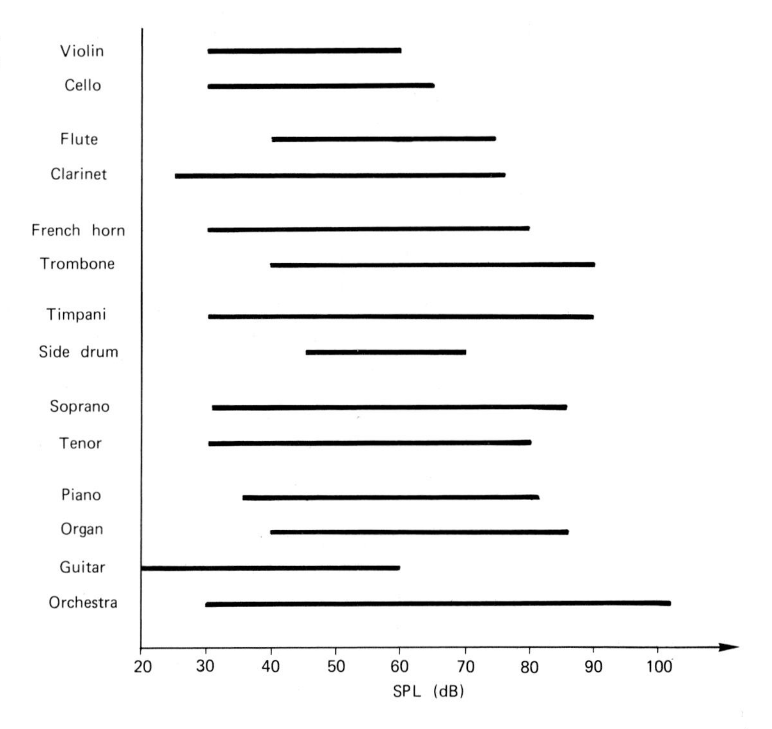

Figure 8.2 Dynamic range of orchestra and various individual instruments and voices

players tend to set the minimum loudness at what they estimate will be just audible to the audience in the given ambient noise conditions. In ensemble playing the musicians strive for the most appropriate and pleasing balance in response to the dynamic markings in the score. Similarly, the composer, arranger and conductor use their knowledge of each instrument's intensity range to map out the most effective balance and the rise and fall of dynamic tension. A distant microphone technique at least has the advantage that it should be able to reproduce the full dynamics and balance without further adjustments.

However, a closer microphone can have the effect of exaggerating the dynamic range and upsetting the internal balance by favouring some instruments at the expense of others. Also, some instruments produce transient peak intensities far above their average levels, and a microphone's inability to handle such peaks without overload distortion may set a limit to the minimum working distance for distortion-free pick-up.

8.2.4 Mechanical noise

Some instruments, and some performers, unfortunately generate a variety of incidental mechanical noises, sniffs or grunts, and microphone placement needs to take note of these possibilities and keep them to a minimum. Their nuisance value is generally greater on reproduced music, when they may be heard repeatedly rather than fleetingly at a live performance. They are also more audible on compact disc, for example, due to the very low level of inherent system noise. Sources of this problem include pedal action on piano, key rattling on wind instruments, page turning and foot stamping by the conductor. Where no microphone technique can be found to alleviate the problem, the musicians may be asked diplomatically to help arrive at a cure.

8.3 String instruments

Stretched strings are used as the tunable vibrating element in many musical instruments. Clearly, however, a vibrating string is a poor radiator of sound waves since it will tend to cut through the air and set very few air particles in motion. An efficient radiator must therefore be coupled to the strings and will usually have an important effect on the instrument's tonal quality and ease of articulation or 'speaking' as well as its intensity. The fundamental resonant frequency of a stretched string is given by the expression:

$$f = \frac{1}{2l}\sqrt{\frac{t}{m}}$$

where l = length of the string,
t = tension, and
m = mass.

This summarizes the observable facts that the frequency, and therefore the musical pitch, can be increased:

1. By reducing its length (e.g. by stopping with the fingers)
2. By increasing the tension (e.g. by tightening the peg)
3. By reducing its mass (e.g. using thinner wire).

8.3.1 Controlling string quality

In theory, a vibrating string produces notes which are rich in overtones with the full harmonic series present and each higher-numbered harmonic having reduced amplitude. In practice, however, these simple conditions seldom apply.

For example, varying the method of attack (which sets the string in vibration) obviously produces a change in timbre. The actual shape of the string prior to release gives a hint as to the brightness or dullness of tone. A guitar string being plucked with a sharp plectrum takes up a more triangular shape than one plucked with a soft finger, and a higher proportion of the radiated energy is therefore thrown into the upper harmonics. The same thing happens if the hardness of the covering on piano hammers is increased to give a brighter tone. In fact, striking the piano keys with more force so that the hammers attack the strings more rapidly makes their covering 'appear' harder. As a result, the initial kink in the string is again narrower, and therefore loud passages are not only higher in average intensity but also in harmonic content. This helps to explain why a recording of the piano sounds disappointing unless it is reproduced at something close to its natural sound-pressure level.

A bowed string, unlike the plucked or struck string, is forced into continuous vibrations by the alternating grip and slip of the resin-loaded bow hairs. This has a tendency to produce overtones which are exact harmonics, despite the effects of string stiffness and yielding of the end supports which normally introduce non-harmonic overtones to the free vibrations which follow plucking or striking.

The point of attack also has a marked effect on string timbre. It obviously inhibits the formation of a node (zero amplitude) at the attack point and can diminish or eliminate completely the harmonics which require a node there. Players normally pluck or bow the string at about one-seventh (0.12) of its length from the bridge so as to silence the dissonant seventh, ninth and eleventh harmonics, whose nodes occur at 0.14, 0.11 and 0.09 from the end. Coincidentally, this is an effective point of attack for maximum efficiency of energy transfer to the string. However, a deliberate shift to nearer the bridge (*sul ponticello*) is sometimes made to produce a brighter quality. More bowing pressure is then needed to achieve the same sound intensity.

8.3.2 The violin

The violin consists of arched top- and backplates joined near their perimeters by short side panels. The four strings are stretched between a tail piece and pegs at the end of the fingerboard which are turned to adjust string tension during tuning. Vibrations of the strings are conveyed to the topplate through a bridge (see Figure 8.3) whose treble foot is supported from inside the body by a soundpost or rod which also acts to transmit vibrations to the backplate. Shaped f-holes further influence the vibration properties of the topplate and allow vibrations of the contained air to contribute to the total sound.

Figure 8.3 Close-up view of violin showing the bridge, f-holes and typical use of a clip-on miniature microphone. (Courtesy Shure)

The complex shape of the resonating system and the frequency-dependent paths from the string to the plates produce uniquely varied timbre from note to note and between the same note played on different strings. Adjustments in bowing pressure, point of attack and vibrato (introduced by rocking the left hand at about 6 Hz) add further colour changes. Clipping a mute on the bridge increases its mass, which has the effect of both reducing the sound intensity and lowering the natural frequencies which favours the fundamental at the expense of the upper harmonics.

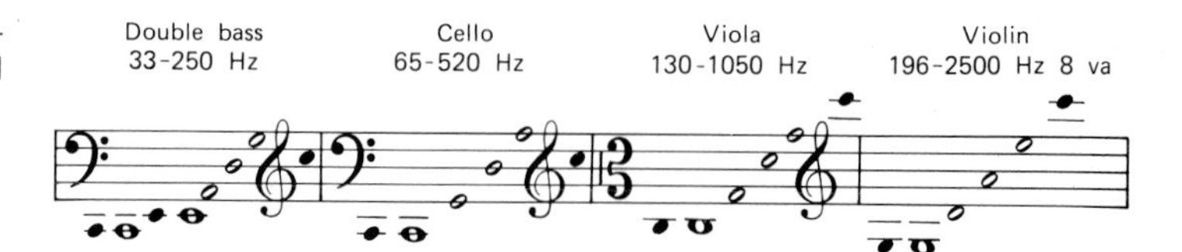

Figure 8.4 String instruments, showing the tuning of individual strings and the range of fundamental notes covered (solid notes)

The violin's four strings are tuned at intervals of a perfect fifth (see Figure 8.4). Shortening the effective string length with the fingers of the left hand (stopping) gives a range of over four octaves, a fundamental frequency range of about 196 Hz to 3136 Hz (wavelengths 175 to 11 cm). Since the resonating violin body is about 32 cm long by 18 cm wide, its radiation pattern is almost spherical for the lowest couple of octaves and then progressively narrower on higher notes and harmonics. As both the top- and backplates act as radiators, the system resembles an acoustic doublet at some frequencies producing a roughly figure-of-eight pattern. At higher frequencies the heavier backplate contributes less energy and the frontal lobes become dominant, partly assisted by the obstacle created by the player's body.

The playing angle directs most of the high-frequency sounds upwards and to the player's right. Microphone placement should ideally be in this region at, say, 2–3 m distance and 2–3 m high, depending on the ambient conditions – the shorter distance in a very reverberant location, a longer one in a very dry studio. If a close position is adopted, trial balances should be carried out to avoid any lobes of markedly uneven frequency response and to check on any problems with performer movement while playing.

Jazz violinists are, of course, adept at holding the instrument within a few centimetres of the microphone. Clip-on and contact microphones (Figure 8.3) can be used for live shows to allow the artiste to move around and yet avoid feedback problems from the reinforcement or foldback loudspeakers.

8.3.3 The viola

The viola is basically a larger-scaled version of the violin, having a more veiled tone, with the body length typically increased from 32 to about 40 cm, a ratio of 4:5. The four strings are tuned a fifth lower than the violin (Figure 8.4), which suggests that the correct dimensional increase to give the viola identical directivity should be 4:6 (see Figure 8.5). This makes the viola less markedly directional than the violin, and less critical as to its position in the orchestra or string quartet. Microphone angle is also less critical as is the viola's orientation with respect to the audience.

8.3.4 The violincello

The cello has a warm velvet-like tone. Its strings are tuned a full octave lower than the viola (Figure 8.4), which would suggest that it should be twice the size to give comparable directional characteristics (see Figure 8.5). In fact, the typical cello body length is about 56 cm, a scaling-up ratio of only about 5:8. Once again, therefore, we have reduced directivity in the lower register and the situation is changed

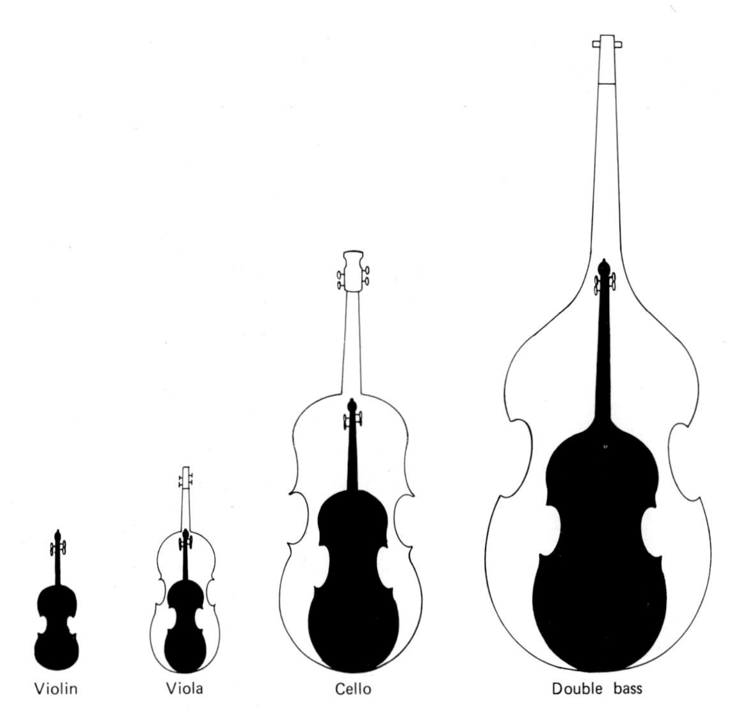

Figure 8.5 String instruments, showing the actual size (shaded) and the increased sizes which would be needed for all instruments to have equivalent directivity

further by the low playing position and the fact that the instrument's spike transmits vibrations to the floor. Image-source reflections from the floor complicate the directivity pattern and can influence the choice of microphone position for best musical balance. A polished wood or stone floor will reflect sounds so efficiently as to create an interference pattern, and a raised box or rostrum, sometimes provided for a cello soloist, may introduce undesirable resonance effects unless heavily damped.

8.3.5 The double bass (contrabass or bass viol)

The fourth and largest member of the orchestral string section has the sloping shoulders and other minor differences which confirm that it is descended from the viol rather than the violin family. Its four strings are tuned in fourths rather than fifths (Figure 8.4), covering frequencies all the way down to 41 Hz (wavelength 8.39 m). The instrument's dimensions, about 137 × 80 cm, are therefore relatively small in relation to the wavelength over much of its fundamental range (see Figure 8.5), though it is just about as large and unwieldy as any practical orchestral instrument can be. Much of the sound energy is radiated over a very wide arc, making the instrument behave as a diffuse rather than a well-defined source, and further diffusion is introduced by floor reflections and vibrations.

Microphone angle is relatively uncritical, though achieving the right degree of definition or bite without moving in so close that perspectives and balance with other instruments are upset can present problems. It helps that orchestral basses are often arranged in a block and so present a more easily captured composite sound source. The double bass in a jazz or pop group is more often plucked than bowed, and, of course, close-miked for optimum separation and attack. A free-standing microphone near the bridge or f-holes works well, or a clip-on or contact type similarly located.

8.3.6 The harp

The harp used in modern orchestras has 48 strings of different lengths stretched on a triangular frame and tuned to cover all the notes of a major scale (like the white notes of a piano) over a range of more than six octaves. Versatile tuning to the intermediate semitones is provided by seven pedals. The soundboard forms the side of the instrument closest to the player and is so small as to be a relatively inefficient and non-directional radiator. One consequence of this is that the notes of the harp die away comparatively slowly. Another is that the microphone position is uncritical, though, for a solo recital where no other instruments are involved, it is worth exploring the region above and even behind the player in search of the optimum tonal balance with rich overtones. Noise from the pedal action makes a very close balance difficult.

8.3.7 The guitar

The classical acoustic guitar has almost flat top and back wooden plates with a round sound-hole forming a warmly resonating structure. The six strings are tuned to the notes E2, A2, D3, G3, B3 and E4. (Other tunings may be used in pop music, such as a G-based one.) They are stretched between a tail piece and tuning pegs over a bridge which rests over an internal sound post, as in the violin. One important difference is that the guitar fingerboard has a series of 22 parallel raised strips called frets. These are spaced so that stopping with the finger on or slightly behind each higher fret has the effect of raising the pitch in exact semitone intervals, whereas the violin has no frets and the player must learn by experience where to place the fingers for each note. The 12–string guitar has six pairs of strings, the top two pairs being tuned in unison and the others in octaves.

Most of the high-frequency energy is radiated in a fairly narrow lobe at right angles to the topplate. An on-axis microphone position will therefore give the most brilliant tone, and this can be modified to increase the resonant lower tones by moving off-axis. A very close balance is often chosen for live performance and somewhere near the bridge is usually best, if noise from the fingers or plectrum (pick) is not excessive. Clip-on or contact microphones (pick-ups) are also common (Figure 8.6) and can be coupled to a wireless transmitter to allow complete freedom of movement around the stage.

Figure 8.6 View of guitar showing possible use of a clip-on microphone. (Courtesy Shure)

Today there are many variations on the traditional mellow-toned gut- or nylon-stringed acoustic guitar construction. Larger steel-stringed guitars, invariably played with a sharp plectrum, produce a much more powerful sound. The Hawaiian guitar has provision for raising the strings above the frets and then a metal bar can be run along the strings to produce the familiar gliding sound.

The *electric guitar* can produce almost unlimited sound levels. Yet it abandons all attempts to radiate efficient or even well-balanced acoustic waves and the instrument body can be reduced to a mere

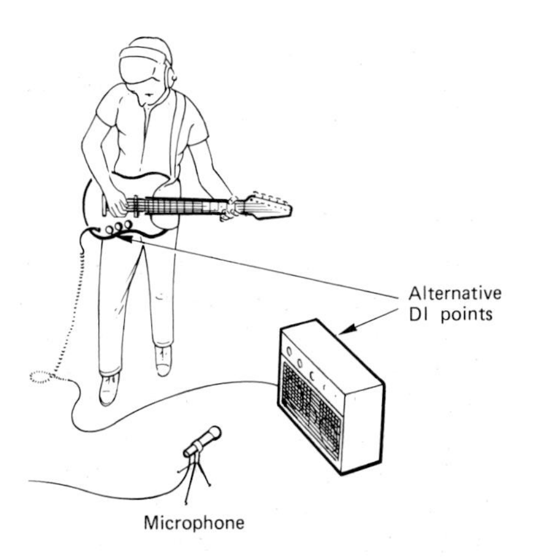

Figure 8.7 Electric guitar amp/loudspeaker, showing alternative microphone (live) or direct injection (DI) techniques

skeleton in a variety of shapes. The solid-body types have the advantage that they are much less prone to acoustic feedback (howl-round) from the loudspeaker than the hybrid acoustic-electric hollow-body guitars. A built-in contact microphone or pick-up transducer is wired to a guitar amplifier. This may be a large on-stage system, but for studio recording it is usually a fairly small amplifier and loudspeaker combined in a single box (Figure 8.7). Early guitar amp units often produced a boomy bass with unwelcome amounts of mains hum and amplifier hiss. Much more sophisticated units are now common, which makes close microphone balance of the speaker sound perfectly feasible. Experiment is needed to seek the best position with respect to the loudspeaker cone – on-axis for brightest tone, off-axis for reduced treble and hiss (almost always present to some degree). Good-quality moving-coil microphones are best for close working, at, say, 15 cm or less, with condenser microphones substituted (or mixed in) at one or more metres distance. The various trick effects need to be checked for microphone balance; e.g. wah-wah pedal, fuzz box, sustain and synthesizer.

The rather coloured sound from an electric guitar amplifier (even with suitable EQ) may be just what is wanted. Often, however, it is disappointing unless it is turned up to such a high volume that it presents separation problems with other instruments and microphones. This has led to the technique of direct injection (DI), in which the electric signal from the guitar pick-up is recorded or sent to line with no microphone involved (as discussed in Section 7.2.1). Amateur connections are not recommended because of risks of electric shocks or, at a minimum, ground loops and consequent hum interference. Instead, most studios provide ready-made DI boxes which make safe connections between the guitar output or a socket on the guitar amp/speaker and the mixing console. Proper impedance matching and conversion from unbalanced to balanced line is included. Of course, the DI sound is acoustically dry, though distortion- and interference-free, and may need some measure of artificial reverberation as well as EQ. Where something closer to the on-stage sound is preferred, the DI signal can be mixed to taste with a microphone picking up the guitar amp live. As an alternative, the player can wear headphones and switch the amp off completely if it is spilling onto other microphones.

The *electric bass guitar* has replaced the double bass in most modern groups. It may be fretted or unfretted, and covers essentially the same pitch range as the double bass, but provides a variety of effects as well as almost unlimited sound-pressure levels. Direct injection is the preferred recording technique, giving maximum control over the sound quality with minimum reliance on the guitar amp's often inadequate response to high-level deep bass signals. As with the electric guitar, however, a microphone balance may also be set up and mixed with the DI or recorded on separate tracks for later mixing – when some compression may be used to iron out any resonant peaks. If the live sound is being used, it may be advisable to reduce any bass-lift the player normally uses and apply suitable boost on the control desk if guitar amplifier buzzes and rattles can be heard. This also allows the DI and live tracks to be treated differently; e.g. bass boost on the DI and treble boost on the microphone channel, reverb on one and a limiter on the other.

8.4 Wind instruments: general principles

Orchestral wind instruments are divided into the woodwind and brass groups even though, to be strictly accurate, some woodwind instruments are now made of metal.

A woodwind instrument consists of a narrow tube or pipe, whose effective length determines the pitch of the note being sounded, coupled to a means of interrupting the steady airstream from the player's lips. The interrupting mechanism may be a thin single reed (as in the clarinet and saxophone) or a double reed (oboe and bassoon) or merely a blowing hole or embouchure (flute and piccolo). The initial jet of air sets up standing waves in the contained air column at its fundamental frequency (and harmonics) and the resulting note is sustained by the continuing airstream. For reed instruments the airstream is broken up into a series of bursts by the throttling or constricting action of the vibrating reed. Blowing such a reed in free air produces a screeching sound rich in overtones but, when the reed is coupled to a resonant air column, its vibrating rate is controlled by the latter which also dictates the frequency and timbre of the emitted note. For embouchure instruments the interruption effect is called air-reed behaviour, and results from an enforced shift in airstream direction alternately above and below the hole edge as the internal air pressure builds up and escapes during alternate half-cycles of the air column's natural frequency – the effect familiar to anyone who has blown across the mouth of a bottle.

The resonant frequency of the air column inside a tube or pipe is directly related to its length, as in the case of a stretched string, though there are important differences, depending on whether the pipe is open at both ends or closed at one end and whether it is straight-sided (cylindrical) or flared.

Figure 8.8 summarizes the production of the vibratory modes for the fundamental and lowest harmonics. It will be seen that an open end must always act as an antinode (maximum displacement). This means that an open pipe has a fundamental wavelength equal to $2l$, and the fundamental frequency is given by:

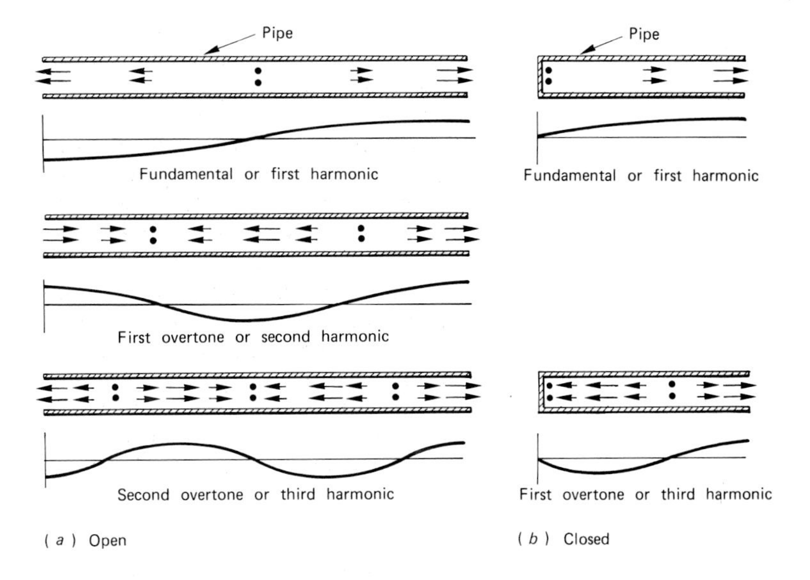

Figure 8.8 Wind instruments, showing vibration modes in (a) open and (b) closed pipes

$$f = \frac{c}{2l}$$

where l = length of the pipe
c = velocity of sound in air

The harmonic series of an open pipe contains all odd and even harmonics: $2f$, $3f$, $4f$, etc.

In the case of a pipe closed at one end, the closed end acts as a node and so the fundamental wavelength is equal to $4l$ and the fundamental frequency is half that of an open pipe of the same length: i.e.

$$f = \frac{c}{4l}$$

The harmonic series of a closed pipe contains only the odd harmonics: $3f$, $5f$, $7f$, etc. The timbre is therefore more hollow and ethereal.

Straight-sided (cylindrical) pipes form the basis of several musical instruments. Pipe organs employ separate pipes cut to appropriate lengths for each note, the pipes being open or closed to suit the timbre required and the space available (closed pipes occupy only half the space needed for open pipes of the same pitch). The woodwind instruments such as the flute, oboe, clarinet and bassoon have a single pipe whose effective length can be varied by uncovering a series of holes cut in the side of the pipe.

As well as straight-sided pipes, with or without a flange at the end, several types of flared pipe or horn are used in which the cross-sectional area expands at a controlled rate. This flare may be linear, giving a conical pipe, or accelerating to give an exponential, parabolic or hyperbolic shaped horn. As a rule, a flared pipe behaves like an open cylindrical pipe of the same length, i.e. its fundamental wavelength is $2l$ and it produces the full harmonic series. It is a feature of flared pipes that, by controlled blowing action, they can be persuaded to resonate and therefore sound the corresponding note at any one of the harmonic series of frequencies. It is in this way that players of posthorns, bugles and cornets can produce a selection of different notes from a single horn of fixed length. The number of playable notes is naturally limited, and widely spaced at the low-frequency end of the scale to the pitch intervals we have come to associate with bugle calls, fanfares and the like.

By a process of evolution the modern family of brass wind instruments has acquired a more versatile means of altering the effective pipe length to produce the full chromatic scale of semitones over several octaves. In the trombone a cylindrical slide is moved to give a telescopic variation in total pipe length covering all possible frequencies between the primary harmonics selected by the player's blowing action. In the trumpet, French horn and tuba the effective length is increased in discrete steps by coupling in short tubes through keyed valves.

8.5 Woodwind instruments

8.5.1 The flute

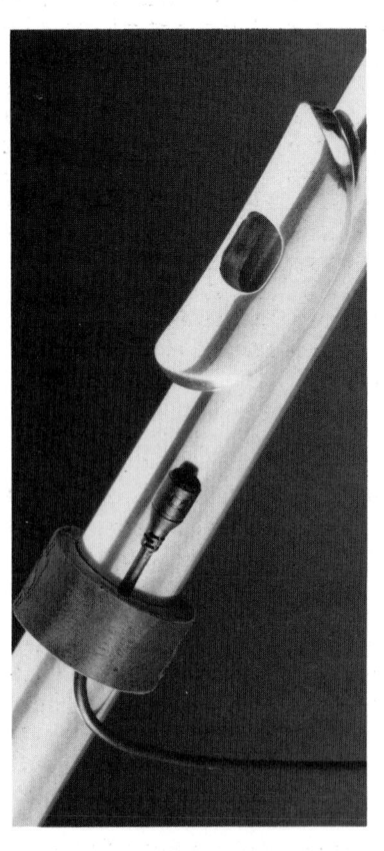

Figure 8.9 Close-up view of flute showing possible use of an attached miniature microphone. (Courtesy Shure)

The transverse flute found in the modern orchestra is regarded as a member of the woodwind family, though it does not possess a reed, as all the others do, and may often be made of silver. It has evolved from a whole range of early instruments mainly held in-line like the recorder and tin whistle. The air-reed principle involves the player in directing a flat jet of air across the blowing-hole or embouchure towards a shaped edge (Figure 8.9). Control of timbre and change of register (a leap to the next octave) depends on the rate and direction of the airstream and the distance of the lips from the far edge. Basically, the flute comprises a straight cylindrical pipe, with a parabolic tapered head and, although the end just beyond the blowing-hole is closed (corked), the system behaves as an open pipe with a fundamental frequency corresponding to $\lambda = 2l$. The usual length is 65.7 cm (overall physical length is 67 cm), making the lowest note middle C, i.e. 261 Hz (see Figure 8.10). Higher notes on a rising scale of an octave are produced by opening the lowest side hole and then successive holes to shorten the resonant pipe length in fixed steps. As suggested above, the next octave or register is substituted by adjusting the blowing technique to align the air-reed action to the air column's second harmonic rather than its fundamental. This process is repeated to cover about three octaves in all. The *piccolo* is a smaller version, measuring about 33 cm overall and tuned about an octave higher.

There are distinct shifts in the radiation pattern for individual notes and overtones between sideways emission and a concentration on the instrument's axis out from the end hole. Close microphone placing at 1 m or less may miss certain partials and boost others. Then, without the reverberation contribution, a distorted tone may be the result. Certainly when the microphone is placed only a few centimetres away (Figure 8.9) intelligent equalizing by someone who knows what the real instrument sounds like may be needed to produce a satisfactory sound.

Figure 8.10 Woodwind instruments: range of fundamental notes covered

8.5.2 The oboe

The oboe is a double-reed instrument with a tapered (conical) tube and slight flaring at the end. Its overall length is 62 cm and it covers a range of about three octaves. Air pressure through the double reed produces a sawtooth waveform containing the full series of harmonics. The fundamental note sounded and the particular pattern of overtones is a function of the resonant length of the air column selected by the opening of side-holes, as described for the flute. There is evidence of a broad formant reinforcement of tone between 600 and 1500 Hz.

The note A4 (440 Hz) played by the oboe is traditionally used for orchestral tuning, though more precision is now sought by means of a 440 Hz electronic tone generator. The oboe tone is rich in harmonics, making it nasal and penetrating but not shrill. As with the flute, there is a smoothing out of timbre when heard at a distance which may be upset at microphone distances of 2 m or less. There is a further complication in that end-fire radiation from the flared end is directed towards the floor and will be reflected up into the listening (microphone) area. Care is needed to avoid oddities in sound balance caused by this reflected wave.

A larger version, the *cor anglais* (English horn), measures 90 cm overall with the reed set into a thin curved pipe and a pear-shaped lower end. The timbre is characteristically reedy and melancholic. It is a transposing instrument, its part being written a fifth higher than it is intended to sound. There is also the *oboe d'Amore,* somewhat intermediate in length (71 cm overall).

8.5.3 The clarinet

The clarinet is an odd man out among orchestral woodwind instruments in that it employs a single reed and behaves as a closed cylindrical pipe, i.e. the register change shifts the pitch upwards by an octave plus a fifth (a twelfth, or to the third harmonic) instead of an octave. This complicates the mechanical construction to incorporate extra keys covering some six more semitones. It also explains the lower pitch of the clarinet compared with the similarly dimensioned oboe, since a closed pipe sounds an octave lower than an open one. Also, the predominance of odd-numbered harmonics (see Figure 8.8) gives the clarinet its characteristically creamy tone – dark in the lowest (chalumeau) register and richly expressive in the upper ones.

The overall length of the most common B-flat clarinet is 64 cm and it covers more than three octaves (Figure 8.10). Note, however, that it is a transposing instrument, i.e. its music is written or notated, for the convenience of the player, a tone higher than the actual note sounded. Hence a written B-flat sounds the note C. There is also an A clarinet (transposing three semitones higher than written) as well as the seldom-used E-flat, D and C models. In addition, a bass clarinet is occasionally used which has a particularly warm, low-pitched tone. To avoid an unwieldy length, the lower end of the tube is bent upwards and bell-shaped, while the top end is bent downwards to present the reed to the player's mouth at a more convenient angle. The bass clarinet is tuned to B-flat but transposes a major ninth below the written notes.

Microphone positioning for the clarinet family follows the same scheme as that for the oboe, there being a certain narrowing of the high-frequency 'window' radiating in the vertical plane in front of the player, complicated slightly by floor reflections. The clarinet has a wider dynamic range than most woodwind instruments. It is very common in jazz ensembles when a close-mike balance will often be suitable, aiming the bell of the instrument to the side of the microphone rather than straight into it.

8.5.4 The bassoon

The bassoon is the bass member of the woodwind family, its tone being plaintive or jocular as required. It has an effective conical tube length of 236 cm which is doubled back on itself to about 123 cm for convenience and fitted with a thin pipe holding the double reed. The wide compass is indicated in Figure 8.10. The radiation becomes increasingly directional at high frequencies with lobes sloping upwards and to the front plus some concentration of overtones on the axis above the open top end.

The *contrabassoon* has approximately twice the effective length of the bassoon, though folded to just 127 cm and having a slightly flared top opening. It sounds an octave below the written notes.

Figure 8.11 Close-up view of saxophone bell with clip-on microphone. (Courtesy Sennheiser)

8.5.5 The saxophone

The saxophone is a hybrid form of instrument having a single reed like the clarinet and a conical flared-end body like the oboe but made of brass. The members of the saxophone family most often met are the B-flat alto, E-flat tenor, E-flat soprano and B-flat baritone – all transposing as their key-names would suggest, and each covering just under three octaves (see range of the alto saxophone in Figure 8.10).

Saxophones have a wide bore which gives them a quick response or attack and produces a smooth singing tone. They are found in wind bands and, of course, jazz music, but only occasionally in the symphony orchestra. In general, the microphone balance gives a more realistic sound from a position looking downwards at the instrument from an angle. If a position near the bell is chosen (Figure 8.11) some EQ will almost certainly be necessary to smooth out the high frequencies.

8.6 Brass instruments

8.6.1 The trumpet

The tube of the modern trumpet has a basic length of about 2 m of which the first part is cylindrical but changes to conical leading to the flared bell. There is a cupped mouthpiece where the player's lips act as a lip-reed to produce a complex airstream which tunes to one of the resonant frequencies of the tube as selected by appropriate lip tension and blowing pressure. This gives the limited range of harmonic notes familiar from bugles and posthorns, but a series of three piston valves can be used to couple in extra tube lengths. This enables the player to choose from eight notes at each playing tension to cover the full chromatic scale over a range of about three octaves (see Figure 8.12). The standard trumpet is in B-flat but other versions exist.

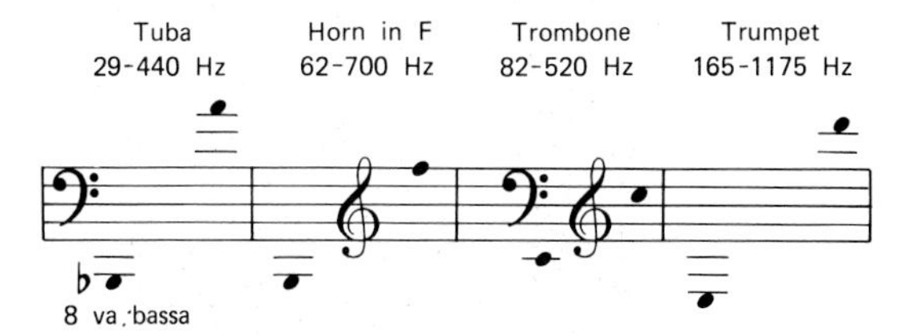

Figure 8.12 Brass instruments. Range of fundamental notes covered

The bell-shaped mouth is the source of most of the sound, which is largely contained within a 90° forward axial cone at high frequencies, tending to become omnidirectional below about 500 Hz. The bright ringing tone, rich in harmonics, is directed fully towards any listeners on the axis, and, of course, to any microphone placed there. However, too close a balance on the axis gives an overbright effect and is liable to fluctuations if the player moves. In such cases an off-axis position may be more consistent, though less liable to cut through the sound of accompanying instruments. Various shapes of mute may be placed in the bell to give a subdued sound or, with harder blowing, a more strident tone. The overall dynamic range of the trumpet is very wide indeed.

8.6.2 The trombone

Two sizes of trombone are found in the modern orchestra, the tenor (Figure 8.12) and the bass covering A1 to Gb4. The cupped mouthpiece is coupled to a cylindrical U-shaped tube via a telescopic slide which provides continuous adjustment of the effective length, down to about 3 m. This allows slides or glissandi as in unfretted string instruments like the violin. The onus is therefore on the player to pitch notes accurately and, except for the valve trombone occasionally seen, requires control of blowing pressure to move up and down the harmonic scale to change register. Directivity is similar to that of the trumpet, the larger bell tending to focus the sound axially down to proportionately lower frequencies. Mutes are also available to provide a choice of special effects.

8.6.3 The tuba

The tuba is the true bass of the brass family, its coiled tube being about 6 m in length and gradually tapered to end in a large bell-shaped mouth pointing upwards. The musical compass is three octaves (Figure 8.12) with four or five valves coupling in extra lengths of tubing to produce the full chromatic scale as in the trumpet.

The large bell gives a fair degree of axial (upwards) directivity, confining the upper harmonics to this region. Placing a microphone above the instrument therefore produces a brighter tone than is generally heard by the audience, and so an off-axis balance is probably safer and more natural. The most common version of the tuba in the orchestra is the bass tuba, but there are sundry close relations such as the Wagner tuba, the euphoniums and the Sousaphone, which has a helical coiled tube terminating in an enormous 60 cm (24 in) diameter bell.

8.6.4 The French horn

The coiled conical tube of the French horn is effectively about 2.7 m in length with a funnel-shaped mouthpiece and widely flared bell. It covers the range shown in Figure 8.12, transposing down a fifth, with valves acting to introduce additional tube lengths or 'crooks' for chromatic tuning. The player can also place a hand inside the bell to raise the pitch by a semitone and simultaneously produce an eerie, muffled quality or use a pear-shaped mute.

As the instrument is pointing backwards when played, much of the bite or attack is directed away from the audience and the main front microphones. This can be ignored in auditorium situations where a proper balance results. If some help is needed, large screens may be set up behind the horn section to reinforce the forward radiation by reflection. Any attempt to achieve a close balance by siting microphones behind the player on the bell axis should be approached with caution as the timbre is breathy and unnatural.

8.7 Percussion instruments

The multitude of instruments making up the percussion family have one thing in common – they involve some form of striking action.

They can be difficult to balance correctly in classical music where the rule about using as few microphones as possible may need to be waived in favour of spot microphones to achieve the best mixture of attack, sonority and perspective. However, such microphones should be used as a last resort, after attempts to produce the desired balance by modifying the player's position or technique have proved ineffective.

Percussion instruments also range from the quietest to the loudest, i.e. they may add the merest subtle pulse to the music at one moment and set the peak sound-pressure level for the whole ensemble the next. It is usual to make a distinction between definite-pitch and indefinite-pitch percussion instruments, though they are generally grouped together physically for the convenience of the players.

8.7.1 Definite-pitch instruments

Even the brief descriptions given here may logically begin with the *tuning fork*, the two-pronged device helpful to piano tuners and others for its ability always to sound the precise note (frequency) for which it has been constructed and labelled. When struck, the fork initially sounds its natural resonance frequency plus a higher 'clang tone', but the latter soon subsides, leaving practically a pure sinewave tone. For better audibility the point of the fork handle is usually rested on a wooden surface or mounted on a resonator box.

Groupings of individually tuned bars of metal or wood form the basis of the *xylophone* and its larger relation, the *marimba*. The bars are laid out horizontally on a frame in two rows broadly in accordance with the black and white notes of a piano and struck with round-topped hard or soft hammers. Each bar is supported at the two nodal points of its fundamental vibration mode and the naturally weak and short-lived sounds are reinforced by tuned resonator pipes located underneath (primitive marimbas traditionally use gourds as resonators). The normal four-octave xylophone covers all the semitones from C3 to E7 and the five-octave marimba from F2 to F7. The *vibraphone* is a modern version of the marimba, whose resonator tubes have rotating lids at the top, kept in constant motion by an electric motor, to produce a sustained pulsating singing tone. The smaller *glockenspiel* (orchestral bells) has a more tinkling sound when struck by dulcimer-type hammers. The *bell lyre* is a vertical form of glockenspiel on a lyre-shaped frame. It is sometimes used in marching bands.

Resonating tubes suspended on strings and directly struck by hammers form the basis of the *tubular bells* (chimes) often used to simulate real bells in the orchestra. Again, the tubes' layout resembles a two-octave piano keyboard and there is a foot-operated damping bar. The *celesta* has an actual keyboard covering the four octaves from C4 to C8. The keys activate individual hammers placed over horizontally laid-out metal bars. Resonator boxes reinforce the sound. There is a damper which mutes each bar when the key is released, unless the sustaining pedal is pressed to release the dampers and allow prolonged vibrations. Natural bells and keyboard-operated groupings of bells called *carillons* are also met occasionally.

The principal definite-pitch percussion instrument is the kettledrum. Groups of two or more *kettledrums* (timpani) in scaled diameters, those most often seen being 58 and 76 cm, consist of a hemispherical bowl or 'kettle' over which is stretched a head of calf-skin or other material. The fundamental frequency of such a circular membrane is proportional to:

$$\frac{1}{D}\sqrt{\frac{T}{d}}$$

where D is the diameter
T is the tension, and
d is the density of the membrane material.

Head screws alter the skin tension for fine tuning, the main two compasses being Bb2 to F3 and F2 to C3. The pedal-timps give preset shifts in tuning to enable the timpanist to play different notes in quick succession and also produce glissandi. Choosing different hammers and positions of striking can give a range of tone qualities: striking at about half-way between the rim and the centre, for example, tends to suppress the fourth partial and weaken some others. Muffled beats or rolls can be produced by placing a cloth on the skin and using soft-headed sticks.

The sound radiation is diffuse rather than directional. When the pick-up on distant main microphones appears to lack definition, perhaps due to wall and floor reflections, a spot microphone (or stereo pair) may be added, but too close a position will run the risk of upsetting tonal balance and peak clipping.

8.7.2 Indefinite-pitch instruments

Percussion instruments which produce sounds with a random noise spectrum rather than identifiable pitch on the musical scale have featured in music making since the earliest times and are still prominent today, adding rhythm and excitement to music of all kinds.

The *triangle* is simply a steel rod bent into a triangular shape which the player suspends from one hand and strikes with a metal beater held in the other. Its high-pitched clang is not particularly loud but it cuts through the general orchestral sound and practically never needs a spot microphone. This is fortunate because the initial transient is ugly in tone when close-miked.

The orchestral *bass drum* consists of a cylindrical shell with a membrane of skin or parchment stretched over one or both ends. It can be anything up to 150 cm in diameter and is usually positioned vertically, the player using large, soft-headed sticks to produce beats or rolls which can range from a thunderous roar down to a gentle tapping. Though the instrument is normally placed edge-on to the audience, it does in fact radiate more efficiently at right angles to the plane of the drumskin in a roughly dipole manner. This has been demonstrated on some Telarc recordings, for example, where the bass drum was turned through 90° to very good effect.

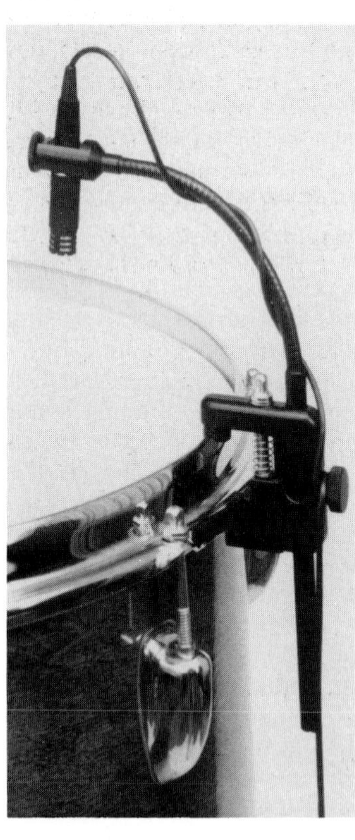

Figure 8.13 Side drum, showing clip-on gooseneck mounting kit. (Courtesy Shure)

Smaller cylindrical drums (Figure 8.13) include the *military drum, tenor drum* and *side drum*. In the special case of the *snare drum* (usually about 36 cm in diameter) cords or 'snares' of catgut are stretched over the lower membrane to add a brilliant rattling effect. The snares can be muted with a handkerchief or wooden wedge (or even taped down) either because the rattling sound is not required or because unwanted sympathetic vibrations are being set up by other instruments in the vicinity. The *tambourine* is a kind of handheld drum with a single membrane and pairs of metal discs or 'jingles' loosely strung on wires within the outer hoop. It can be played by shaking or tapping or rubbing with the thumb to produce a continuous jingling effect.

The *cymbal* is basically a circular metal plate with a sunken rounded portion at the centre from which it is suspended or held by a strap. Orchestral cymbals produce a wide range of brilliant effects, notably when the player holds one cymbal in each hand and either rattles the two edges against each other or clashes them together at some orchestral climax. They can also be struck with hard or soft drumsticks or foot-operated by mounting one fixed cymbal pointing upwards on a stand and bringing another down onto it by pressing a pedal. The *Chinese gong* (Tam Tam) resembles a cymbal turned over at the edge into a dish shape. It is beaten with soft drumsticks to produce single crashes or a sustained roaring drumroll. Tiny *finger cymbals* can be tapped together to give an Oriental tinkling sound. *Castanets* are distinctly Spanish in character and comprise pairs of hollow wood pieces attached to the finger and thumb and rhythmically clacked together. For convenience, orchestral castanets are fastened to the end of a stick and tapped or shaken.

The modern *drumkit* dates back only about 60 years to the early dance-band era but has evolved into a key component in all types of popular music. The individual drums and cymbals have sometimes been stylized almost out of recognition and even replaced by electronic equivalents or samplers. The basic make-up (Figure 8.14) includes a pedal-operated bass drum (kick drum) in front with an array of snare drum, one or more high and low tom-toms, cymbals and hi-hat (pedal-operated cymbals). Each drummer puts together a personalized kit and has usually developed a careful setting-up procedure of tensioning the skins for a given compromise between

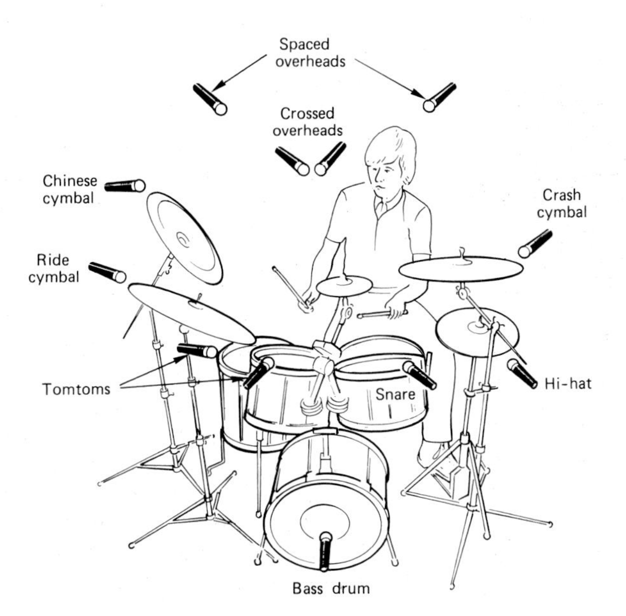

Figure 8.14 Drumkit, showing typical set-up with close-miking and over-head 'ambience' microphones

sharp attack and a more open projection of sound. The procedure for establishing the right microphone balance should begin with a period of listening to the live sound as the player completes the tuning process, and locating moderately distant overhead or 'ambience' microphones (1 m or more) in an attempt to reproduce the drumkit's natural balance and stereo spread. In an ideal world this open balance would become the final one if the acoustic environment and considerations of pick-up of other instruments allowed. It has the advantages of low distortion, avoidance of crushed transients due to microphone overload, low mechanical noise intrusion and a pleasingly integrated sound.

More often than not, these 'ideal' conditions cannot apply and it becomes necessary to adopt an ultra-close-miking technique with separate microphones dedicated to each instrument and placed as near as possible to ensure good separation while avoiding accidental bumps from the instrument or sticks. The front skin is usually removed from the bass drum and a cushion or blanket placed against the back skin. The microphone is then placed inside, using bass-cut if necessary to reduce any bass tip-up and produce the popular tight bass-drum sound. The snare drum very often needs to be taped and the snare tension adjusted to reduce ringing. It must always be borne in mind that deafening SPLs of up to 150 dBA can exist close to percussion instruments, which could easily damage some sensitive condenser or ribbon microphones. Various vintage dynamic microphones have become favourites for this application because of their robustness and smooth overload characteristics. However, some newer microphone designs are also favoured where sharp attack is the prime requirement or small dimensions capable of being clamped near cymbal rims, etc.

8.8 The pianoforte

The piano is at the same time a stringed and a percussion instrument. In terms of tonal colour and dynamic range it is almost orchestral in its scope and we should not be surprised, therefore, if the modern grand piano presents a unique set of problems to the balance engineer.

The development of the piano arose from a general need to produce a keyboard instrument combining the expressive powers of the faint-toned clavichord with the brilliance and force of the harpsichord. The invention is generally credited to Cristofori in Florence, who produced in 1709 what he called a *gravicembalo col piano e forté* (a harpsichord with soft and loud). The main feature was the use of a hammer action whereby the initial loudness could be controlled over a wide range by the force with which the finger-key was struck. An escapement action immediately returned the hammer, leaving the string to vibrate freely until the key was released and a damper moved against the string to suppress the vibration. Numerous developments followed in Germany, England and elsewhere to change the shape and detailed construction.

The modern grand piano commonly has a 1.2 m wide keyboard of 88 keys covering more than seven octaves from A3 to C8 (see Figure 2.9), though some makers, notably Bösendorfer, aim for additional sonority and add up to nine further bass semitones. The strings are stretched across a steel frame and coupled through a bridge to a massive soundboard forming the floor of the entire instrument. Unlike the hollow body of a violin, for example, the piano soundboard is a single wooden sheet reinforced with carefully positioned bars and acts as an efficient means of reinforcing the radiation at all frequencies without introducing specific self-resonances. The strings are graduated in length, somewhat like those of a harp, with extra mass introduced on the lower strings by wrapping wire around them. For the upper five octaves there are three strings tuned to each note and struck simultaneously by the appropriate hammer. The next two lower octaves have two wrapped strings per note and the remaining bass notes have one wrapped string each. The overall width of the piano is about 1.5 m and lengths vary from about 1.5 to 2.7 m (5 to 9 ft).

Three pedals are usually provided. The right pedal holds off the dampers from all strings and is called the sustaining or 'loud' pedal. Not only does this prolong the sound and enable many more notes to sound together than two hands could normally encompass, but it introduces sympathetic vibrations from all the other strings to reinforce fundamentals and harmonics alike. The centre pedal acts as an optional bass-sustaining pedal affecting the dampers of the lowest two octaves only or, in superior designs, giving a sostenuto effect by holding only the dampers of played notes clear of the strings, leaving others to play unsustained if required. The left pedal is called the 'soft' pedal, and acts to reduce the sound level by one of the following methods: (1) allowing the dampers to fall against the strings without waiting for the key to be released, (2) interposing a strip of felt between the hammers and strings, (3) reducing the length of travel of the hammers (in uprights) or (4) shifting the action sideways so that the hammers strike only one string of each note (*una corda*).

In fact the piano cannot produce a truly sustained note in the way that a bowed violin or blown clarinet can, and its notes start to decay from the instant of striking. The timbre also alters as the note dies away, the higher harmonics falling silent more rapidly. Initial timbre varies very much with the playing force applied (though a player's 'touch' is a complex amalgam of holding down and overlapping of notes plus pedalling). For the loudest passages, the felt-covered hammers arrive at the strings at speeds which make them appear harder and produce a kink in the string, throwing more energy into the higher harmonics. In soft passages, by contrast, the lower harmonics predominate. It is for this reason that experienced listeners can always tell when a piano is being played loudly or softly, even from distance away or from a recording reproduced at the wrong level.

Dynamic range can be very wide, up to 50 dB or more (see Figure 8.2), and this makes it almost impossible to reproduce the true range with a close microphone balance – which exaggerates dynamics – though many engineers prefer close-miking. Polar distribution in the horizontal plane of the soundboard or a little above is very variable with frequency. All that can be said is that moving round the piano, at some 'safe' distance like 3–5 m, in an arc from the line of the keyboard to the top end, picks up a tonal balance which starts off as

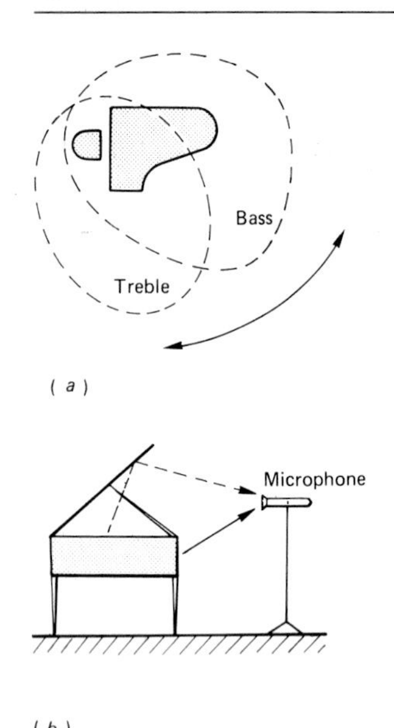

Figure 8.15 Piano directivity. (a) Plan view showing change in treble/bass emphasis; (b) end view showing possible interference from lid reflections

slightly treble-heavy and ends bass-heavy (see Figure 8.15(a)). The position of the lid is critical. With the lid on its short stick, sounds radiate much as just described and the microphone height can be altered with impunity. Opening the lid wide, on the longer stick, gives best musical results to an audience but introduces strong lid reflections which interfere with the direct wave (Figure 8.15(b)). The effect is worst in the plane bisecting the angle between the lid and the soundboard, and may make this particular microphone height unusable.

A sensible balance procedure for a classical piano recital in the studio is to set up the microphone (or stereo pair) about 3–5 m from the instrument, depending on the length and quality of the studio reverberation, and pointing down towards the centre of the soundboard from a height of about 3 m, which seems to avoid problems with lid and floor reflections. Moving in the arc from keyboard to tail will then modify the bass/treble balance to taste. In a live recital, standard presentation usually sets the piano sideways on to the audience but a suitable microphone balance can usually be obtained by adjusting the four parameters of distance, height, angle of tilt and lateral position across the hall on the lines described above.

Close balancing for jazz or rock piano is fraught with problems of dynamics, tonal imbalance and mechanical noise, as we have said, but a working compromise can be obtained with a pair of microphones aimed at the upper and lower strings or frame holes from a few centimetres away. Oddly enough, the acoustic effect is not so dead as might be expected because the soundboard and lid add a degree of 'bloom' to the sound.

Recording other keyboard instruments such as the *upright piano, clavichord, harpsichord* and *spinet* calls for similar sensitivity in microphone technique. Mechanical noise can be more of a nuisance with the softer-voiced instruments, but, in general, a commonsense identification of the main radiating surfaces and openings, plus some careful listening at various angles, will indicate the most promising microphone locations.

Electronic keyboards and synthesizers are now very sophisticated, some being touch-sensitive and therefore able to simulate a natural piano as well as provide numerous effects such as phasing, echo, chorus and tone control/filtering. Traditional reed and pipe organs were joined by electronic types from about 1934 and their evolution has been very rapid. The original Hammond used a motor turning 95 tone-wheels to which was added the Leslie tone cabinet having a two-horned loudspeaker mounted on a revolving spindle – needing special care with microphone balance. All-electronic keyboards will, of course, be recorded using DI or live pick-up of the loudspeaker sound, or a mixture of both, in the same way as was discussed for electric guitars in Section 7.2.1.

9 The human voice

9.1 How the voice works

The mechanism of the human voice may be compared to that of the brass wind instruments but with the addition of a complex control system capable of almost unlimited and subtle variations in pitch, timbre and intensity growth and decay.

The lungs produce a controlled steady flow of air which passes up the windpipe and through the gap between two flap-like folds of muscle known as the vocal cords. These act as a double reed in the same way as a trumpeter's lips. The vocal cords lie in the horizontal plane and are about 1.25 cm or 2 cm long for women and men respectively. They open widely during breathing and narrowly during whispering to produce a rushing or hissing sound akin to white noise. Their edges move together for speech or singing to set up the double-reed vibratory action generating a pulsed waveform at the chosen fundamental frequency and rich in harmonics. This superimposition of a triangular wave on a steady airstream passes through the vocal tract to the head cavities and emerges from the mouth and nose. The physical dimensions of the vocal tract and head cavities will introduce natural resonances or formants but these are subject to extensive timbre modification by muscular action and movements of the tongue, jaw and lips. Vowel sounds are produced in this way with the average fundamental frequency in speech occurring at about 270 Hz for women, 145 Hz for men (see Figure 9.1). The typical ranges for singing voices are also shown in the figure and cover a total compass of about four octaves, each voice having a range of about two octaves. The particular sound quality corresponding to each vowel sound is a function of two (or more) broad resonance regions or formants (quite independent of the fundamental frequency, as is evident from our ability to sing any vowel on any chosen note of the scale). These formants arise principally from tongue movement dividing the mouth cavity into two portions. They introduce a boosting of vocal cord partials in the specific frequency bands for all voices at any pitch: e.g. around 400 and 800 Hz for the 'oo' in pool; 825 and 1200 Hz for the 'ah' in father; 375 and 2400 Hz

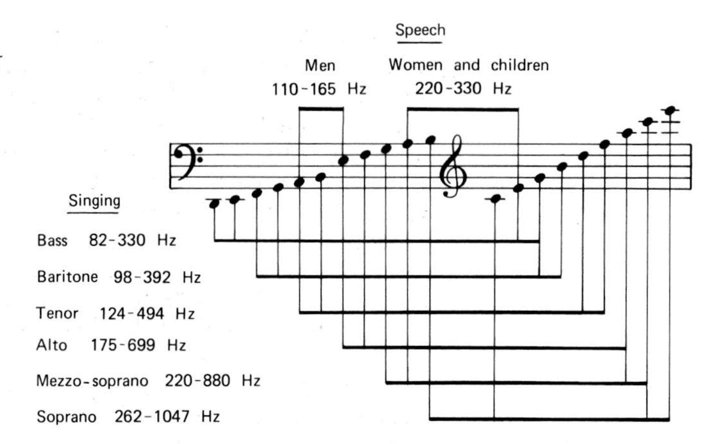

Figure 9.1 The human voice, showing fundamental frequency ranges for speech and singing

for the 'ee' in team. As an aside, it may be noted that the relatively low formant frequencies for some vowels such as 'oo' and 'ee' make them difficult to sing on high notes, sopranos and altos having the greater difficulty. Also in passing, the bands of frequencies sometimes causing coloration in loudspeakers can often be identified by recognizing the characteristic vowel sound they produce.

The unvoiced consonants such as 's', 'p' or 't' do not use the vocal cord vibration but are produced by sustained or interrupted edge effects as air passes between the tongue, teeth and lips. The voiced consonants include elements of both the vowel and unvoiced mechanisms.

9.2 The voice and the microphone

Voice intensity is controlled by the air pressure and tends to increase with pitch, rising to as much as 1 W (120 dBA SPL) on the loudest high notes. Dynamic range can therefore be remarkably wide, particularly with sopranos (see Figure 8.2), making microphone balance uncertain unless all a singer's effects in a given song or aria are rehearsed at 'full voice'. Harmonic content is high but capable of great variety and shifts during growth and decay. As a rule, the voice character changes significantly between soft and loud speech. This is illustrated in Figure 9.2, which shows a considerable boost for loud

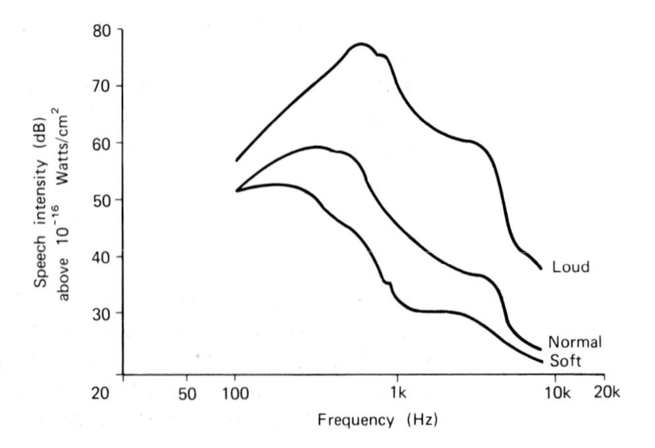

Figure 9.2 Speech spectrum, showing the changes which take place for soft, normal and loud speech (after Hilleard)

voices in the 500–700 Hz region. Most of the sound is naturally emitted through the mouth, though the nose and chest also contribute. Air motion can be a problem with close microphone technique, leading to 'popping' or 'blasting' on plosives such as 'p' or 'b'. Singers and commentators often develop a technique of directing their voices across rather than straight into the microphone, which reduces the problem. However, for such close working generally there is a need to choose a microphone type known to have minimum 'popping' troubles or to fit a windshield. At distances beyond about 0.5 m (20 in) this problem largely disappears.

Except with omnidirectional microphones, there is still the need to consider the 'proximity effect' (see Section 4.1) or bass tip-up experienced when pressure-gradient microphones (bidirectional, cardioid, etc.) are used at distances less than about 1 m. Again, many microphones are designed for close working and have appropriate bass roll-off built-in, sometimes via an on/off switch. Otherwise, bass-cut can be introduced on the mixer console. Some cardioid (and noise-cancelling) microphones make a virtue of the proximity effect and the fact that it can help to discriminate against low-frequency random noise or feedback from PA loudspeakers. Figure 9.3 shows the published response curves for a handheld vocal microphone at 6, 76 and 610 mm working distance. Clearly, close-miking will greatly reduce feedback problems since distant sources are not subject to the bass tip-up.

Figure 9.3 Use of proximity effect in a vocal microphone to boost low frequencies on the voice but not on distant sources such as PA loudspeakers (Electro-Voice BK1)

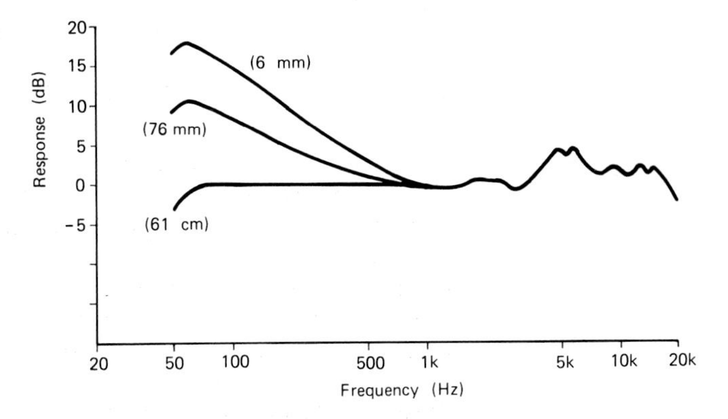

In terms of directivity in the horizontal plane, all frequencies are fairly evenly radiated within a 90 arc°, though high frequencies are rapidly attenuated at wider angles to the side or back. A similar situation occurs in the vertical plane, with a slight upwards bias due to the baffle effect of the body.

9.3 Microphone technique – the spoken word

At one level it may be admitted that the speaking voice occupies a narrower frequency and dynamic range than most musical ensembles and therefore it demands less of the recording or broadcasting medium. However, against this must be weighed the infinitely more acute sensitivity of human hearing to subtleties in speech timbres. In short, a narrow bandwidth and compressed dynamic range may provide adequate speech intelligibility for telephone or communications systems. Yet in applications where true fidelity to individual voice characteristics and inflections are important, both the microphone and its associated electronics should be up to the highest standards available. In the same way, microphone technique as exercised both by the speaker and the balance engineer can have a considerable influence on the results.

9.3.1 Telephones, intercoms and public address

The ubiquitous telephone system has evolved with a frequency range not much better than 300–3000 Hz, peaking somewhat at the higher end. This gives adequate speech intelligibility, but only just, and the speaker must avoid obvious faults if results are not to deteriorate. Talking loudly straight into the mouthpiece from a distance of only a few centimetres can aggravate popping or overload distortion which is easily avoided by lowering the mouthpiece to speak across it. At the same time, at least on older installations, the inherent system noise is often so high that reasonably loud and clear speech is essential if some words are not to be missed.

Many office or hotel intercommunication systems have a considerably better technical specification than the ordinary telephone but are let down by the failure of users to address the microphone correctly. Where ambient conditions allow, too close a speaking position should be avoided and the tone adopted should be that of addressing a small group of people across a table with no microphone present. In very noisy surroundings it will be necessary to speak closer, but again a moderate conversational approach avoiding microphone popping will be best.

In public-address systems for lecture halls or auditoria there may be powerful amplification and arrays of sizeable loudspeakers. Engineers in charge therefore have a heavy responsibility to choose the best microphones for the job and position them securely. They must also ensure that all speakers understand how to use the microphones for best results, or all their work will be wasted. The main problems are:

1. To maintain adequate and consistent voice level and intelligibility over the entire audience area;
2. To avoid spurious noises from the microphone, lectern or table thumps, script rustling, etc.;
3. To keep below intensity levels which run the risk of annoying the audience or causing uncontrolled feedback (howl-round) from loudspeakers to microphone.

The choice of microphone fixing will be between a floor or table stand, possibly on a gooseneck, and a pin or lapel (lavalier) type. Stand mounting is likely to give the best voice quality, but too close a balance will make the speech intensity vary considerably as the speaker looks up or down onto his or her script, or turns aside to write on a board or operate a slide projector. Gain riding by the control engineer will only marginally overcome these difficulties, and so a middle-distance microphone placement might be tried to reduce the fluctuations in balance with movement. A cardioid microphone

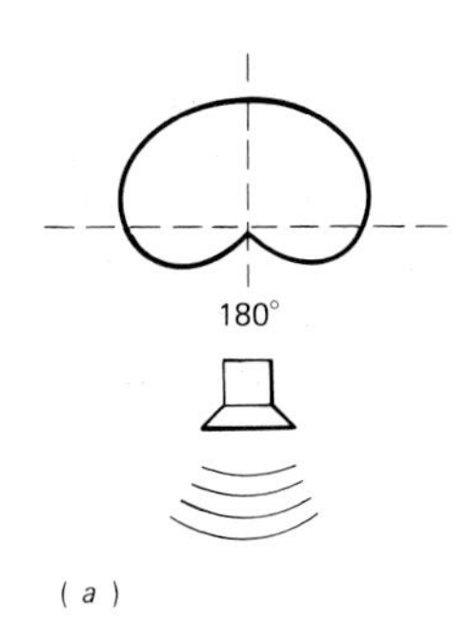

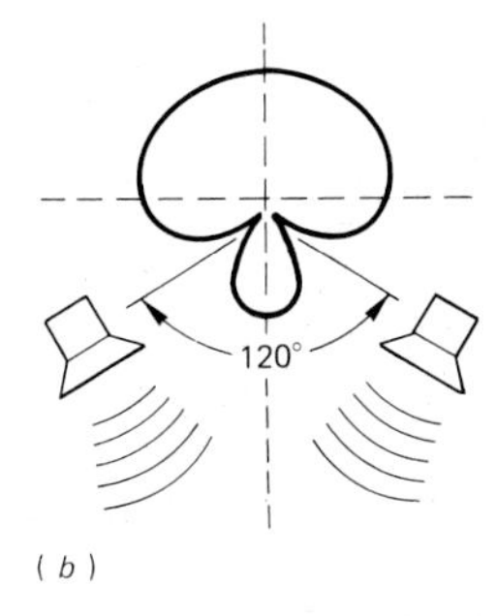

Figure 9.4 Avoiding PA feedback by using (a) a cardioid with centrally placed loudspeakers and (b) a supercardioid with loudspeakers located at the sides

pointing towards the speaker is the most popular choice since it has the twin advantages of a wide frontal pick-up angle and good rejection of loudspeaker feedback from the audience area (Figure 9.4(a)). Note, however, that a supercardioid might be better if the loudspeakers are at the sides (Figure 9.4(b)). The alternative to stand-mounting is a pin-on or lavalier microphone, which will at least have the advantage of moving around with the wearer, though the balance will still vary as he or she looks up, down and sideways. The disadvantages are rustling noises due to clothing friction and some loss in voice fidelity. In telephone switchboard and communications work, headphone-attached microphones are often used.

9.3.2 Studio talks, interviews and discussions

At every hour of the day and night, countless broadcasters are speaking into microphones all over the world. Whatever their message, the medium can play its part if it succeeds in reproducing natural sound quality free of distractions, as if the speaker were in the listener's room. The best broadcasters adopt just the right conversational tone and, when this is combined with a well-chosen microphone at a sensible 0.5 m or so distance with high-quality electronics and transmission, communication is complete. Poor broadcasters hector the microphone or get too close: poor studios produce boomy bass, aggravated by the proximity effect, or buzzing air conditioning, etc.

Some of these problems may be difficult to cure, but a straightforward talk, interview or discussion around a table can usually be tackled easily enough. A slung or boom-suspended microphone will avoid table knocks. Alternatively, special studio tables can be used having a non-reflecting top with a hole in the centre where the microphone can be placed on a spring-suspended platform. If the reader tends to look down, causing interfering reflections (comb filter distortion) from the script or table-top, an angled script-rest or lectern may help, or the reader can be encouraged to hold the script up alongside the microphone.

Two-way interviews in mono are comfortably set up with the speakers facing each other and the microphone in the centre. It should be just below eye level, when either a bidirectional or omnidirectional type will be suitable, and moved slightly to favour the weaker voice (usually that of the less-experienced interviewee). Alternatively, a cardioid could be hung centrally, looking down or up from a position above or below eye level as appropriate.

The same techniques work well for three or four voices, but a larger group will probably need extra microphones. This brings the same problems which always arise when several microphones have to be used in a confined space. Each microphone inevitably picks up 'unwanted' voices as well as those for which it is intended, and odd cancellation effects are audible when the microphone signals are mixed together. This problem of phasing and interference was discussed in Chapter 4, and the solution is to choose a layout which keeps all 'unwanted' pick-up levels to a minimum. Figure 9.5 illustrates one approach to a (mono) table discussion with five speakers. The chairman is placed at one end with a cardioid

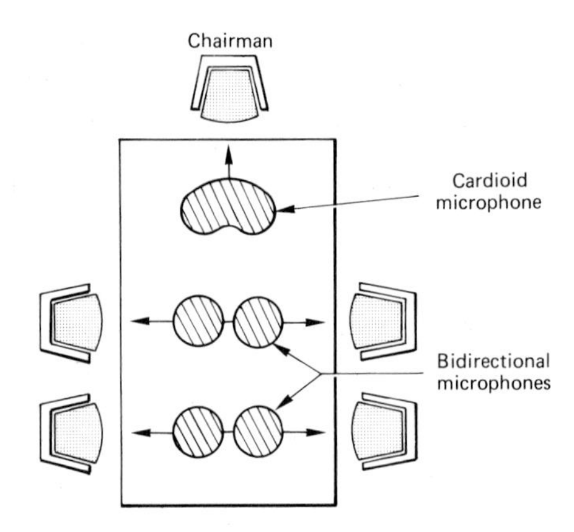

Figure 9.5 Round-table discussions. Typical use of a cardioid for the chairman and figure-of-eights for each facing pair of contributors

microphone to reduce pick-up of the other speakers, while they are grouped in pairs on either side of two bidirectional microphones.

Stereo is only occasionally used for this type of programme and its projection of different voices from points spread over the arc between the stereo loudspeakers, with no visual component, adds very little to the broadcast's effectiveness. It also becomes very difficult to maintain mono compatibility, so vital when the majority of listeners to the programme are probably using mono equipment. However, when applied sensitively and not in a 'ping-pong' manner, stereo speech programmes can work well enough. Where audio fidelity is important, the speakers should be arranged in a line or arc facing one of the coincident pair or spaced pair microphone arrays described in detail in Chapter 5. A trial balance should be arranged to check that each voice is picked up well and in a believable stereo spread. In a less purist approach, several mono microphones can be used and panned as necessary to position the voices across an acceptable stereo stage.

9.3.3 Studio drama

Broadcast plays and their equivalents on 'spoken-word' commercial recordings have a unique power to capture the listener's imagination, as witness the often-quoted comment of a child asked whether she preferred plays on radio or television, who replied 'I like radio plays best because the scenery is much better'. Given proper presentation, sound-only actors are free to move into any real or imaginary location, or no location at all for dream or space-travel sequences, and the listener's imagination does the rest. Very elaborate drama studios have been built in West Germany, Britain and elsewhere to produce a wide range of acoustic environments from totally dead or 'anechoic' (best for simulating the open air) to very reverberant or 'live' (for church or courtroom scenes). Sometimes a single live-end, dead-end studio with dividing screens can produce various apparent environments within one set of four walls between which actors can move quickly. Added realism relies on artificial echo or reverberation assisted by concrete slabs and gravel pits for footsteps, a bath for water effects, dummy doors, telephone handsets, etc. These live or 'spot' effects can be operated by the actors themselves or, more usually, by special-effects operators. Of course, they are reinforced by prerecorded effects of all kinds, from animals to crowds, and vehicles to weather.

Traditionally, such effects have needed to give only an approximation to the real sound, but advancing broadcast and recording quality, including stereo, have made high-quality and realistic sound effects essential. Therefore, as in film work, special prerecording and assembly of individual effects will often take place for each new production and most drama studios have built up a library of tapes and discs – ideally, recorded digitally for split-second cueing and avoidance of wear and noise at each playing.

On receipt of a drama script, these effects should all be assembled and the studio laid out to suit each scene with the microphones in their approximate positions before the actors arrive. An acoustic contrast between scenes is often effective and should be introduced even when a truly realistic ambience is not possible. When movable screens are used, or microphones have to be positioned near a wall, the risk of coloration due to reflections must be checked and avoided by suitable angling. The best studio screens are large and heavy. Anything less than 1 m wide will become a poor reflector for frequencies below about 340 Hz. In the best designs, one side is made reflective and the other absorptive, and sometimes a clear plastic window is built in at eye level to assist in visible cueing.

Stereo drama presents special problems compared with mono, both technically and artistically, but when these are understood and tackled properly the result can be very rewarding. One basic difficulty is that only a minority of radio listeners can be assumed to be sitting centrally in front of a good-quality stereo system (commercial discs and tapes have a better chance of being auditioned properly). Therefore the balance must be fully mono/stereo compatible. The stereo element may be used to heighten the dramatic impact (for the stereo listeners) but it must not be essential to an understanding of the action. There is also a danger that individual voices or sound effects which can be distinguished well enough in stereo (helped by the 'cocktail party' effect, which enables us to listen selectively by using two ears) may overlap and mask each other in mono. It is also very important to keep antiphase sounds to a minimum, since these will cancel, and simply disappear, when the left and right channels are multiplexed in stereo broadcasting. This problem is largely avoided if pure (Blumlein) intensity stereo is used (see Section 5.2) and so a crossed pair of figure-of-eight microphones is often preferred. This gives a 90° frontal acceptance angle in which actors can move, but the side and rear quadrants must be avoided as they produce out-of-phase effects, unpredictable stereo imaging and poor mono.

Wider acceptance angles of 130° and even 180° can be obtained from crossed hypercardioids and cardioids, respectively (see Figure 5.7) and so variable-directivity microphones are useful to provide a degree of flexibility in studio layout. With all microphone types, the entrances, exits, approaches and recedes of actors, as well as the location of effects doors, telephones, etc., need to be rehearsed and checked for naturalness. Ideally, the balance engineer should enlist the services of an assistant and check all these parameters before the actors assemble. Microphone identification and phasing should be checked as described in Section 7.3.1 and then the assistant should walk in a slow arc, speaking continuously, to check the recording angle as heard over the monitoring loudspeakers for effective changes in location and perspective as may be demanded by the script. Panpotting can be used to help move things around, especially prerecorded mono sound effects, but at some loss in realism. Where time allows, a film technique gives best results, recording individual scenes with real-time actor movements followed by an assembly/editing session at which authentic stereo sound-effects and music are added.

Though still only of minority interest, experiments in binaural and 'surround sound' drama have proved artistically successful and may become popular in the future.

9.3.4 Live theatre

Until a few years ago the sound installation in traditional theatres was primitive. Its function was merely to reproduce occasional recorded music and such sound effects as could not be simulated by backstage thundersheets and the like. Now all that is changed. Most theatres have installed comprehensive amplification and fast-cue playback systems and all new theatre designs give audio requirements a prominent place.

Part of the reason for this is the need to accommodate a wider range of productions, from Shakespeare to musicals; part is the greater public expectation for realism based on their experience with film and television; and part is the tendency for playwrights to introduce more exotic sound and lighting effects. Amplification of on-stage voices or effects calls for all the techniques discussed earlier in this chapter with the additional requirement that the microphones should be invisible. This can be achieved by concealing them in flowerpots or furniture, or providing actors with hidden radio microphones.

When a theatre performance is to be recorded or broadcast, the existence of some degree of in-house amplification can be a nuisance. Normal coverage of the stage action is traditionally tackled as for a live concert or live opera (see Section 10.6). This may mean using slung microphones, above the audience's line of sight if possible, or edge-of-stage microphones – possibly boundary types – in the footlights. However, any amplified sound will emerge from loudspeakers which are normally placed for optimum radiation into the audience area. These may be located above or on each side of the proscenium arch, on the side walls or even concealed on- or off-stage. Microphone balance must take account of this anomaly and the only answer may be to mike the loudspeakers separately or ask theatre technicians for line feeds from their microphones to be mixed with the signals from the main microphones. In some auditoria the sound system may have a complex system of time-delayed feed to the more remote loudspeakers. This is introduced to present the audience with a forward-biased sound even when they are nearer to one of the loudspeakers than they are to the stage (using the Haas effect). This can complicate microphone balance for a live recording and may make the use of on-stage microphones obligatory.

9.3.5 Reportage

Proprietary handheld microphones now provide a good measure of insulation against hand or cable noise. They also have built-in or push-on windshields and switchable bass-cut to make them suitable for outdoor use or close talking. There remains the problem of setting and maintaining the best balance between the voices and the ambient noise or reverberation. This is largely a matter of technique and reaching the right decision on the relative importance of the ambient background noise – at a sporting event, for example – for the given programme.

In all but the noisiest surroundings it is best to hold the microphone about 15–25 cm (6–10 in) away from and a little below the mouth. Holding a small loop of cable in the microphone hand, with a minimum of hand movement, will help to avoid cable noise. In an interview or multiple-commentator situation it is generally a mistake to push the microphone close to each speaker in turn. Apart from handling noises, this causes unexpected changes in loudness and perspective, and can go hopelessly wrong in unscripted situations. It is better to hold the microphone steady at some central position which will pick up direct speech from all contributors, or slightly favour the interviewee.

At a sporting event or in a machine shop the ambient noise may swamp the voices unless a very close talking technique is adopted. Where possible, it will then be advantageous to provide each speaker with an individual microphone either held close to the mouth or pinned on clothing. A two-track recorder can be used to allow later rebalancing of voices. Otherwise a mixer will be needed. Sporting commentators round the world have adopted special lip-ribbon or other noise-cancelling microphones, as described in Section 4.6.7. These can provide considerable rejection of surrounding noise, so much so that it often becomes necessary to put out separate 'atmosphere' microphones mixed in at a suitably lower level to reproduce the noise of the crowd or other activity which the commentator is describing. Control of levels is often difficult as the dynamic range in a noisy situation may reach unexpected proportions and the commentator may whisper one minute and shout the next.

9.4 The singing voice

The fundamental frequency range of singing voices was shown in Figure 9.1, notionally divided into the usual choral divisions of soprano, alto, tenor and bass. Of course, a much wider range should be reproduced for true fidelity as overtones exist up to the limits of audibility. Again, as with speech, our hearing is specially critical of singing quality, due to familiarity, and the slightest imperfection in the recording and reproducing chain can cause major losses in fidelity to the individual voice characteristics. A singer not only has considerable control over the subtle features of each note but can, and does, vary the sound quality within a note. The soprano voice can make the greatest demands on microphone performance, with a smooth and extended frequency response essential as well as a good ability to handle high-level peaks without distortion.

Opera singers in particular have learned to make themselves audible over the orchestral accompaniment in a number of ways. For example, male singers are able to put more energy into the upper partials by controlling the air flow from the larynx. They improve audibility still further by introducing a 'singer's formant' at around 2500–3000 Hz, where human hearing is most sensitive. Finally, they use vibrato, at around the usual vibrato rate of 6 Hz, as a further means of increasing the listener's ability to recognize the sinewave components in each note. A certain amount of artistry is needed to vary the emphasis of the 'singer's formant' if a harsh obtrusiveness is to be avoided. The amount and rate of vibrato should similarly be varied and all efforts made to subdue the random vibrato always present due to inherent muscular unsteadiness in the vocal cords. Sopranos are additionally able to 'tune' particular formants to match the voice frequencies in the music. This gives a gain in loudness with little extra effort, increases the available dynamic range and imparts smoothness and fullness of tone.

Most of the microphone techniques described for speech apply equally to singing. Dynamic range is naturally wider and can be a problem which is aggravated at close distances.

9.4.1 Solo voice – classical

For classical music, singers should preferably never get nearer than about 1 m (3 ft) from the microphone and, in conditions where the acoustics and the nature of the accompaniment allow, this distance can be increased considerably. Placing the microphone exactly at the height of the singer's mouth (position A in Figure 9.6) will give maximum direct pick-up and works very well if there is no tendency to emphasize sibilants or breath noises. However, considerations of the particular form of accompaniment to be included in the balance, or obscuring the singer's face from cameras or the audience may make a higher or lower position desirable. There will be some loss of treble at the higher position B but slightly less loss at the lower position C. This has the distinct advantage in a live performance that the microphone can be virtually invisible to the audience and, when the performers are raised up on a stage, this lower position is partly screened from the piano or orchestral accompaniment, perhaps making an ideal balance easier to achieve. In general, songs with orchestra may be treated in the same ways as described for concertos in Section 10.4.

Some classical singers like to 'work to the microphone', especially during studio recordings, by which they mean moving back and forth for loud and soft passages. This should be tactfully discouraged, as it almost always introduces such unwanted side effects as variable acoustic perspective and a change in balance with the other musicians.

9.4.2 Solo voice – pop

Pop vocalists rely very much on the properties of particular microphones and the way they use them (or, in a few cases, misuse them). In the most usual close-mike technique the essential requirements are ruggedness, good windshielding and suppression of

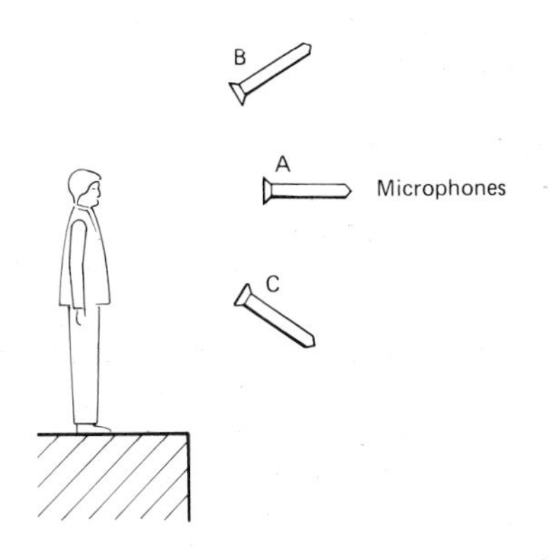

Figure 9.6 Solo voice microphone technique. (A) gives maximum direct pick-up; (B) gives some treble loss; (C) is less obvious to the audience and is partly screened from the accompanying musicians

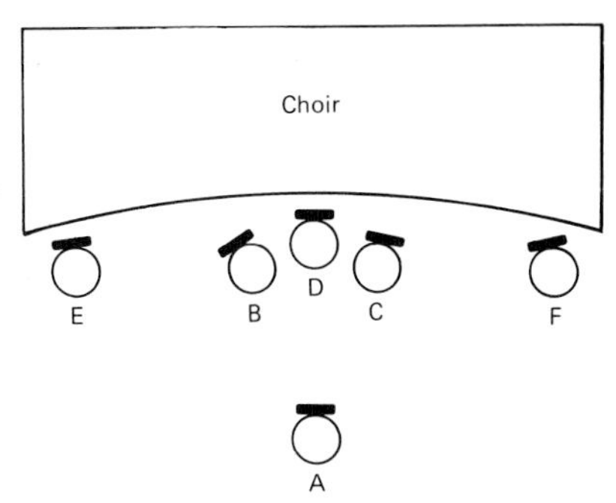

Figure 9.7 Choir microphone technique. (A) provides an overall natural balance using a mono microphone or stereo coincident pair; (B and C) represent the alternative spaced pair stereo technique; (D) may be added to improve centre imaging; (E and F) are outriders for a very wide choir

handling noise. High sensitivity is not important; indeed, attenuation as well as bass-cut will often be switched in, but low distortion at high sound-pressure levels is very necessary. Care is always needed to avoid problems with sibilance or popping. The former can be reduced by careful microphone technique, using a de-esser unit or asking the vocalist to soften 'S' sounds. The latter may be avoided by using a special pop-shield but this will sometimes introduce unwanted treble loss: some studios therefore rely on a stocking stretched over a wire coat-hanger placed a few inches in front of the microphone. Incidentally, this is quite a good way to prevent the vocalist from getting too close to the microphone. During live shows the vocalist may hold the microphone for a while and then clip it onto a floor stand to leave both hands free. This needs careful watching by the balance engineer or, as a back-up, it is sometimes possible to use a suitably concealed radio microphone to follow the singer more consistently.

Extra problems arise when singers accompany themselves on a guitar or keyboard, and it is usually better to balance these separately with fixed or contact microphones or direct injection of the electrical output. These topics and the use of feedback loudspeakers on-stage are discussed in Section 11.3. In a studio recording the singer will generally work to a fixed microphone and record to a separate track on multitrack tape. It is the usual practice to keep all vocal takes and then choose the best sections from each and dub or 'bounce' them onto another track to assemble the final edited version. Another possibility is to record simultaneously from two or three microphones to separate tracks, perhaps with compression or EQ on one, and decide later which gives best results.

9.4.3 Choirs

Choral ensembles invariably sound at their best in a fairly reverberant acoustic which will help them to produce and sustain full tone and blend the vocal lines into a satisfyingly integrated whole. This applies to church music, of course, and much of what we have decided to call classical music – from the Renaissance to the present day – was evidently written to be performed in reverberant surroundings. On the other hand, a more intimate sound may be appropriate for madrigals and chamber choirs, and operatic choruses are often heard in relatively dry theatre acoustics.

Decisions on microphone balance need to take account of these historical and practical listener expectations. Where the music is being performed in nearly ideal conditions, the microphone balance should retain this aspect: where ambient conditions are less than ideal, or the existence of odd echoes or proximity to organ pipes or orchestral instruments necessitates a closer balance, supplementary 'atmosphere' microphones or artificial reverberation should be added. It may be mentioned here that various tricks can be devised to overcome balance problems. For instance, where sibilants appear excessive due to enforced microphone closeness it is possible to ask a number of singers not to voice the sibilants and so produce a softer, more natural result.

Ideally, a choir will have a balanced number of sopranos, altos, tenors and basses which should make a fairly distant frontal microphone arrangement sufficient. However, many amateur choirs have a predominance of sopranos and too few male singers, so that some rebalancing may be needed. The simplest technique is to use a single microphone for mono or any of the coincident pair or spaced microphone arrays described in Chapter 5 for stereo. This basic approach is illustrated in plan view in Figure 9.7, where A is a mono microphone or coincident stereo pair, B and C represent the alternative spaced pair technique, D is an optional centre microphone to improve centre imaging with B and C (or focus on a centre soloist) and E and F are supplementary outrider microphones which may be needed to cover a very wide choir. Clearly, the output from D will be sent equally to the left and right stereo channels, whereas E and F will be panned to left and right, respectively.

Microphone height should be chosen for best direct-to-reverberant sound balance with the angle of tilt perhaps slightly favouring the back-row singers. This is indicated in the side view of Figure 9.8, which also illustrates the benefits of raised rostra or tiers allowing all voices a clear view of the audience and microphones alike. Very large

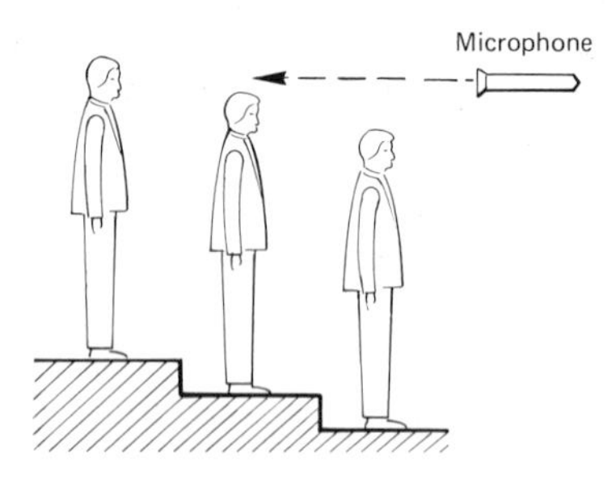

Figure 9.8 A choir, showing advantage of using a tiered stage and directing the microphone to favour the back rows

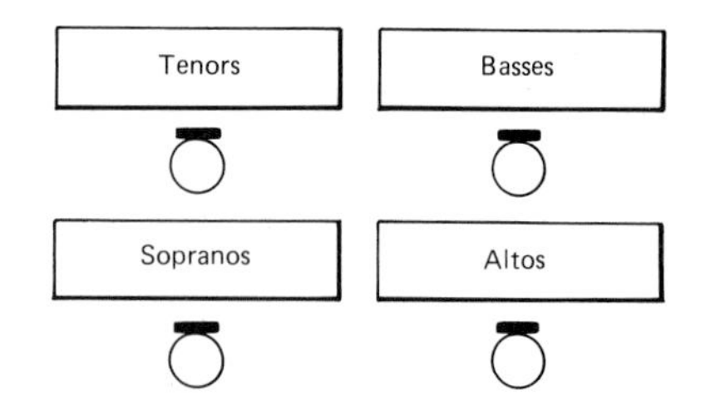

Figure 9.9 Large choirs may need extra microphones to give good intelligibility on all voices

choirs may be difficult to encompass with this basic microphone set-up and then a multimicrophone technique may be the only answer (see Figure 9.9). Cardioids will obviously give the widest frontal coverage with a possibly useful attenuation of sounds at the rear. Otherwise suitably tilted figure-of-eights give good rejection of brass and percussion instruments (for example, when the choir is located behind an orchestra).

Solo singers within a choir should already be in correct musical balance and need no special microphone, but this should be checked in advance and rehearsed for best effect. Featured soloists will take up a forward position and a trial balance may again be needed to see if the main microphones produce the desired effect or whether one or more soloist microphones are needed. A general rule applies here, that fading up any additional microphone increases the total pick-up of random reverberant sound in the overall mix. Therefore any such 'spot' microphones should preferably be placed fairly close to their respective sources and faded up only to an extent which leaves the overall ambience virtually unchanged between choir-plus-soloist and choir-only items. This suggests directional microphones in preference to omnis, since they pick up only about one-third of the reverberant sound for any given working distance.

9.4.4 Vocal groups

Small groups of singers often feature in pop music. They may stand close together or in twos or threes, or be scattered throughout a band and play various instruments as well as sing. Microphone technique at a live show has to rely on ultra-close balance because so much is going on at once (see Section 11.3). However, in a studio, multimicrophone recording separation is much less of a problem. Each singer can be given a separate microphone or, if they like to keep together for close harmony, they can share one or more microphones. A cardioid will easily accommodate two or even three singers on its live side, or the singers can face each other from opposite sides of a figure-of-eight microphone.

Occasionally, if the studio acoustics are suitable, a true stereo balance can be set up, but it is more usual to work with mono microphones and pan them across the stereo stage for best effect. A wide left/right spacing for the group voices may be best to leave the lead vocalist clear at the centre. However, there are other ways of achieving this, apart from simply making the lead vocalist louder. These include the use of compression or double-tracking for the lead voice or slightly attenuating the mid-frequency band on the group vocal tracks.

10 Ensembles – classical

The characteristics of solo instruments and voices which have an affect on microphone balance have been described in Chapters 8 and 9, respectively. These considerations still apply when groups of musicians perform together but it then becomes necessary to reproduce the ensemble in as realistic, believable and effective way as possible following the guidelines given in Chapter 7.

For classical music the starting assumption must be that the ensemble will take up playing positions which both provide all the players with the best visual and audible contact and present a correct balance to the audience or, in the case of a non-audience recording, to the microphones. There is no single solution to the problem of choosing the best microphone types and positions but the examples which follow are indicative of balance layouts and procedures which have worked well over the years. They should point the way to achieving acceptable results for the ensembles illustrated and, by extension, to any other ensemble met in practice. For simplicity, a single microphone is usually shown at each location, as if for mono, but in almost all cases this should also be interpreted as a stereo microphone pair using any of the coincident or spaced microphone arrangements outlined in Chapter 5.

10.1 Soloist with piano

The main difficulties presented by a grand piano are mentioned in Section 8.8. Adding just one further performer, singer or instrumentalist certainly makes the capturing of an ideal piano sound much more difficult, and some degree of compromise will usually be necessary. In the live recital situation the soloist will naturally face the audience and require a central balance (see Figure 10.1). Microphone distance and height will be chosen for best fidelity on the soloist, as shown by the alternative positions M1 and M2. The basic question of relative loudness balance between soloist and piano is less of a problem with a fairly distant microphone (say, D2 = 10 m or more), and this will also give about the same direct-to-reverberant

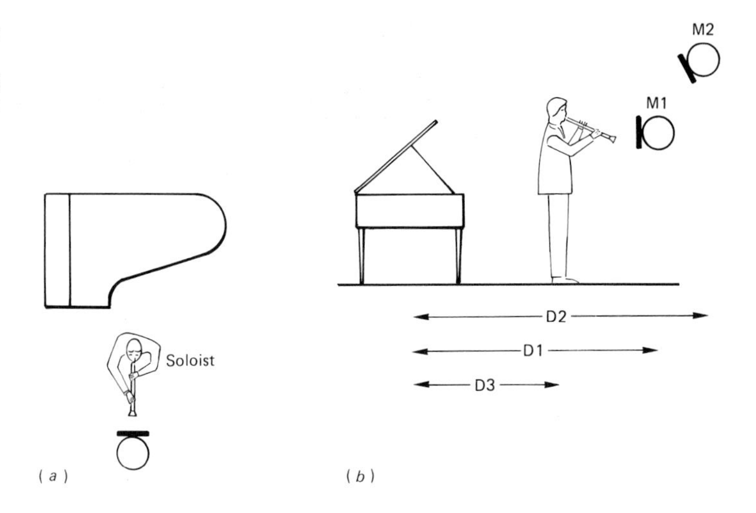

Figure 10.1 Soloist with piano. (a) In a live recital the soloist will normally be placed in front of the piano, facing the audience; (b) side view showing how microphone height and soloist/piano distance can be adjusted for correct balance

sound balance for soloist and piano. However, at a closer distance D1, the soloist will tend to be relatively too loud and have less acoustic bloom. This imbalance can sometimes be corrected by tilting the microphone to favour the piano, or it may be possible to vary the distance D3 between the soloist and the piano while keeping D1 fixed – by asking the soloist to take up a suitably marked position on the stage. However, this latter manoeuvre is limited by the performers' preferences in respect of eye-contact, etc.

Positioning one or more separate microphones for the piano may be the only answer, as illustrated in Figure 10.2. Of course, this affords an opportunity to restore any tonal losses to the piano balance resulting from pick-up on the main microphone M1 which was placed primarily for optimum soloist balance. At a position such as M2 it will help to use a directional microphone or stereo array which picks up a minimum of soloist sound. For example, M2 could be a single figure-of-eight microphone with its dead side towards the soloist. As usual when additional microphones are introduced in this way, they should be faded up only to the extent necessary to restore correct balance. Often the main microphone M1 may need to be moved slightly closer because the extra microphone contributes a significant amount of random reverberant sound.

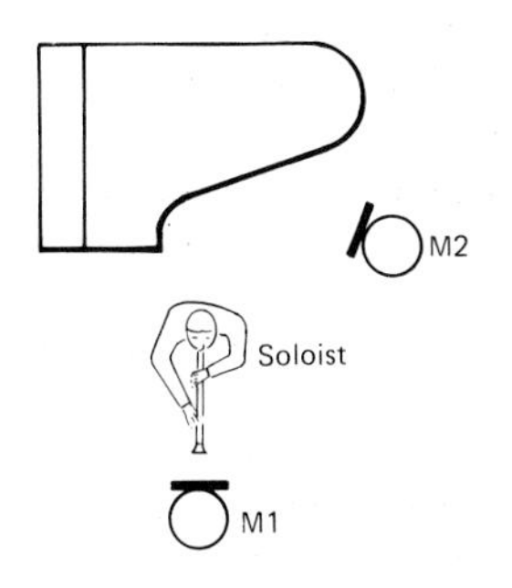

Figure 10.2 Soloist with piano. Adding a separate piano microphone or stereo pair to improve piano sound

When no audience is present, various non-standard layouts may be tried for improved balance and performer comfort – but only if the performers agree. Three possibilities are shown in Figure 10.3, with the soloist (a) facing the piano; (b) positioned alongside the keyboard; (c) fairly close to the piano and separately miked.

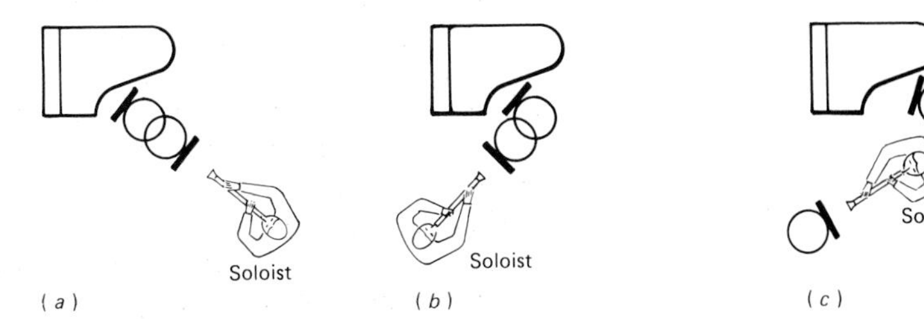

Figure 10.3 Soloist with piano. In a studio recording the soloist may take up various non-concert hall positions

Whenever the normal platform layout is abandoned, however, the engineer takes on the responsibility of reproducing a believable stereo 'stage' by sensitive microphone panning. It is an established part of listener expectation that the soloist should be located at the centre with the piano also broadly central – and the reverberant sound providing the rest of the stereo image. However, a slightly split balance can occasionally work quite well, with a cello soloist, for example, just right of centre and the piano given a compensating left-hand bias. As a rule, however, any balance which in total produces an off-centre effect sounds wrong or at least unintentional.

10.2 Chamber music

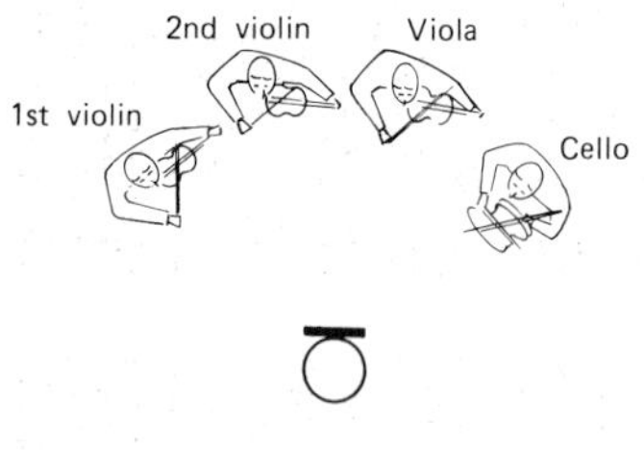

The preferred platform layout used by a chamber ensemble in live performance is also a first choice for studio recording. To capture the true internal balance and perspectives the aim should be to use a simple microphone array placed as far away as the acoustics and ambient noise level will allow. This is illustrated in Figure 10.4(a) for a string quartet but will apply equally for a string trio, wind quintet, etc. The microphone distance may be anything from 3 to 10 m.

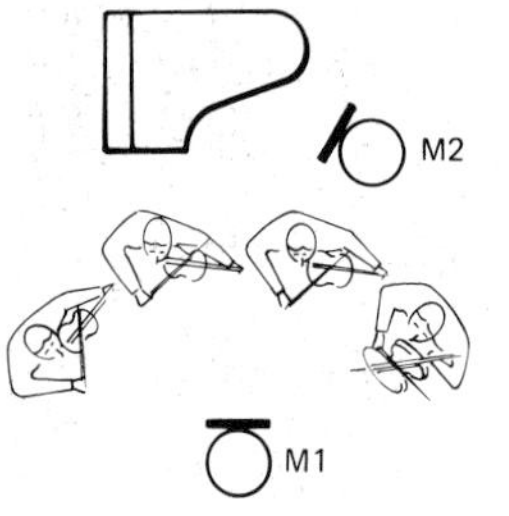

Departure from this ideal, and relatively simple, microphone scheme may become necessary at an acoustically unsatisfactory venue or where one or more of the instruments needs special treatment. This can sometimes arise with period instruments. The lute and harpsichord often need careful close-miking with, again, the problem of restoring stero imaging and believable ambience which should appear consistent for all performers to provide a properly integrated effect.

The same is true when a chamber group includes a grand piano which can so easily appear too loud – or disproportionately reverberant if pushed further back in an attempt to reduce it in intensity. This problem is less likely in the typical, fairly compact, piano quintet layout shown in Figure 10.4(b), with the strings closer to the microphone but the piano just behind them. Indeed, conversely, an additional piano microphone M2 may be needed to enhance the presence on piano, as already suggested for the soloist balance in Figure 10.2.

Figure 10.4 Chamber music. (a) Typical central balance of string quartet; (b) a chamber group including piano may need a separate microphone M2

10.3 Orchestral music

Orchestras come in all shapes and sizes, and indeed recent years have seen a broadening out in attitudes to the orchestra with numerous ensembles of 'period' or 'authentic' instruments and playing styles competing with the conventional modern symphony orchestra for public interest in the concert hall as well as on records. Figure 10.5 shows one very common layout for a full orchestra of a size ideally suited to early classical compositions. The larger string numbers shown in Figure 10.6 – about 20 first violins, 18 second violins, 14 violas, 12 cellos and eight double basses – is about a maximum which will usually be scaled down to suit the repertoire or the given resources. Note, too, that the strings layout in both photographs is the modern so-called American seating arrangement, with the first

Figure 10.5 Typical layout of classical/chamber orchestra (Academy of St Martin-in-the-Fields with conductor Sir Neville Marriner: Courtesy Philips)

Figure 10.6 Full-sized symphony orchestra showing frontal array of microphones with many spot microphones (Cleveland Orchestra with conductor Lorin Maazel: Courtesy CBS)

and second violins placed together on the conductor's left. This helps them to keep in strict time synchronism in 'difficult' music. The right-hand placing of the cellos and basses incidentally gives a pleasant balance in stereo recordings or broadcasts, and is very occasionally modified, as shown in Figure 10.7(a), to bring the violas to the edge of the platform. Another popular layout is the German or European plan shown in Figure 10.7(b). This has the first and second violins at the edge of the platform facing each other. Advantages of this layout include close contact between the conductor and the principal cellist, and good contrapuntal differentiation between the two violin parts. These layout variations are of no more than academic interest in a mono recording or broadcast but they create quite different effects in stereo.

Again, as was suggested for chamber ensembles in the previous section, the microphone technique which stands the best chance of reproducing the true internal balance of the orchestra with the real-life perspectives and left/right locations is a simple frontal microphone array as far removed from the musicians as ambient

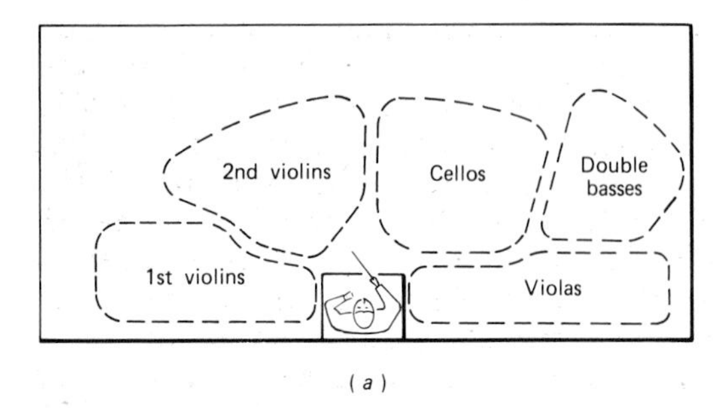

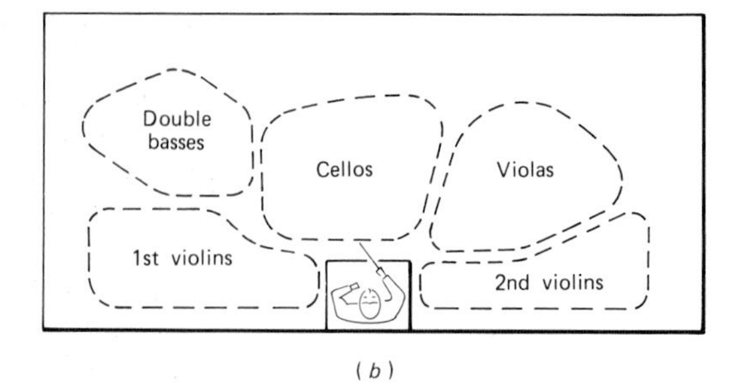

Figure 10.7 Symphony orchestra: showing alternative groupings of the strings with (a) violas on the right; (b) second violins on the right

conditions will allow. The precise choice between the various stereo microphone options discussed in Chapter 5 will largely depend on personal taste:

1. Basic Blumlein coincident pair for most accurate pinpoint stereo imaging, though with the risk of introducing unwanted phase effects at the sides and rear;
2. Spaced pair for a more 'open' effect, though at some loss in precise imaging;
3. Near-coincident or M–S, etc. as conditions or tastes dictate.

Figure 10.8 illustrates the use of crossed bidirectional microphones M1 and M2 and shows clearly that this system's narrow 90° frontal recording angle can be an embarrassment unless good-quality acoustics and low ambient noise permit a fairly distant balance. Of course, outrider microphones at M3 and M4 may be introduced to pick up extreme left and right musicians, panned L/R to create a smooth stereo arc embracing the full orchestra. However, this will, to some extent, dilute the stereo precision and inevitably destroy the phase coherence which is basic Blumlein's greatest attribute. A change to crossed cardioids or hypercardioids or M–S may be a better solution.

Figure 10.9 illustrates the use of a spaced pair of omni microphones M1 and M2. Omnidirectional microphones allow closer positioning

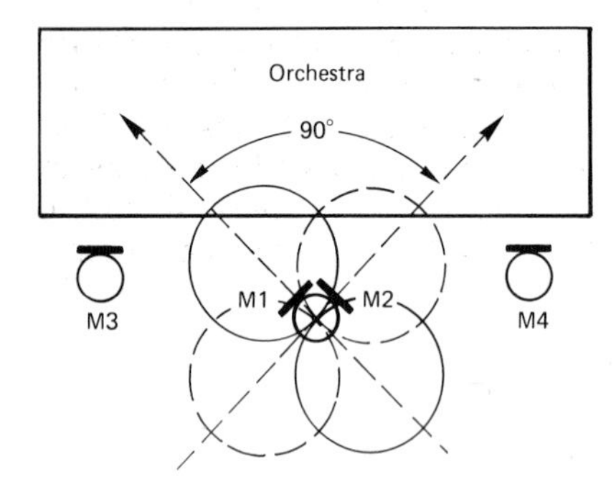

Figure 10.8 Basic Blumlein arrangement of crossed bidirectional microphones M1 and M2 Outrider microphones M3 and M4 are added only if the basic 90° recording angle proves too narrow

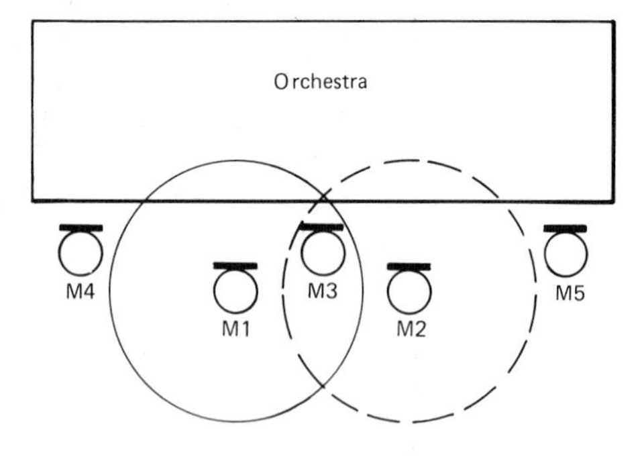

Figure 10.9 Stereo pair of spaced microphones M1 and M2 with a central microphone M3 added for possible enhancement of the central image. Again the outriders M4 and M5 may be introduced for wider coverage

to the musicians for a given direct-to-reverberant sound ratio and, of course, give a wide stage coverage. However, cardioids or hypercardioids could be substituted at a greater working distance. Typical M1/M2 spacings are 1–5 m at a height of 3 m or more and some 3–10 m out from the front of the orchestra. A third identical microphone M3 at the centre and about 0.5 m forwards is frequently added and faded up until any 'hole-in-the-middle' tendency disappears, as discussed in Chapter 5. Raising the level of M3 too far reduces the stereo width, and careful monitoring from a central position between the loudspeakers is very important during all such adjustments. With a very wide orchestra, or enforced close balance, outrider microphones M4 and M5, suitably panned L/R, can help to ensure full orchestral coverage.

In the hands of an experienced engineer any of these main microphone arrays can produce excitingly natural-sounding recordings in a hall or studio with helpful acoustics. However, circumstances often demand additional 'spot' microphones to pick up individual instruments or groups which sound too distant or lacking in presence on the main microphones. This can be done sensitively by placing the spot microphones as close as fidelity to instrumental timbres and avoidance of too forward a sound will allow, so that they need to be faded up only slightly. Note, however, that any close-microphone technique is liable to produce brighter tonal balance than is normally heard at a distance. This is because of the progressive loss in high frequencies with distance due to air attenuation. The exact position, type and angle for spot microphones must be chosen to ensure maximum rejection of neighbouring, possibly louder, instruments. This need for 'separation' clearly influences the choice of directivity pattern and, as a last resort, may mean asking some musicians to move. A typical problem case would be miking a celeste in the vicinity of timpani or horns; or the intrusion of brass instruments onto a woodwind microphone.

Whenever a spot microphone is used, it is naturally standard practice to pan this close-up image so that it takes up the same position across the stereo spread as it does on the main microphone balance, and adjust its level for best effect. This still leaves the problem that there is a time difference when spot microphones are mixed with more distant main microphones. For example, in Figure 10.10 a spot microphone M2 is shown positioned 1 m in front of a clarinet which is 11 m away from the main microphones M1. This difference in distance will result in two arrival times which are 29 ms apart (10 ÷ 344 = 29 ms). The resulting double image will lead to phase-related distortion (comb filter effect) and further emphasizes

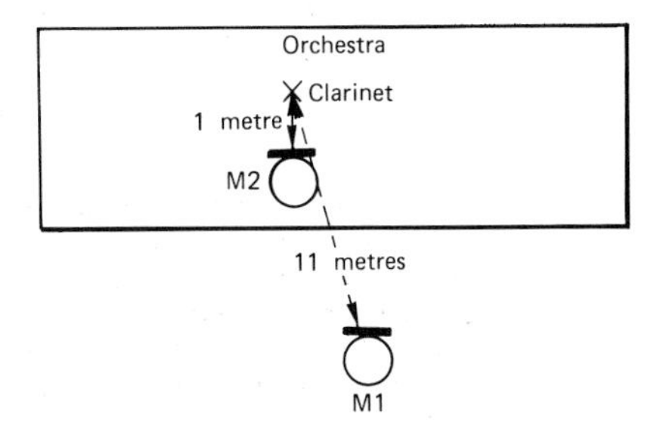

Figure 10.10 Using a spot microphone introduces a time difference between sounds reaching microphones M1 and M2 respectively

the need to work for good separation. The recent introduction of high-quality digital delay processors, sometimes built into digital consoles, offers a solution. It becomes possible to calculate, or adjust by ear, the appropriate time delay to insert in each spot microphone channel to make the two images coincide. Note, however, that any pick-up of the brass (for example, on the clarinet's spot microphone) will also be subjected to this fixed delay in addition to the time lapse from their location to the spot microphone. This compounding of the timing aberrations introduced by spot microphones generally emphasizes the need to choose directivity patterns which maximize the pick-up of the 'wanted' instrument and reject everything else.

There is clearly a limit to this technique of adding spot microphones when mixing becomes so complex that the integrity of the main microphones is largely lost. The alternative is then to work out a multimicrophone balance at the outset, abandoning any attempt to produce a single overall perspective. The comprehensive multimicrophone approach illustrated in Figure 10.11 provides one or

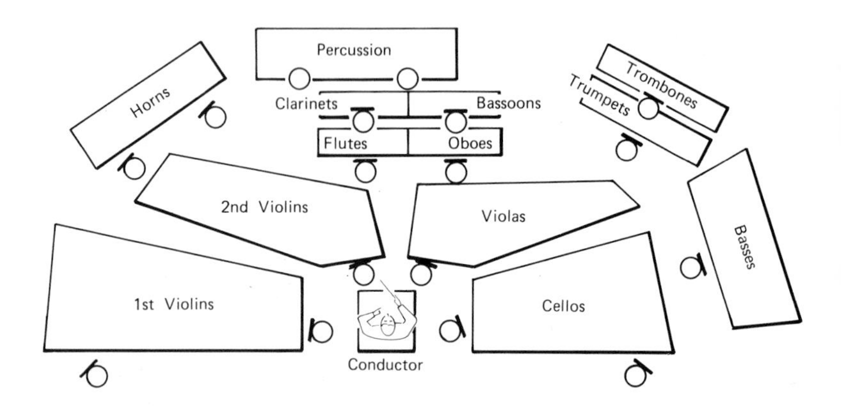

Figure 10.11 The multimicrophone technique

more (mono) microphones for each section of instruments, or even each desk. The responsibility for final musical balance no longer rests solely with the conductor. The engineer must assemble or synthesize the orchestral balance from all these separate channels and recreate a believable left/right spread with appropriate front-to-back perspective and overall acoustic ambience. The task is a formidable one and, it must be admitted, rarely succeeds even when the outputs from the microphones are sent to separate tracks on a multitrack recorder and a final balance is worked out at a protracted mixdown session aided by EQ, reverberation and other processors.

Any residual phase distortions or oddities of balance are much more audible on digital playback (e.g. compact disc), and this has led to a decline in the multimicrophone approach. It remains an option for light music, show bands, pit orchestras and ensembles appearing on-camera in film or TV when the musicians may be laid out in blocks for visual effect and camera accessibility – perhaps in the excessively dry acoustics of a television studio, where noisy camera trollies make it impossible to set up main frontal microphones at a reasonable distance.

10.4 Concertos

It is now regarded as old-fashioned to highlight a concerto soloist to the extent that used to be common when domestic playback equipment was of very restricted quality. However, there is still a need to favour the soloist slightly, if only to compensate for the lack of a visual element, while preserving the feeling of a natural concert layout. At a studio recording all parameters should be capable of adjustment to produce a natural concerto balance using any of the techniques described for orchestral music in the previous section.

In the best acoustic situations it may be possible to place the soloist centrally or a little to the conductor's left (the normal position at a live concert), as shown at S1 in Figure 10.12, and use the same main

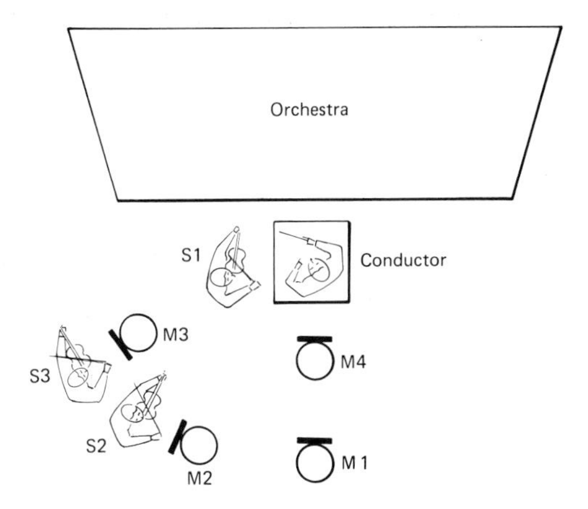

Figure 10.12 Concerto. Soloist position S1 may not need a separate microphone; alternative positions S2 and S3 will need a spot microphone and then the main microphone M1 may have to be moved forward to position M4

microphones M1 as for an orchestral item. This does not compromise the orchestra's internal balance, which will be particularly important in a classical concerto with an extended orchestral introduction. Of course, the soloist's position will be highly critical and, once established at rehearsal, must be carefully marked and adhered to thereafter.

All too often this ideal layout will be impossible, for instance in a piano concerto where the solo instrument is both too bulky and too powerful to fit conveniently in front of the main microphones M1 – unless the studio acoustics have allowed M1 to be fairly distant. Then it may be best to move the soloist away from the orchestra to S2 and use a spot microphone M2. This compromise layout should at least make it possible to optimize the pick-up quality on the soloist following the guidelines in Chapter 8 and, in this example, M2 could be a figure-of-eight type with its dead side towards the orchestra to facilitate separation. Yet another possibility is shown at S3, where the soloist looks directly at the conductor/orchestra and the solo microphone M3 is a cardioid, again minimizing orchestral pick-up.

These various alternative soloist positions will work best if the main microphones are fairly high and aimed at a point well back in the orchestra. Again it should be noted that M2 or M3 will contribute further diffuse ambient sound and probably require a slightly more forward position M4 for the main microphones to give the same

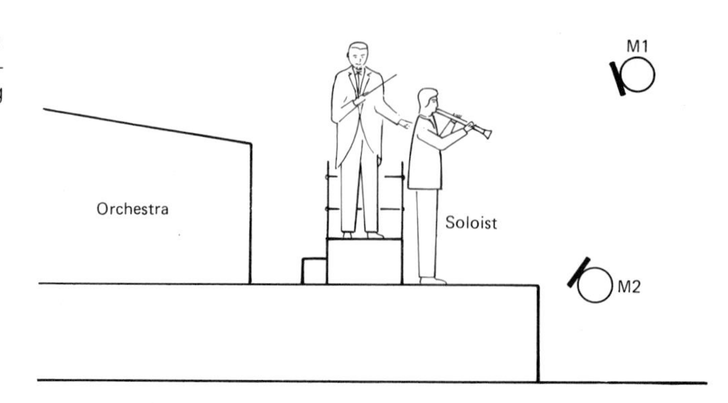

Figure 10.13 Concerto or voice with orchestra. Side view shows low position for soloist microphone M2 giving reduced pick-up of orchestra

overall ambience as would be obtained in a purely orchestral item using M1 alone with the solo microphone faded out. In a concerto with a long orchestral introduction it then becomes a fine artistic decision whether to use the mixture of solo microphone plus M4 throughout or begin with the optimum (cleaner and more purist) M1-only orchestral balance, and crossfade to the solo microphone plus M4 mixture as the soloist enters.

At a live concert the platform layout will be dictated by convention, with the soloist playing directly to the audience. There will then be a strong case for keeping to the main frontal balance using M1 alone if at all possible. When this does not produce the ideal soloist/orchestra balance, any solo microphone will need careful positioning for minimum pick-up of the orchestra. One possibility is to choose a low position for the solo microphone, as was suggested in Figure 9.6 for solo voice (see Figure 10.13).

4.5 Voices with orchestra

Basic guidelines for microphone technique with solo voices and choirs were given in Chapter 9. These can be carried over to the orchestral situation, though some compromises are usually inevitable. One or more solo voices can be treated in the same way as concerto soloists in the previous section, ideally relying on a main microphone array but introducing extra solo microphones when absolutely necessary.

For choir with orchestra, the usual layout at a live concert places the choir behind the orchestra and raised on stepped rostra or tiers. This arrangement is also best for studio recording, as it gives the conductor (and the microphones) the most direct visual communication. Any of the main microphone techniques used for orchestra alone should again form the basis for choir-with-orchestra music (see Figure 10.14). It might work out, for example, that raising the main microphone (or stereo pair) M1 and moving it to a slightly more forward position than that arrived at for orchestra-only will produce a suitably balanced and in-focus choir sound. More often, unfortunately, this counsel of perfection either upsets the orchestral balance or leaves the choir lacking in presence and vocal clarity (intelligibility). Then extra choir microphones will be needed, as indicated at M2 to M5.

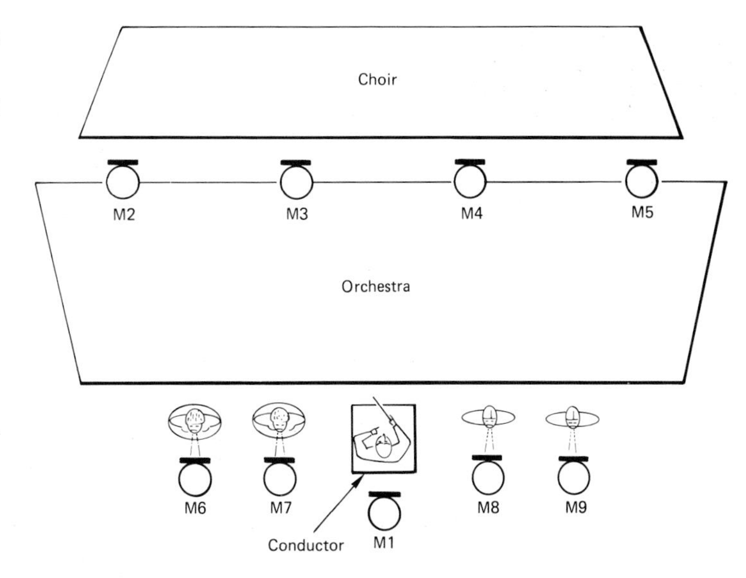

Figure 10.14 Choir and soloists with orchestra. The main microphone M1 may have to be reinforced by choir microphones M2 to M5. Soloists will almost certainly need individual microphones M6 to M9

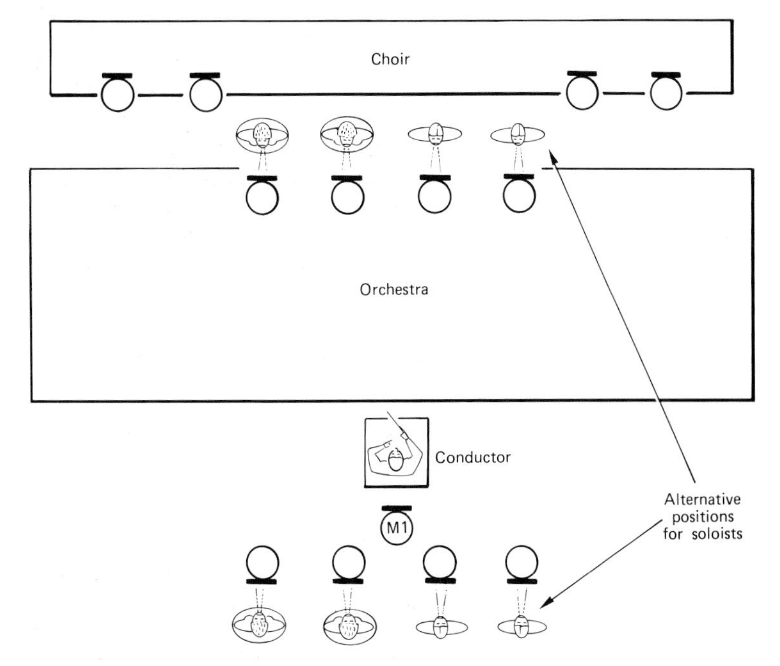

Figure 10.15 Voices with orchestra. In a studio recording the soloists may be placed behind or in front of the orchestra

When solo singers are also present, as in an oratorio, the soloists may be arranged on either side of the conductor as shown in Figure 10.14 or, when the singers prefer it, all grouped together on the left or right. Except in the most favourable acoustic environments, each soloist will be given a separate microphone as shown at M6 to M9, despite the known phase and separation problems, as the only sure way of maintaining optimum balance throughout.

At studio sessions it becomes possible to choose an unconventional layout with the soloists placed in front of the choir (also occasionally done at live performances) or even behind the conductor, with microphones having polar characteristics which provide the best degree of separation (Figure 10.15).

10.6 Operas and musicals

Everything said in the previous section about recording voices with orchestra can be taken to apply to studio recordings of operas and musicals. However, there are extra requirements including the need to give the impression of stage movements, sound effects, off-stage voices or instruments and perhaps spoken dialogue.

Before stereo came along in the late 1950s, productions were inevitably static, relying on some front-to-back perspective contrasts to convey the feeling of action taking place in real space, and using a minimum of approaches and recedes. With the advent of stereo, however, and notably in John Culshaw's productions of Wagner's *Ring* cycle during the period 1958–1965, a measure of heightened realism became possible. In the best examples this style of production can add considerably to the listener's feeling of involvement. However, as for stereo drama (Section 9.3.3), each move and effect has to be carefully rehearsed to make sure that its conversion to the standard 60° arc of domestic stereo listening preserves as much of the realism as possible. Some producers mark a chessboard pattern on the studio floor, or at least provide cues to each singer on how they should move between microphones from scene to scene. Given time for proper rehearsal, this can often produce a better feeling of real live action than simply relying on panpots plus artificial reverberation.

In the live theatre, the microphone balance can at best attempt to capture the series of impressions received by a well-placed member of the audience. Even this modest objective is notoriously difficult to achieve in practice. If it were possible to deal with the pit orchestra in isolation, then one or other of the microphone arrays discussed in Section 10.3 should, in theory, work well enough. Unfortunately, however, most theatres and opera houses are acoustically less reverberant than the ideal concert hall, while inevitable noise from the audience almost certainly rules out the kind of distant main microphone placement which might produce a pleasing direct-to-reverberant sound balance. In addition, the usually cramped layout of the orchestral players makes an ideal balance almost impossible. When the need to pick up singers' voices throughout the entire stage area is added, the problems are seen to be severe.

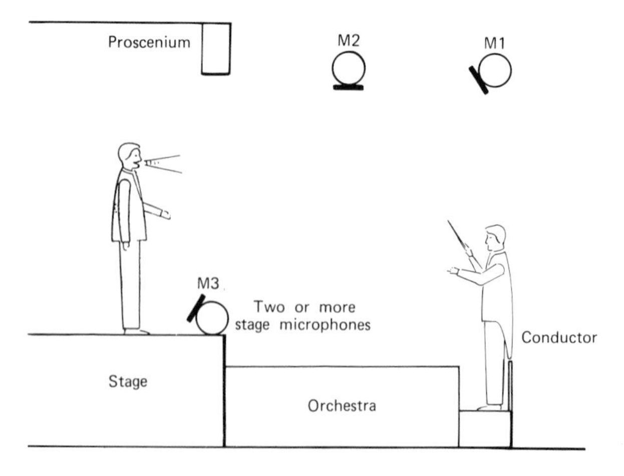

Figure 10.16 Operas and musicals. Main microphones M1 or M2 will usually be reinforced by two or more stage microphones (plus radio microphones in a modern musical)

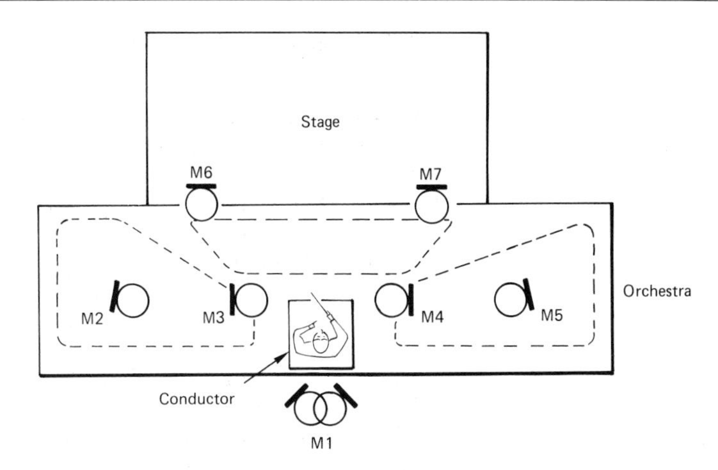

Figure 10.17 Operas and musicals. Example of closer balance of the pit orchestra supplemented by stage microphones M6 and M7

It might be thought that a conventional main microphone array at M1 in Figure 10.16 would be a good starting point, tilted to favour orchestra or singers as necessary. This may be feasible at a non-audience recording in the theatre but generally runs into difficulties at a live performance, because it gets in the way of spotlights and is too visible to the audience. A high-position M2 in front of the proscenium arch is a possibility, mainly directed down at the orchestra and reinforced by two or more stage microphones M3 set in the footlights. If this does not provide a satisfactory degree of presence and separation on the orchestra, then a hybrid orchestral balance may be better, as shown in Figure 10.17. Here a coincident pair M1 provides general coverage and is supplemented by various spot microphones M2 to M5. Two footlight microphones M6 and M7 are shown and these should normally have cardoid characteristics to minimize pick-up of the orchestra. Alternatively, this is seen to be an appropriate application for PZM boundary microphones which would combine the qualities of wide stage coverage and inconspicuousness. Whatever type of microphone is used, foam pads or other shock-resistant mounts are desirable to cut down the sound of footsteps and other stage noises. Windshields are often needed because of the constant flow of air resulting from the temperature gradient between the warm auditorium and the cooler backstage area.

11
Ensembles – light music and pop

The very large spread of music repertoire which can be called non-classical, and was therefore left out of the discussions in the previous chapter, is so varied that only a few clearly identified categories will be discussed. This omits folk music, country and western, brass and military bands, synthesized and electronic music. However, it should need very little effort to see how the suggestions made in this chapter and elsewhere throughout the book can be helpfully applied to any other musical situation. The two categories singled out for separate treatment are light music and pop groups, with some general notes on the handling of live concerts.

11.1 Light music

Bridging the gap between classical music and the pop and rock bands is an important category which relies very much on 'straight' musical instruments and ensembles but adds a 'microphone music' freedom in regard to spotlighting, amplifying and processing plus a busy and prominent rhythmic element. This kind of music is much in demand for easy listening or middle-of-the-road albums, film, TV and commercials. It can be approached in the same way as classical music to varying degrees; indeed, some established symphony orchestras and soloists earn a good income from this type of 'crossover' or semi-classical recording session, and so we may assume that their classical sound is what is wanted. Therefore the studio or hall should preferably have appropriate open acoustics to make sure that the classical players will find it easy to produce and sustain their natural tone and blend. When the music is being dubbed direct to visuals, there is the added space requirement to accommodate a large film screen. Yet when the arranger calls for a full rhythm section with drumkit, electric bass, bongos, etc. and wants vocal or instrumental solos to be close-miked, the classic coincident or spaced pair of microphones suspended out in front of the orchestra ceases to be viable.

In most cases it is as well to make a complete move towards a multimicrophone layout and balance. It then makes sense at the same time to record to multitrack – with a trial stereo or mono mix available to the producer and artistes at the session. The final balance and 'sound' will be engineered at a later mixdown stage. The standard orchestral layout is particularly at risk from problems with rhythm and percussion, which lose their edge in a reverberant acoustic, even with close-miking, and unwanted mike-spill with poor separation is inevitable.

At the very least it becomes necessary to place the different instrumental sections – strings, woodwind, horns, brass and percussion – some distance further apart than they usually prefer, and perhaps ask the players within each section to move closer together or play to the microphone. Large screens will help with separation, the type having built-in windows being best for providing visual communication with the conductor. Booth-type isolation enclosures are sometimes constructed for the loudest and quietest musicians – drums, brass, acoustic guitar, vocals, etc. These booths must be well designed to provide sufficient screening and yet remain free of box-like resonance effects.

Figure 11.1 is typical of this light-orchestra approach but, of course, is only one of many possible layouts. Today's conductors seem happy to work in this way, though they cannot see or hear

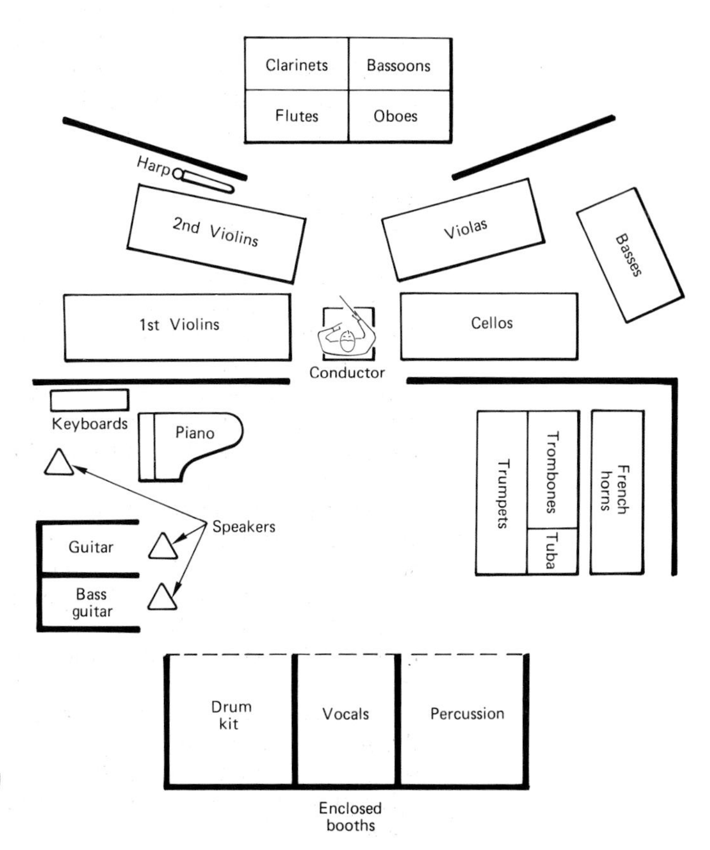

Figure 11.1 Light music orchestra. Typical multimicrophone set-up using screens and spaced layout to improve separation

everyone properly. Headphones may help, fed with a suitably pre-mixed foldback signal having the vocals or selected instruments emphasized as necessary. Studios dedicated to this kind of orchestral pop should ideally have a live reflective end to reinforce string and wind tone, plus a dead or absorbent one where separation is all-important. Judicious addition of artificial reverberation can then be used to match the sound from all sections.

Although a trial stereo mix is usually set up on the monitor channels, it is common practice to record the output from each microphone or related group of microphones to a separate track at fixed level. Peak level is essential in the case of analogue recording because of the cumulative effect of tape hiss at each copy/transfer or mixdown stage. It is perhaps less critical in digital recording where the noise floor is lower and not degraded to the same extent on copying. Nevertheless, most engineers keep to the tradition of peaking each channel close to the indicated limits on the level meters.

Overdubbing of extra musical lines or individual artiste retakes is obviously possible with multitrack, and can be used to correct for situations where ideal separation or balance could not be achieved with the full orchestra playing. By an extension of this idea, the final multitrack master can be assembled from recordings made at different times or even in different studios or countries. For example, a particular string section sound, free of microphone spillage problems, might be recorded using specially chosen musicians and acoustics at one venue, and the rest overdubbed at one or more studios with state-of-the-art facilities but less ideal acoustics. Clearly, the microphone techniques used for this light-music category can extend all the way from the purist classical approach to the latest pop methods, which will be outlined next.

11.2 Pop groups

Much of the music featuring in the pop record charts in recent years has been performed by groups of just three to five artistes. However, to increase the scope for interesting sounds, they will often use electronic effects pedals and a variety of keyboards or synthesizers. Their musical arrangements may also bring in sound effects, brass sections, a string quartet, woodwind solos or even a full symphony orchestra (using the real thing for the album but substituting a synthesizer when performing live). For these various reasons, pop recordings are usually assembled or overdubbed track by track, often at sessions extending over several days or weeks. In the extreme examples, only one vocalist or instrumentalist performs at any one time, this new track being laid alongside previous tracks to which the artiste listens on headphones or a foldback monitor loudspeaker. This technique has evolved over the years because it offers a number of production benefits:

1. It provides a unique ability to capture each voice or instrument in total isolation. The absence of other instruments means that a more distant microphone technique can be used, for example to avoid picking up key or wind noises, and a purist stereo microphone pair may be easier to position for best effect.

2. It allows each balance to be fine tuned to everyone's satisfaction, musically as well as technically, without the complication (or embarrassment) of asking all the performers to play through a sequence many times.
3. It enables the producer, artiste and engineer to experiment with new ideas away from the business of ensemble performance and balance.
4. It allows the music to be recorded in short sections using a 'drop-in' procedure in which the musician first records, say, 16 bars of music, then listens to playback on headphones and starts to play at the beginning of bar 17, when the record button is pressed.
5. In a one-man-band extension, of course, it allows one artiste to play more than one instrument or sing more than one line.

Microphone technique for recording only one performer at a time will take as its starting point the basic ideas suggested for solo instruments and voices in Chapters 8 and 9, respectively. The microphone balance for small groups of players or vocalists can similarly be approached initially using the classical ensemble techniques outlined in Chapter 10. From that point on there are absolutely no rules regarding the extent to which special microphone effects and electronic processing can be employed in the search for the desired individual 'sound'. Bearing in mind the limitless possibilities for modifying existing sounds and creating new ones which modern consoles, add-on processors and synthesizers can provide, the prime ingredient supplied by the microphone should be clean, undistorted signals with minimum noise or interference.

Freed from the need to achieve separation, omnidirectional microphones can more often be chosen with benefits in eliminating the proximity bass tip-up effect and securing the smoother frequency response shown by many omnis. The artiste/microphone relationship can also be optimized, taking account of any microphone preference which the performer may have cultivated but also evolving the ideal working distance and angle for each recording. Some tracks will be produced without a microphone, namely those using direct injection of electric guitar, keyboards or synthesizers – perhaps physically set up and played in the control room instead of the studio. This gives optimum artiste/engineer communication and allows the performer to monitor everything on the main control room loudspeakers.

11.2.1 Multitrack planning

Engineers will evolve their own favourite sequences for allocating voices and instruments to tape tracks and, provided that these are suitably logged for later identification, any track-numbering sequence is possible. The edge tracks (1 and 24 on a 24-track, for instance) are naturally at most risk from tape-head azimuth misalignment or physical damage. They are therefore kept as spares, used for less critical instruments, or perhaps a timecode for synchronization with other machines. One or two tracks may be needed to store data in some computer-assisted mixing systems.

Establishing the right drumkit sound can often be so time consuming, and use up so many tracks, that it is usually recorded at an early stage, as is a guide vocal track. The latter is simply recorded loud and clear as an important part of the headphones' foldback signal, helping all subsequent tracks to be performed with a good feeling for the words and timing of the song. The final recording of lead and backing vocal tracks will supersede the guide vocal, and they are of such importance as to merit special care and attention. When the drumkit sound has been captured on tape, and a rough mix of its tracks established on the monitor controls, the other rhythm instruments can be added, followed by the melody instruments, each artiste being provided with an updated pre-mix of the previous tracks as required. Indeed, the artistes can be given their own individual foldback mixer units to adjust the playback balance to taste.

An important key to these overdub procedures is the provision of a digital metronome, or 'click-track'. This gives the musicians an exactly timed count-in so that they can begin playing at the right moment and tempo. For film soundtrack recording, this fixed-tempo click-track may assume extra importance to ensure that a given musical sequence exactly fits the required number of frames. The click-track will therefore be made available on headphones throughout the take and may need to be faded down during quiet passages if there is any risk of its sound leaking into the microphones. Note that sound spilling from too-loud headphones has to be guarded against whenever close-miking is being used.

11.2.2 Sound processing and mixdown

Presented with individual, ideally balanced tracks, the engineer can finalize the sound processing or sweetening operations for each track or combination of tracks as necessary, and then move onto the concluding stage of mixdown to produce the two-track stereo master. Though these processes are beyond the scope of this book, a brief listing of the possibilities would include the following:

1. *Frequency-domain processing*: using simple bass, midrange and treble EQ (equalizers), graphic narrow-band equalizers, parametric equalizers giving variable frequency and width EQ, low-pass and high-pass filters or de-essers. These can be used creatively to enhance the frequency balance and tonal quality, or to correct and compensate for any deficiencies in the signal.
2. *Dynamic range domain processing*: using compressors, limiters, expanders, noise gates, noise-reduction systems (in analogue recording) plus manual fader adjustments. Again, these can be used creatively to make best use of the available dynamic range for artistic effect or realism or both. They also have a corrective role. For example, a noise-gate will shut down a flute or drum microphone to avoid pick-up of unwanted instruments or reverberation, and fade it up only when the instrument is played.
3. *Space-domain processing*: using panpots or Middle–Side width controls for adjusting left/right spread, plus mid-frequency 'presence' EQ perhaps combined with simple level adjustment for front/back movement. In surround sound recording, two-dimensional processing will be needed to provide both left/right and front/back control over the 360° soundfield.
4. *Time-domain processing*: using digital delay devices, reverberation chambers, plates or springs, phasing or flanging, pitch shifting

or time-squeezing. These can be used for simple enhancement of the acoustic ambience or for ADT (automatic double-tracking), echo or chorus effects, phasey space-travel effects, etc.

A comparatively new application of time-domain processing is known as sampling. This simply locks a musical sound, often a drum rhythm, into a digital memory or sampler unit in conjunction with a sequencer. An edited portion of the stored sample can then be triggered to play only when initiated by an unwanted drum sound. At the mixdown stage only the sampled drum sound is heard, and can, of course, be tighter and more consistent than is humanly possible on the live sound. When sampling is used for track-laying of the actual musician at a session, or one of the new generation of drum machines, it provides a degree of certainty and can also save a good deal of time when it is adapted for the process of 'spinning in', using the sampler to repeat a whole chorus, for example, between verses. Most sampler manufacturers offer computer software containing a selection of pre-sampled musical instruments and sound effects. These can be modified or completely changed by signal processing and then triggered via a keyboard or drums to play back in any key or any tempo. Taking samples from another artiste's albums without permission is strictly unethical.

11.2.3 A return to ensemble recording

Though overdubbing of single performers on multitrack continues to be the principal technique for studio pop recording, some groups have always preferred to perform in ensemble despite the difficulties of getting everything right at once. Also there are signs that some producers are returning to the earlier practice of recording all or most of the musicians playing together. They are also coming to prefer a more open ensemble sound than the enclosed claustrophobic effect heard on some heavily overdubbed recordings. This ensemble technique does not get any easier as group arrangements continue to look for more extravagant effects, but it has been welcomed by some older engineers who can remember mixing and recording big bands and jazz ensembles 'straight to stereo'. At the same time, it has forced some younger engineers brought up solely on overdubbing to learn the art of mixing and multimicrophone balancing from scratch.

It is very important to isolate some of the weaker instruments from powerful drum or bass guitar sounds, so that screens or isolating booths may be even more necessary than in the light-music situation outlined in Figure 11.1. This produces a studio layout something like that shown in Figure 11.2. Microphones will be chosen for their inherent quality, ability to handle high sound-pressure levels and, above all, their directivity pattern. Direct injection will be used for most of the electronic instruments, with their individual loudspeakers possibly set at a lower volume than for a full on-stage performance, unless the musicians insist otherwise or miking of the loudspeakers is considered essential.

The network of foldback feeds to the various performers' headphones, or sometimes carefully sited loudspeakers, can become very complex when everyone wants a different mix. In any case, all microphones will normally be recorded onto separate tracks for later

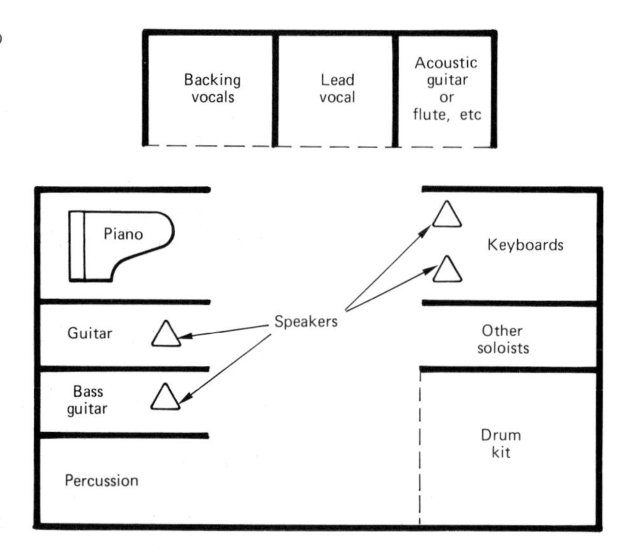

Figure 11.2 Pop group. Typical studio layout giving maximum separation

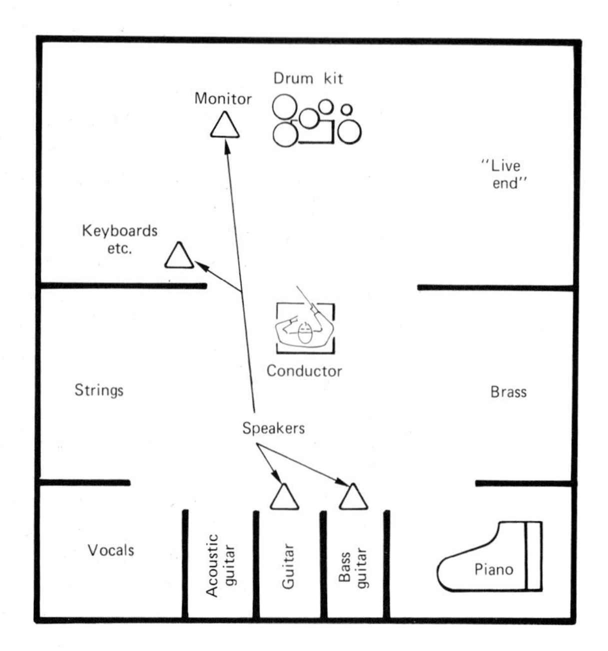

Figure 11.3 Live-end/dead-end (LEDE) studio layout giving a more open sound for drums, strings and brass

processing and mixdown – though there is less scope for individual rebalancing from this ensemble recording than when voices and instruments are recorded at different times.

As a further element in this return to group recording there has been a trend to open out the drumkit sound and produce a more powerful and acoustically integrated effect. For this the drumkit is taken out of its isolation booth and set up in the live studio area, yet with screening from the other instruments as necessary (Figure 11.3).

11.3 Live concerts

There are two quite distinct microphone balance assignments which can arise at a live concert:

1. *Sound reinforcement*, i.e. placing microphones, mixing their signals, amplifying these to just the right levels and routing them to arrays of loudspeakers positioned for optimum coverage of the audience area;
2. *Sound recording or broadcasting*, i.e. placing microphones specifically for optimum recorded sound quality, perhaps on multitrack tape for later mixdown to stereo.

The sound-reinforcement (public address) situation calls for detailed knowledge of the correct acoustic gain needed for remote parts of the auditorium and what this demands in terms of amplifier power, loudspeaker efficiency and directivity. There may be a need to introduce time delays to the more distant loudspeakers so that the audience still hears the direct stage sound first, and retains the natural impression that all sounds are coming from there. Most important, too, is the avoidance of acoustic feedback between loudspeakers and microphones (howl-round). This calls for careful choice of microphone and loudspeaker directivity patterns plus equalization or other electronic processing such as frequency shifting or phase reversal if the required acoustic gain is to be achieved without running into feedback instability. This is no easy task at a rock band concert, for example, where sound-pressure levels of around 115 dBA are demanded. These special aspects of public-address and sound-reinforcement engineering are beyond the scope of this book, but the techniques of microphone balance given here are equally valid for effective PA installations.

When a live concert or indeed any PA-assisted presentation is to be simultaneously recorded for issuing as a 'live' album or relayed live on radio or television, the two different balance operations come together and two independent teams of engineers may find themselves working on the same stage. The PA team may comprise the local house personnel or road engineers accompanying the artistes.

The recording/broadcasting team will have arrived solely to cover this one event. As far as they are concerned, planning and carrying out location recordings away from the comparative comfort and security of the studio will follow the broad principles outlined in Section 7.3.6. Their difficulties are further compounded by the fact that halls, arenas and other venues may be booked on only a 24–hour basis. This gives very little time to move in all the gear, set it up, cover the show, de-rig and get back on the road. Well-equipped mobile recording vehicles at least reduce the amount of heavy equipment that needs to be carried in and out, but there is still a major microphone, loudspeakers and cables installation to set up. Care will be needed to dovetail and coordinate all these activities with the team rigging the main show. The Rolling Stones, for example, have been known to arrive in a convoy of 32 trucks with a staff of 150.

In most cases the need to put on the best possible show for the audience is the first priority. This will determine the layout and movements of artistes as well as the choice and disposition of

Figure 11.4 Typical live group concert. Note TV camera and large loudspeaker arrays on each side of the proscenium, fill-in loudspeakers on the ceiling and foldback/monitor loudspeakers on-stage. (Courtesy Electro-Voice)

microphones, auditorium loudspeakers, foldback monitors, etc. (see Figure 11.4). Relaying the performance to the tape machines (or to radio or landline links in the case of a live transmission) is therefore more than usually difficult. Hardly any of the final balance subtleties or the effects of audience reaction can be exactly predicted or rehearsed in advance. The most that can be done is to set out enough microphones (up to 50 in some cases) to cover all possible eventualities, and anticipate as far as possible the types and ranges of mixing, level control (manual and automatic) and sound processing which may become necessary in the heat of the performance.

It is probably a waste of effort trying to sling overhead microphones specifically to capture an overall musical balance. The situation clearly calls for a multimicrophone approach from the outset, plus playing in of prerecorded tracks or effects. Some virtuoso console manipulation will be required if a straight mixdown to two-track stereo is the aim. A better solution is to record to multitrack for later mixdown back at base.

However, some thought should be given to the rigging of special atmosphere or audience reaction microphones. Audience and hall sounds picked up solely on the stage microphones will often sound thin, distant or even distorted. When these off-stage sounds are seen to be an important element in the production they should be deliberately balanced on one or more stereo pairs. These are probably best hung high over the audience, clear of lights, cameras and the risk of accidental damage. It then becomes possible to mix in this sound sufficiently to mask the poor atmosphere sound picked up on the stage microphones and produce a more controlled live-show sound picture. Indeed, it is not unknown for an engineer to keep a collection of audience/applause tapes and mix these into the master to get the special feeling of a live performance. Such tracks of atmosphere sound can also be used to smooth over edits when some parts of the performance or announcements are being cut out or put in a different running order. At the concert, the acoustic effect should be carefully noted with a view to reproducing this as accurately as possible on the final mix. For example, a show in a small acoustically dead club will sound very different from a major event in a concert hall or open-air stadium.

To avoid a tangle of microphone stands and cables, which would also make life difficult for the musicians, it is simpler to make key vocal and solo instrumental microphones on-stage do double duty, feeding them both to the public-address system and to the recording mixer via a splitter box (see Figure 6.6), bearing in mind the numerous problems with earthing and the provision of phantom power to condenser microphones which this splitting technique can introduce. Level checking and musical balance can then be experimented with during any preliminary sound-checks or rehearsals. An even better idea of what to expect during the show itself will be obtained if there is a 'warm-up' run-through of one or more musical numbers with the audience present. Close liaison with the PA personnel is vital, and lines of communication both by telephone and by a fast runner armed with spare microphones and gaffer tape should be arranged to cover all possible problems.

The increasing use of radio microphones, handheld, pinned to clothing or even concealed in the artiste's hairline, often calls for a backstage radio-microphone assistant in a large production. He or she will monitor the various receivers, switch between the limited number of frequencies legally available and even take a transmitter from one artiste and put it on another as the show proceeds.

Further reading

AES Anthology, *Microphones*, New York: Audio Engineering Society (1979).

AES Anthology, *Sound Reinforcement*, New York: AES (1978).

AES Anthology, *Stereophonic Techniques*, New York: AES (1986).

Alkin, G., *Sound Recording and Reproduction*, London: Focal Press (1981).

Amos, S.W. (ed.), *Radio, TV and Audio Technical Reference Book*, London: Butterworths (1977).

Anazawa, T. *et al.*, 'Digital time-coherent recording technique', Presented at 83rd AES Convention (New York), preprint 2493 (H-2).

Bartlett, B.A., 'Tonal effects of close microphone placement', *J. Audio Eng. Soc.*, **29**, 726–738 (1981).

Bauer, B.B., 'A century of microphones', *J. Audio Eng. Soc*, **35**, 246–258 (1967).

Benade, A.H., *Fundamentals of Musical Acoustics*, New York: Oxford University Press (1976).

Boré, G., *Microphones for Professional and Semi-professional Applications* (trans. S.F. Temmer), Berlin: Neumann (1978).

Borwick, J. (ed.), *Loudspeaker and Headphone Handbook*, London: Butterworths (1988).

Borwick, J. (ed.), *Sound Recording Practice*, Oxford: Oxford University Press (3rd edn 1987).

Bruel & Kjaer Application Books, *Acoustic Noise Measurements* (1978), *Architectural Acoustics* (1978), *Frequency Analysis* (1977), Denmark: B & K.

Burroughs, L., *Microphones: Design and Application*, New York: Sagamore (1974).

Cabot, R.C., Genter, C.R. and Lucke, T., 'Sound levels and spectra of rock music', Presented at 60th AES Convention (Los Angeles), preprint 1358 (G-6).

Campbell, M. and Greated, C., *The Musician's Guide to Acoustics*, London: Dent (1987).

Clifford, M., *Microphones*, Blue Ridge Summit, PA: Tab Books (3rd edn 1986).

Dickreiter, M., *Handbuch der Tonstudiotechnik*, Munich: Saur Verlag KG (5th edn 1987).

Dooley, W.L. and Streicher, R.D., 'M–S stereo: a powerful technique for working in stereo', *J. Audio Eng. Soc.*, **30**, 707–718 (1982).

Dooley, W.L. and Streicher, R.D., 'Basic stereo microphone perspectives – a review', *J. Audio Eng. Soc.*, **33**, 548–566 (1985).

Eargle, J., *The Microphone Handbook*, New York: Elar Publishing (1981).

Franssen, N.V., *Stereophony*, Eindhoven: Philips Technical Library (1964).

Gayford, M.L., *Acoustical Techniques and Transducers*, London: Macdonald and Evans (1961).

Hilliard, J.K., 'A brief history of early motion picture sound recording and reproducing practices', *J. Audio Eng. Soc.*, **33**, 271–278 (1985).

Huber, D.M., *Microphone Manual*, Indianapolis: Howard W. Sams (1988).

Jeans, Sir J., *The Growth of Physical Science*, Cambridge: Cambridge University Press (2nd edn 1951).

Kinsler, L.E. and Frey, A.R., *Fundamentals of Acoustics*, New York: John Wiley (3rd edn 1982).

Lipshitz, S.P., 'Stereo microphone techniques . . . are the purists wrong?' *J. Audio Eng. Soc.*, **34**, 716–744 (1986).

Martin, G. (ed.), *Making Music*, London: Frederick Muller (1983).

Mazda, F. (ed.), *Electronics Engineer's Reference Book*, London: Butterworths (6th edn 1989).

Meyer, J. (Trans. J.M. Bowsher), *Acoustics and the Performance of Music*, Frankfurt: Verlag das Musikinstrument (1978).

Nisbett, A., *The Technique of the Sound Studio*, London: Focal Press (4th edn 1979).

Nisbett, A., *The Use of Microphones*, London: Focal Press (3rd edn 1989).

Olson, H.F., *Music, Physics and Engineering*, New York: Dover Publications (2nd edn 1967).

Paquette, B., 'Mikes of the pre-war era', *Audio* (December 1974).

Rumsey, F., *Stereo Sound for Television*, London: Focal Press (1989)

Sank, J.R., 'Microphones', *J. Audio Eng Soc.*, **33**, 514–546 (1985).

Scroggie, M.G. and Amos, S.W., *Foundations of Wireless and Electronics*, London: Newnes (10th edn 1984).

Streicher, R.D., see Dooley.

Sunier, J., *The Story of Stereo: 1881–*, New York: Gernsback Library (1960).
Williams, M., 'Unified theory of microphone systems for stereophonic sound recording', Presented at 82nd AES Convention (London), preprint 2466 (H-6).
Wood, A. (revised J.M. Bowsher), *The Physics of Music*, London: Chapman and Hall (7th edn 1975).
Woram, J.M., *Sound Recording Handbook*, Indianapolis: Howard W. Sams (1989).

Index